FRONTIERS IN COMPUTER EDUCATION

ICT in Education

ISSN: 2377-1119
eISSN: 2377-1127

VOLUME 1

2ND INTERNATIONAL CONFERENCE ON FRONTIERS IN COMPUTER EDUCATION (ICFCE 2014), DECEMBER 24–25, 2014, WUHAN, CHINA

# Frontiers in Computer Education

*Editor*

Yuanzhi Wang
*Huazhong Normal University, China*

CRC Press
Taylor & Francis Group
Boca Raton London New York Leiden

CRC Press is an imprint of the
Taylor & Francis Group, an **informa** business
A BALKEMA BOOK

Published by:
CRC Press/Balkema
P.O. Box 447, 2300 AK Leiden, The Netherlands
e-mail: Pub.NL@taylorandfrancis.com
www.crcpress.com – www.taylorandfrancis.com



ISBN 13: 978-1-138-02797-8 (hbk)
ISBN 13: 978-1-315-68658-5 (ebk)

Typeset by MPS Limited, Chennai, India

*Frontiers in Computer Education – Wang (ed.)*
*© 2015 Taylor & Francis Group, London, ISBN 978-1-138-02797-8*

# Table of contents

## *Communication and intelligent control*

*Frontiers in Computer Education – Wang (ed.)*
*© 2015 Taylor & Francis Group, London, ISBN 978-1-138-02797-8*

# Preface

The 2014 2nd International Conference on Frontiers in Computer Education (ICFCE 2014) was held December 24–25, 2014, in Wuhan, China.

The 2014 2nd International Conference on Frontiers in Computer Education mainly focused on Frontiers in Computer Education. The objective of this conference was to provide a forum for different researchers in different fields especially in Frontiers in Computer Education as well as Information Technology to exchange their different findings. The conference brought together researchers from all over the world to share their new findings, and thus to promote academic exchanges.

ICFCE 2014 features various topics including Computer Science, Big Data, Cloud Computing, Green Communication, Robotics, Internet of Things, Recommendation Systems, Web Applications, Telemedicine, Artificial Intelligence, Software Engineering, Database Technology, Computer Graphics, Computer Education, Engineering Education, Engineering Education Mode, Curricula and Courseware Design, Education and Teaching Reform, Vocational Education, Educational Information, Public Education Policy Management, Communication and Intelligent Control, Circuits and Systems, Communication Systems, Control Theory and Applications, Power Systems and other related topics.

All the papers accepted have passed a strict peer-review process before publication by CRC Press/Balkema (Taylor & Francis Group). Here, we should show our sincere thanks to all authors for their high-quality papers and their careful presentations at the conference. At the same time, all reviewers should be thanked for their careful work. In addition, all sponsors and all team members of CRC Press/Balkema (Taylor & Francis Group) should be especially acknowledged for their hard work and careful editing of ICFCE2014, their joint efforts helping to make our conference successful.

We would like to express our sincere gratitude to all the TPC members and organizers for their hard work, precious time and endeavor to prepare for the conference. Our deepest thanks also go to the volunteers and staffs for their long hours of work and generosity they have shown to the conference. Last but not least, we would like to thank each and every one of the authors, speakers and participants for their great contributions to the success of ICFCE 2014.

*Frontiers in Computer Education – Wang (ed.)*
*© 2015 Taylor & Francis Group, London, ISBN 978-1-138-02797-8*

# ICFCE 2014 Organizing Committee

*Computer science*

*Frontiers in Computer Education – Wang (ed.)*
*© 2015 Taylor & Francis Group, London, ISBN 978-1-138-02797-8*

# The implementation of a campus information management system design in a vocational school

Bin Han & Xiaobin Bi
*Tianjin Bohai Vocational Technology College, TianJin, China*

ABSTRACT: Campus information management is the foundation of education informatization construction. So the Internet construction of campus information management systems has become an important part of school information construction based on so much attention. In order to achieve the school of information sharing, enhance the links between parents and schools, improve the efficiency of the management of schools at an educational level, according to the requirements of campus information management, VisualStudio2005 + ASP.NET2.0, as the design platform, puts forward the overall design scheme of the system, the analysis and implementation process of the system's detailed design, and the key techniques to realize the system are researched in the system.

## 1 THE BACKGROUND AND SIGNIFICANCE OF RESEARCH WORK

With the rapid development of Internet, the Internet has become an important channel for people to obtain, release, and transfer information. It plays an important role in many people's work, life, study, etc. Publishing information on the Internet is mainly achieved through websites, and access to information is the Internet 'sea' in a certain way to retrieve the information needed to browse. So the Internet construction of a campus information management system has become an important part of school information construction. In order to achieve a school of information sharing, enhancing the links between parents and the schools, improving school management efficiency and the managerial level, a special construction of a campus information management system for national vocational schools is needed.

## 2 NET TECHNOLOGY

The NET is a new generation of Web application development platforms of Microsoft. The NET framework is a multi language environment set up, configured, and run by Web services and applications, NET components and applications, and provides the best development environment for writing ASP.NET2.0 applications.

The Web service is the basic building block of distributed applications based on the network. The distributed application platform is independent and is usually created using the object template and a variety of languages. A Web service based on HTTP and XML, open Internet standards, and the formation in Microsoft of a programmable network theory. The Web service can receive a request and XML data by HTTP. The Web service is not specific to a certain component and object calling conventions, so it can be accessed in any language, or operating system component model. The Web service is an application, it can use the standard internet protocol to embody the function of being online in the Internet and also an internal enterprise. The Web service has a Web component programming.

The quality of a good selection of development tools and the speed of the system's development and the system's future maintenance and upgrading is very important. Based on the current system development tools on the Internet, such as ASP, JSP, and NET, according to the system characteristics and the practical development, finally I chose ASP.NET as the system development tool, version 2.

Select the ASP.NET 2 as a development tool, mainly based on the previous version adds new features a convenient, practical. Mainly in the following areas:

### 2.1 *An increase of more than new server controls*

According to the control's characteristics, they can be divided into: data control, login control, Web control, site navigation control, and other controls.

### 2.2 *The ASP.NET page only need one after the compiler does not need to be recompiled*

The ASP.NET page only need one after the compiler and does not need to be recompiled, until the page has been modified or the Web application is restarted and the multiple access speed has been greatly improved.

### 2.3 *GridView*

GridView, DataDroplist and other database component ASP.NET 2.o provided by ADO.NET can have a direct database connection.

### 2.4 *Code behind*

ASP.NET 2 'code behind' is a way to write the code making the code easier to write, the structure clearer, reducing the complexity and cost of the system's development and maintenance.

## 3 SQL RELATIONAL DATABASE

This system uses SQL Server2000. It has the following characteristics:

### 3.1 *Internet integration*

To provide a complete XML support for the SQL Server2000 database engine. It also has the scalability, availability, and security features required for data storage components, constituting the largest web site. The SQL Server2000 programming model and Windows DNA framework integration, developed web applications, and SQL Server2000 English Query and Microsoft search service and other functions, in the Web application contains a user-friendly query and powerful search function.

### 3.2 *Scalability and availability*

The same database engine can be used in different platforms, from the portable computer running Microsoft Windows 98, to large multi processor servers running Microsoft Windows 2000, data centre edition. The SQL Serve 2000 Enterprise Edition supports the Federation server, with indexed views and large memory support and other functions, so that it can be upgraded to the level of the performance required for a maximum Web site.

### 3.3 *Enterprise class database function*

The SQL Server 2000 database engine supports the demanding data processing environment required function. The database engine fully protects data integrity, and the management of thousands of concurrent databases and keeps user overheads to a minimum. For any query, it allows you to reference data from different data sources in SQL Serve 2000 distributed, as if these data are part of the SQL Server 2000 database, and the distributed transaction supports the integrity and the full protection of any distributed data updating. Replication also allows you to maintain multiple copies of information, at the same time it ensures synchronous replica alone. It can copy a set of data to multiple mobile offline users, allow the user independent operation, and then merge the changes back to the server.

### 3.4 *Easy to install, deploy and use*

SQL Server 2000 includes a series of management and development tools, these tools can improve process at multiple sites on the installation, deployment, management and use of SQL Server 2000. SQL Server 2000 also supports programming model standards and integration with Windows, based on DNA, using a SQL Server database and a data warehouse, it becomes a seamless part of a generated scalable powerful system. These features enable you to quickly deliver a SQL Server application, while the client only needs to install it and the management overhead can at least achieve these applications.

### 3.5 *Data warehouse*

SQL Server 2000 includes extraction and summary data analysis and On Line Analytical Processing (OLAP) tools. Some tools include a SQL Server and can be used to visually design databases and analyse data by English Query.

## 4 SYSTEM ARCHITECTURE

The system uses the B/S architecture design. The Web application is an application distributed by the server, the Web server and the client browser to cooperate with each other to complete the function, so the structure is also known as B/S (Browser/Server) structure. With the traditional C/S (Client/Server), the structure is different, most of the functions in the B/S structure are implemented on the server. According to the user's browser server sends request,.Therefore, control of the client interface is all Web applications must realize the function. In addition, Web applications generally use the database, so the function module and database interaction are necessary for the application.

According to the characteristics of Web application and ASP.NET technology, the system uses a three layer of Web application development and design model, based on ASP.NET technology. This model is the Web application and is divided into three basic levels: user interface layer, business logic layer, and data access layer.

### 4.1 *Software architecture*

#### 4.1.1 *The page data display technology*

##### 4.1.1.1 Based on the data source control data processing technology

In order to solve the problem of data processing of the ASP.NET1.x version, ASP.NET 2 with a number of improvements. To improve the data access model, the improvement of ADO technology, one of the most attractive is the introduction of a variety of data source controls. The original code to achieve the function of data processing is encapsulated in the data source control, and the specific data to these controls to call. Such as the use of select, update, insert and delete and other methods to achieve the database query, update, insert and delete records operation record. ASP.NET 2 also supports data bound controls which are many, such as: DropDownlist, RadioButtonlist, CheckBoxlist, ListBox, and GridView server controls.

### 4.2 *The database design*

A reasonable design of the database structure can improve the efficiency of data storage, to ensure the integrity and consistency of data. After fully understanding the functional requirements of the system, the general system of the database, according to the module, consists of an Information Management System Data Base (IMSDB), a student database (Student DB) and a message database (Guest Book DB).

### 4.3 *The design of system function module*

The main task of this system is to realize the automation of teaching management in the school, parent school contact information, and the campus information management system is an important guarantee of educational information. The main goal of the system is the realization of:

Campus news publishing and the introduction into the school of;

The informatization of teaching management and resource sharing;

Parents and students to achieve a good interaction;

Student achievement query and daily behaviour.

According to the target of the system, taking into account the different functions, module independence is strong, the characteristics of data set, system versatility, reliability, openness, from the function of the system is divided into seven modules.

#### 4.3.1 *The campus news management module*

The campus information management module mainly includes news and the news to be added, modified, deleted, and other basic management.

#### 4.3.2 *Student management module*

Student management module includes student, class information query and student, class information management, the students, the class information management to the school class, the student information to add, modify, delete and other basic management.

#### 4.3.3 *The daily student attendance and school performance management module*

The daily student attendance and school performance management module includes student daily evaluation queries and daily evaluation situations to be edited, features added, the main student daily evaluation with queries and teachers daily evaluation, with regularly published, management updates, etc.

#### 4.3.4 *A parents, school contact module*

Parents, school contact module includes messages, replies and message management, the system for parents and school (class, the teacher) constructs a communication message platform.

#### 4.3.5 *An on-line recommendation vote module*

The online voting module includes comments for online voting and appraisals of project management, online voting and a view real-time vote view function, an evaluation of project management to manage the candidate's projects.

#### 4.3.6 *In the examinee on-line query results*

The examinee on-line result inquiry module includes query results and performance management, the examinee enters the ticket number and is given online query test scores.

#### 4.3.7 *Professional skills display module*

The professional skills display module includes professional skills on demand and resource management, the professional skills on demand to realize online audio, video on demand, professional skills resource management resource upload, download.

## 5 THE SYSTEM HAS REALIZED THE FUNCTION

With the development of the Internet, each link of educational activities will be gradually replaced by informatization. The school affairs information should be preferred as the first step in educational informatization. This process greatly improves the efficiency of school educational activities, deepens the interaction and relationship between the school, the students, and the parents, so as to improve the school's running level.

The work in this paper is based on the Internet, according to information technology education, research and the design of a campus information management system and an implementation method. The purpose of the effective management of campus information is to improve the level of running a school.

According to the campus information management system design requirements, the system has realized the following function:

1. *School news release, additions, modifications, deletions.*
2. *The student information, class information query, add, modify, delete.*
3. *Students submit online timely, and then teachers, parents can understand the situation of students.*
4. *Parents, students, teachers, interactive message communication.*
5. *Online voting behaviour.*
6. *The examinee can enter online the examination number and name, and the results are online.*
7. *The professional skills of students are submitted online, by parents and students on demand communication and they can upload and download files.*

## REFERENCES

"THE CONCEPTS OF DATABASE MANAGEMENT, 2E", (database management system second version) Philip J. Pratt, Joseph J. Adamski

Ma. Proficient in ASP.NET2.0 network application system development of the posts and Telecommunications Press 2006.6
"THE CONCEPTS OF DATABASE MANAGEMENT, 2E", (database management system second version) Philip J. Pratt, Joseph J. Adamski 2006.8
"Network oriented information: database and search engine" Liu Yanzhang Northwestern Polytechnical University press in 2007 April
Miao Xuelan. Technology and application of database. Beijing: Mechanical Industry Press, 2006
Li Yingjun, Bi and.ASP.NET dynamic web development tutorials, Tsinghua University press, 2004.12
Wang Huazhang practical. ASP.NETZ.O network system development course. China Railway Publishing House 2006.6
Zhou Xu, Lina, Bai hypo.SQLserver2000 Chinese version of the introduction and improvement of Tsinghua University press, 2001.5
Cheng Wenjuan, Chen Dongfang. Information network technology to promote modern education function [j]. Journal of Anhui Agriculture University: 2001 (2).

*Frontiers in Computer Education – Wang (ed.)*
*© 2015 Taylor & Francis Group, London, ISBN 978-1-138-02797-8*

# A study in the design and production of TV titles techniques in modern digital media

Qi Liu & Xiao Wang
*Tianjin Bohai Vocational Technology College, TianJin, China*

ABSTRACT: Drama is a professional communication arts technology, culture and art in one, and is an important part of a television program packaging and image design. With the wide application of digital media technology in the field of television production, planning, creative, design TV titles get tangible reproducing, TV also works greatly to meet the aesthetic needs of the masses of the people. In this paper, from the general rules of TV titles design creation, the themes of creative drama creation, colour, and music are explored, with several problems of digital media technology and other elements of harmony and unity.

## 1 EFFECT OF TV TITLES IN THE TV

With the development of the media industry, the channel has already become the basic forms of existence of TV. Because of television media's demand for ratings and the advertising market, TV and TV channel images are being targeted with more and more attention and this gradually becomes a new profession. How to construct the television and television channel brand image, enhance the brand value, brand utility achieved by TV media's interest, has become an important task, for each big television station. So, major television stations have to pay attention to image management, and have introduced the concept of television publicity, and at the same time more pay attention to the overall channel packaging, in addition to television TV titles. The channel film advertisement interval has been used as an essential element in a day full of communication broadcast television in the process. It also appeared in the new topic of TV title design.

## 2 TECHNICAL INDICATORS SHOULD PAY ATTENTION TO THE TV TITLE CREATION

Drama is an important part of program content and it plays an effective interpretation of the program image, packaging, and positioning of the role. TV titles, like the title of an article, are summarized and hint at the program's content. This generalization and prompt strictly within a certain period of time, at the same time, these are used the methods of TV image to conceive and design. This approach makes the audience's understanding, identify left the impression, and also allows the audience to accept the entire program.

According to the State Administration of radio, TV program technical quality awards (Jin Fanjiang), "recording technique quality assessment approach" of "video graphic evaluation method" requirement, generally speaking, drama produced by its success or failure, should, from the following technical indicators, be measured.

### 2.1 *On the theme creative evaluation*

The drama theme is clear, the originality is novel, and the form is clever.

### 2.2 *The evaluation of production skills*

TV title composition and modelling is pretty standard, colours and materials use is reasonable, with exquisite light, clip is consistent, with smooth motion rhythm; technology is abundant, and can accurately express the creative intention.

### 2.3 *The overall effect on the overall visual effect evaluation*

The drama is prominent and sublimates the program theme; the special nature and the style of TV music whether it reflects the program picture and music, coordinates, and has strong appeal.

The above elements, through the technical means of production, accurately express TV's creative intention, and is the key to evaluating the success of the drama's production.

## 3 DESIGN AND PRODUCTION OF TV TITLES

### 3.1 *The theme and the creation of TV titles*

#### 3.1.1 *The theme of the drama*

Drama is the theme of the central idea of drama performance, reflected in the program orientation and program style. A good TV title must make clear its theme with prominent, rich and deep meaning, positive, to withstand the test of practice, to show the spirit of the times, but also can be understood and accepted by the audience.

3.1.2 *The creative drama*

TV titles of creativity use very simple TV language, vivid, with an orderly combination by means of technology, creating a specific mood, so that the viewer recognizes and resonates with its impressive connotation. Originality, as the whole drama packaging activities of the soul and life, is the core of the whole packaging program, only to create novel, unique perspective, the one and only, let a person gifted with an extraordinary retentive memory, can talent showing itself in numerous television channel brand program, to be recognized by the audience.

3.1.3 *The relationship between creative theme and TV titles*

If creativity is from the soul, then, the theme is from the brain, and the two complement each other, theme determines creativity; creativity the theme. Drama idea is to express the theme in an audio-visual language program used to visualize the process to create, through technical means, a television special language form, so that the viewer has recognition and resonance from its connotation, which impresses. Program Program positioning accuracy is the premise of creativity, and grasping the drama theme is the foundation for accurately positioning. Originality, as the whole drama packaging activity of the soul and life, is the core of TV titles. To show itself in numerous TV titles, for the audience to memory, just do the new in order to be different, unique perspective, out of the ordinary, and the photographic memory.

Different types of programs with different requirements of content and form, which formed the individual characteristics of different program titles, is a typical representation of the characteristics of personality. Therefore, the design thought of drama creation cannot do without the personalized, personalized design ideas to the subject; and full of individual creativity through advanced technology, to shock the audience, the purpose of emergent themes. For example, "law and economics" and "today" and other TV titles are the reality of material processing technology for the performance of the theme, is a true form for the audience watching. CCTV's publicity film is a kind of freehand brushwork form. It is used in ink digital art forms to express, through several abstract expression highlighted titles on the theme 'believe in the power of the brand', the title character is outstanding, the connotation is rich, so there was a kind of 'find everything fresh and new' feeling.

### 3.2 *Using the TV title creation colour technology*

Television is a visual art, first without light and colour. In the artistic creation of TV titles, how to make the colour technology and colour art together to create the fascinating TV titles is very worthy of our consideration. Colour plays a rendering of the atmosphere, deepens the theme, emotion is an expression of the theme and is an important factor on the TV screen. The TV title manufacture person has to experience the colour to express the importance of artistic creation, and the colour also has a very strong design sense. If the particularity in order to emphasize the picture colour sense from the application of colour in TV titles, this colour design is a failure.

3.2.1 *Colour reflecting the visual effect*

Elements of TV pictures are dynamic, television titles in colour are moving, and the titles show more visual effects that comprehensively reflect a dynamic art. Therefore, only reflect the program content specific colour, in order to better reflect the visual effect.

3.2.2 *The colour of a psychological impact*

According to people's visual habits, blues have contractive feeling on the vision. Therefore, in television program packaging blue is the most used colour, blue is also known as the eternal colour, bright, blue, free and other various feelings can be expressed. TV news titles, mostly use blue colours, combined with the 3D effect of light and a strong sense of space, to reflect the news program clear, timeliness, modernity and the atmosphere.

3.2.3 *Application of gradient colour*

With a lively sense of colour, feature rich layers. Rational use of gradient colour, can make the picture more variability with TV titles and fluidity. As a CCTV news titles are made of blue colours, blue with a calm, objective, atmosphere, dignified temperament. Its flagship program "news broadcast", "news 30", "evening news", "focus interview" and other programs, packaging colour their titles reflects the program style, reflecting the CCTV news channel in the matter.

3.2.4 *Colour and a sense of rhythm*

In the first use of colour, the same or a similar colour, on the screen more than two reproductions can create the most basic visual rhythm. For example: colour image duplication, alternate, gradual change, mutation, rhythmic colour, if the formation of linear relationships between specific objects on the picture, can give a person to feel full of vigour colour themes. The film title manufacture of clever use of colour point, line, surface and their spacing, transfer, change and other technical means, can produce the vitality of a colour rhythm.

Therefore, as a drama producer, should consider the special nature of colour art, according to the program orientation and style characteristics, reasonable use of colour, the colour is more compact, orderly, organized, so that television titles give more visual impact, and more imaginative and expressive force in creating more excellent works of art.

### 3.3 *The design concept of TV music*

TV music has special characteristics, it because the program orientation and program style is different. Generally speaking, television news music should adhere to the objective, the atmosphere, solemn, clear, strong sense of rhythm design concept; children's program of music is relaxed, lively, lively rhythm characteristics; while the other TV themes according to the characteristics of this program. But no matter what kind of program music must have a common characteristic: that is music must be closely combined with the screen, a painting of harmony and unity, so that the visual image and listening constitute an indivisible whole, with enhanced visual effects.

Significantly different, as everyone knows, the CCTV "news broadcast" and "weather forecast" will be in the music style. In 'News broadcast' the music gives people a solemn atmosphere, the feeling, and the music has been experienced for more than ten years, bears the national news music image, and expresses the in-depth audience's bone marrow. The 'weather forecast' is to give a person a kind of relaxed, lively feeling of the music. The positioning or feature visible TV program determines the choice of music, music must be closely combined with the screen, to create visual harmony.

### 3.4 *Application of TV film title manufacture technology*

Drama creation regardless of its form as realistic or freehand, creative ideas and thoughts, will eventually need to be realized by digital media technology. With the wide application of information network technology, digital compression technology in the field of television, television program manufacture technology has undergone revolutionary changes. From the computer caption machine being widely used to the birth and the introduction of a non-linear editing system of 3D animation technology, it brought a technological leap to television titles in artistic creation.

#### 3.4.1 *Application of computer subtitles*

In the late 1980s, with computer graphics technology combined with the television signal processing subtitle editing method, can from the computer up to dozens, even hundreds of fonts in the select program character font, text design, and can collect an appropriate picture, from computer rich art gallery in the pattern and active material, as the ornament design background, provides a broad space for the drama creation.

#### 3.4.2 *The application of 3D animation*

The 3D animation is the combination of computer and high technology and art, it makes use of graphics and the principle of computer aided design into painting, photography and other art categories with professional skill, from the computer to create a three-dimensional sense of space of the virtual objects and scenes. This application realize the creation effect which cannot be achieved by the traditional complex method, and enhances the artistic appeal for the TV title creation.

#### 3.4.3 *The application of a nonlinear editing system*

A nonlinear editing system is a computer operated platform television post production system, is a set of images, sounds, graphics, subtitles, working in an environment, making the system can easily to realize the whole function of the traditional complex system, which brought great convenience for the drama creation. As in conventional picture editing, tricks, superposition and transformation of multi screens multichannel 2D, 3D captions stunt effects, chroma key, high quality audio editing and synthesis, combined with the technology of computer 3D animation and virtual reality technology to finish the film art expression, the titles of the creativity and ideas are most incisively displayed. Real digital media technology plays an indispensable role in the production of TV program.

The total above, design and manufacture of visible TV titles mainly uses the actual shooting elements, 3D elements and plane elements, and through the late synthesis software will these materials and elements for processing and synthesis of special effects, resulting in television titles. Drama production often use software including graphic design software (such as Photoshop, Illustrator etc.), 2/3D animation software (such as Flash, Max, Maya etc.) and post production software (such as After Effect, Digital Fusion, Combustion, Shake), the software is not constant, the general is to choose to use according to the needs of producers and creative design habits. At the same time, the use of digital media technology, television titles shrouded in mystery, with brilliant lustre, it show the unrecognizable change, marks China's TV industry with international standards.

## REFERENCES

Hao Gang. On the TV program titles of personalized [J]. Contemporary television, 2004(3):44–45.

Tang Yu. Development of TV program titles created [J]. Journal of Fujian Radio & TV University, 2007(5):60–61.

Yang Chun-rong, Zhu Bao-bin. program titles packaging [M]. Hanghzou: Zhejiang people's fine arts publishing house, 2010.

Zeng Yu. Thinking about the TV program titles created [J]. Advanced Television Engineering, 2006(1):132–135.

*Frontiers in Computer Education – Wang (ed.)*
*© 2015 Taylor & Francis Group, London, ISBN 978-1-138-02797-8*

# A fuzzy time series forecasting model based on percentages

HongXu Wang & JianChun Guo
*Tourism Management College, Qiongzhou University, Sanya, China*

Hao Feng
*Science and Technology College, Qiongzhou University, Sanya, China*

HaiLong Jin
*Tourism Management College, Qiongzhou University, Sanya, China*

ABSTRACT: Song & Chissom founded a fuzzy time series forecasting theory in 1993. But the existing fuzzy time series forecasting model can only study the simulation and forecast of historical data. In this paper we have presented a forecasting model of fuzzy time series based on percentages. It can predict the data about unknown years.

## 1 INTRODUCTION

Classical time series forecasting theory was founded in the middle of the last century, and is widely applied. 'But the classical time series methods cannot deal with the forecasting problems in wich the values of time series are linguistic terms represented by fuzzy sets [1, 2]'. Song & Chissom had founded a fuzzy time series forecasting theory in 1993, and first studies the forecasting problem of the enrolments in the University of Alabama. It is called 'the enrolment forecasting problem' in that the values of time series are the problem of linguistic terms represented by fuzzy sets. Many scholars have put forward many fuzzy time series forecasting models, and have study 'the enrolment forecasting problem'. But the existing fuzzy time series forecasting model can only study the simulation and forecasting of historical data. Stevenson & Porter [4] and Saxena, Sharma & Easo [2] used the percentage changes of year to year enrolments. Jilani, Burney & Ardil [1] and Stevenson & Porter [4] and Saxena, Sharma & Easo [2] established and applied the concept of the inverse fuzzy numbers. This paper has presented a Fuzzy Time Series Forecasting Model Based on Percentage (FTSFMBP). Because FTSFMBP can predict the data of the unknown years, so the range of possible applications of the FTSFMBP is much wider.

## 2 BUILDING THE FORECASTING FORMULA OF THE FTSFMBP

The enrolments ei in the University of Alabama in 1971–1992 are found in the Table 1, as given by Song and Chissom [3] in 1993. From table 1 we may get the enrolment universe $E = \{e_{1971} = 13055, \ldots, e_{1992} = 18876\}$. Using the formula

$$P_i = \{(e_i - e_{i-1}) \div e_{i-1}\} \times 100\%$$

to compute the percentage changes in year to year enrolments, and obtained the percentage universe $P = \{P_{1972} = 3.89,\ P_{1973} = 2.24, \ldots, D_{1992} = -2.38\}$, and write in the Table 1. In the percentage universe $P$, $P^{\min} = -5.83, P^{\max} = 7.66$. we take

$$T_1 = \frac{1}{3}\max\{|P_{\min}|, |P_{\max}|\} = \frac{7.66}{3} = 2.55$$

as the unit segment, and obtained the segmented universe of percentage as:

$$T = \{T_{-3} = -7.66, T_{-2} = -5.11, T_{-1} = -2.55,$$
$$T_0 = 0, T_1 = 2.55, T_2 = 5.11, T_3 = 7.66\}.$$

On the universes $e, P, T$, to set up the forecasting formula of the FTSFMBP as

$$W_i = \frac{0.0001 + 1}{\dfrac{0.0001}{P_{i-1}} + \dfrac{1}{P_{i-1} + T_j}},$$
$$F_i = e_{i-1} \times (1 + W_i\%).$$

## 3 A SIMULATION AND FORECASTING OF HISTORICAL DATA

We use the enrolments of the University of Alabama in 1971–1992 as a historical data as a forecasting function

Table 1. The enrolments and the percentages in the University of Alabama [3].

| Year | Enrollments $e_i$ | Percentage $P_i$ (%) | year | Enrollments $e_i$ | Percentage $P_i$ (%) |
|---|---|---|---|---|---|
| 1971 | 13055 | – | 1982 | 15433 | P1982 = −5.83 |
| 1972 | 13563 | P1972 = 3.89 | 1983 | 15497 | P1983 = 0.41 |
| 1973 | 13867 | P1973 = 2.24 | 1984 | 15145 | P1984 = −2.27 |
| 1974 | 14696 | P1974 = 5.98 | 1985 | 15163 | P1985 = 0.12 |
| 1975 | 15460 | P1975 = 5.20 | 1986 | 15984 | P1986 = 5.41 |
| 1976 | 15311 | P1976 = −0.96 | 1987 | 16859 | P1987 = 5.47 |
| 1977 | 15603 | P1977 = 1.91 | 1988 | 18150 | P1988 = 7.66 |
| 1978 | 15861 | P1978 = 1.65 | 1989 | 18970 | P1989 = 4.52 |
| 1979 | 16807 | P1979 = 5.96 | 1990 | 19328 | P1990 = 1.89 |
| 1980 | 16919 | P1980 = 0.67 | 1991 | 19337 | P1991 = 0.05 |
| 1981 | 16388 | P1981 = −3.14 | 1992 | 18876 | P1992 = −2.38 |

Table 2. The prediction values of the FTSFMBP.

| Year | Enrollments $e_i$ | $P_i$ (%) | $P_{i-1} + T_j$ | prediction $F_i$ | $E_i - F_i$ | $(E_i - F_i)^2$ | $\lvert E_i - F_i\rvert/E_i$ |
|---|---|---|---|---|---|---|---|
| 1971 | 13055 | | | | | | |
| 1972 | 13563 | 3.89 | | | | | |
| 1973 | 13867 | 2.24 | P1972+T−1 | 13745 | 122 | 14884 | 0.008798 |
| 1974 | 14696 | 5.98 | P1973+T1 | 14531 | 165 | 27225 | 0.011228 |
| 1975 | 15460 | 5.20 | P1974+T0 | 15575 | −115 | 13225 | 0.007439 |
| 1976 | 15311 | −0.96 | P1975+T−2 | 15446 | −135 | 18225 | 0.008817 |
| 1977 | 15603 | 1.91 | P1976+T1 | 15555 | 48 | 2304 | 0.003076 |
| 1978 | 15861 | 1.65 | P1977+T0 | 15901 | −40 | 1600 | 0.002522 |
| 1979 | 16807 | 5.96 | P1978+T2 | 16933 | −126 | 15876 | 0.007497 |
| 1980 | 16919 | 0.67 | P1979+T−2 | 16950 | −31 | 961 | 0.001832 |
| 1981 | 16388 | −3.14 | P1980+T−1 | 16601 | 213 | 45369 | 0.012997 |
| 1982 | 15433 | −5.83 | P1981+T−1 | 15456 | −23 | 529 | 0.001490 |
| 1983 | 15497 | 0.41 | P1982+T2 | 15544 | −47 | 2209 | 0.003033 |
| 1984 | 15145 | −2.27 | P1983+T−1 | 15165 | −20 | 400 | 0.001321 |
| 1985 | 15163 | 0.12 | P1984+T1 | 15187 | −24 | 576 | 0.001583 |
| 1986 | 15984 | 5.41 | P1985+T2 | 15953 | 31 | 961 | 0.001939 |
| 1987 | 16859 | 5.47 | P1986+T0 | 16849 | 10 | 100 | 0.000593 |
| 1988 | 18150 | 7.66 | P1987+T1 | 18211 | −61 | 3721 | 0.003361 |
| 1989 | 18970 | 4.52 | P1988+T−1 | 19077 | −107 | 11449 | 0.005640 |
| 1990 | 19328 | 1.89 | P1989+T−1 | 19344 | −16 | 256 | 0.000828 |
| 1991 | 19337 | 0.05 | P1990+T−1 | 19200 | 137 | 18769 | 0.007085 |
| 1992 | 18876 | −2.38 | P1991+T−1 | 18851 | 25 | 625 | 0.001324 |
| AFER | | | | | | | 0.462015% |
| MSE | | | | | | 8963 | |
| EPER | | | | | | | 1.2997% |

to test the FTSFMBP. When the application of formula (1), can be reached and the number of forecasting values should be seven for each year. We introduce

**Definition1** The forecasting value of the closest to a true data is called prediction value in the forecasting values of a certain year.

We need to fill every year prediction values in Table 2.

## 4 COMPARED WITH EXISTING METHODS

Table 3 and Table 4 are the forecast values of the FTSFMBP and are compared with the existing models. From Table 3 and Table 4, one sees that FTSFMBP provides the lesser AFER and MSE.

## 5 THE FORECASTING FUNCTION OF AN UNKNOWN YEAR 1993

We take parameters $j = -3, -2, -1, 0, 1, 2, 3, 4, 5$. Here the segmented universe of percentage as:

$$T = \{T_{-3} = -7.66, T_{-2} = -5.11,$$
$$T_{-1} = -2.55, T_0 = 0, T_1 = 2.55,$$
$$T_2 = 5.11, T_3 = 7.66, T_4 = 10.21, T_5 = 12.76\}.$$

Table 3. Compared with the existing models [2], [1].

| Year | Enrollments | Jilani, Burney, Ardil [1] | Saxena, Sharma, Easo [2] | Song, Chissom [3] | Stevenson, Porter [4] | Jilani, Burney [5] | Jilani, Burney, Ardil [6] |
|---|---|---|---|---|---|---|---|
| 1971 | 13055 | 13579 | – | – | – | – | 14464 |
| 1972 | 13563 | 13798 | 13486 | 14000 | 13410 | – | 14464 |
| 1973 | 13867 | 13798 | 13896 | 14000 | 13932 | – | 14464 |
| 1974 | 14696 | 14452 | 14698 | 14000 | 14664 | 14730 | 14710 |
| 1975 | 15460 | 15373 | 15454 | 15500 | 15423 | 15615 | 15606 |
| 1976 | 15311 | 15373 | 15595 | 16000 | 15847 | 15614 | 15606 |
| 1977 | 15603 | 15623 | 15600 | 16000 | 15580 | 15611 | 15606 |
| 1978 | 15861 | 15883 | 15844 | 16000 | 15877 | 15611 | 15606 |
| 1979 | 16807 | 17079 | 16811 | 16000 | 16773 | 16484 | 16470 |
| 1980 | 16919 | 17079 | 16916 | 16813 | 16897 | 16476 | 16470 |
| 1981 | 16388 | 16497 | 16425 | 16813 | 16341 | 16469 | 16470 |
| 1982 | 15433 | 15737 | 15657 | 16789 | 15671 | 15609 | 15606 |
| 1983 | 15497 | 15737 | 15480 | 16000 | 15507 | 15614 | 15606 |
| 1984 | 15145 | 15024 | 15214 | 16000 | 15200 | 15612 | 15606 |
| 1985 | 15163 | 15024 | 15184 | 16000 | 15218 | 15609 | 15606 |
| 1986 | 15984 | 15883 | 15995 | 16000 | 16035 | 15606 | 15606 |
| 1987 | 16859 | 17079 | 16861 | 16813 | 16903 | 16477 | 16470 |
| 1988 | 18150 | 17991 | 17965 | 19000 | 17953 | 18482 | 18473 |
| 1989 | 18970 | 18802 | 18964 | 19000 | 18879 | 18481 | 18473 |
| 1990 | 19328 | 18994 | 19329 | 19000 | 19303 | 19158 | 19155 |
| 1991 | 19337 | 18994 | 19378 | 19000 | 19432 | 19155 | 19155 |
| 1992 | 18876 | 18916 | 18984 | – | 18966 | 18475 | 18473 |
| AFER | | 1.02% | 0.34% | 4.38% | 0.57% | 1.40% | 2.38% |
| MSE | | 41426 | 9169 | 775687 | 21575 | 82269 | 227194 |

Table 4. Compared with the existing models [2].

| Year | Enrollments | Hwang, Chen, Lee [7] | Chen [8] | Huarng [9] | Song, Chissom [10] | Chen [11] | FTSFMBP |
|---|---|---|---|---|---|---|---|
| 1971 | 13055 | – | – | – | – | – | |
| 1972 | 13563 | – | – | 14000 | – | 14000 | |
| 1973 | 13867 | – | – | 14000 | – | 14000 | 0.008798 |
| 1974 | 14696 | – | 14500 | 14000 | – | 14000 | 0.011228 |
| 1975 | 15460 | – | 15500 | 15500 | 14700 | 15500 | 0.007439 |
| 1976 | 15311 | 16260 | 15500 | 15500 | 14800 | 16000 | 0.008817 |
| 1977 | 15603 | 15511 | 15500 | 16000 | 15400 | 16000 | 0.003076 |
| 1978 | 15861 | 16003 | 15500 | 16000 | 15500 | 16000 | 0.002522 |
| 1979 | 16807 | 16261 | 16500 | 16000 | 15500 | 16000 | 0.007497 |
| 1980 | 16919 | 17407 | 16500 | 17500 | 16800 | 16833 | 0.001832 |
| 1981 | 16388 | 17119 | 16500 | 16000 | 16200 | 16833 | 0.012997 |
| 1982 | 15433 | 16188 | 15500 | 16000 | 16400 | 16833 | 0.001490 |
| 1983 | 15497 | 14833 | 15500 | 16000 | 16800 | 16000 | 0.003033 |
| 1984 | 15145 | 15497 | 15500 | 15500 | 16400 | 16000 | 0.001321 |
| 1985 | 15163 | 14745 | 15500 | 16000 | 15500 | 16000 | 0.001583 |
| 1986 | 15984 | 15163 | 15500 | 16000 | 15500 | 16000 | 0.001939 |
| 1987 | 16859 | 16384 | 16500 | 16000 | 15500 | 16000 | 0.000593 |
| 1988 | 18150 | 17659 | 18500 | 17500 | 16800 | 16833 | 0.003361 |
| 1989 | 18970 | 19150 | 18500 | 19000 | 19300 | 19000 | 0.005640 |
| 1990 | 19328 | 19770 | 19500 | 19000 | 17800 | 19000 | 0.000828 |
| 1991 | 19337 | 19928 | 19500 | 19500 | 19300 | 19000 | 0.007085 |
| 1992 | 18876 | 15837 | 18500 | 19000 | 19600 | 19000 | 0.001324 |
| AFER | | 2.44% | 1.52% | 1.529% | 3.11% | 3.11% | 0.462015% |
| MSE | | 226611 | 86694 | 86694 | 407507 | 321418 | 8963 |

Table 5. By the FTSFMBP to calculate the forecasting values of 1993 year.

| Year | $P_{1992} + T_j$ | $F^j_{1993}$ | year | $P_{1992} + T_j$ | $F^j_{1993}$ |
|---|---|---|---|---|---|
| 1993-3 | P1992+T-3 | 16981 | 19932 | P 1992+T2 | 19397 |
| 1993-2 | P 1992+T-2 | 17462 | 19933 | P 1992+T3 | 19873 |
| 1993-1 | P1992+T-1 | 18047 | 19934 | P 1992+T4 | 20355 |
| 19930 | P1992+T0 | 18427 | 19935 | P 1992+T5 | 20836 |
| 19931 | P1992+T1 | 18908 | EPER | | 1.2997% |

Application of forecasting formula (1), in the 1993 year can show that the number of forecasting values should be nine, as shown in Table 4.

**Discussion.** In Table 2 we note that the enrolments during the three years before the year 1993 (1990–1992) are between 18876–19337.

When there is nothing unusual happening, we seek the forecasting values of year 1993 between 18876–19337. If there is only one value, it is a forecasting value for 1993. If there are some other values, the mean of these values is the forecasting value of 1993.

Here the forecasting value for 1993 is 18908

When there is good information, we need to seek the forecasting values those greater than 19337. Here we take 19397 or even 19873 as a forecasting value of the 1993.

When there is collected information, the need is to look at the forecasting value of smaller figures than 18876. Here we take 18427 or even 18047 as a forecasting value for 1993.

Table 2 also conveys an important message: in all the forecasting of FTSFMBP values, the maximum forecasting error rate is 1.2997%. The greatest forecasting error rate write into GFER. We hope there are an imagine.

**The greatest forecast error rate.** The forecast error rates of unknown year 1993 is not more than GFER.

Here the forecast error rate of an unknown year1993 is not more than GFER = 1.2997%.

Of course, this is only a kind of irresponsible estimates.

## 6 CONCLUSIONS

The forecasting formula for FTSFMBP is easy to calculate, and simple to structure and it is 'the enrolment forecasting problem for short-term fuzzy time series forecasting models. For the simulation and forecasting of historical data, FTSFMBP provide the lesser AFER and MSE. FTSFMBP can predict the data of an unknown year. The potential for applications of this model could be very extensive.

## REFERENCES


T A Jilani, S M A Burney, C Ardil. Fuzzy metric approach for fuzzy time series forecasting based on frequency density based partitioning. Proceedings of World Academy of Science, Engineering and Technology, Vol. 34, pp. 1–6, 2007

P Saxena, K Sharma, S Easo. Forecasting enrollments based on fuzzy time series with higher forecast accuracy rate. Int. J. Computer Technology & Applications, Vol. 3, No. 3, pp. 957–961, 2012

Q Song, B S Chissom. Forecasting enrollments with fuzzy time series– part 1. Fuzzy Set and Systems, Vol. 54, pp. 1–9, 1993

M Stevenson and J E Porter. Fuzzy time series forecasting using percentage change as the universe of discourse. Proceedings of World Academy of Science, Engineering and Technology, Vol. 55, pp. 154–157, 2009

T A Jilani, S M A Burney. M-factor high order fuzzy time series forecasting for road accident data [M]. In IEEE-IFSA 2007, World Congress, Cancun, Mexico, June 18–21, Forthcoming in Book series Advances in Soft Computing, Springer-Verlag, 2007

T A Jilani, S M A Burney, C Ardil. Multivariate high order fuzzy time series forecasting for car road accidents. International Journal of Computational Intelligence, Vol. 4, No. 1, pp. 15–20, 2007

J R Hwang, S M Chen, C H Lee. Handling forecasting problems using fuzzy time series. Fuzzy Sets and Systems, Vol. 100, pp. 217–228, 1998

S M Chen. Forecasting enrollments based on high-order fuzzy time series. Cybernetics and Systems: An International Journal, Vol. 33, pp. 1–16, 2002

K Huarng. Henuristic models of fuzzy time series for forecasting, Fuzzy Sets and Systems, Vol. 123, pp. 369–386, 2002

Q Song, B S Chissom. Forecasting enrollments with fuzzy time series–part 2. Fuzzy Set and Systems, Vol. 62, pp. 1–8, 1994

S M Chen. Forecasting enrollments based on fuzzy time series. Fuzzy Sets and Systems, Vol. 81, pp. 311–319, 1996

*Frontiers in Computer Education – Wang (ed.)*
*© 2015 Taylor & Francis Group, London, ISBN 978-1-138-02797-8*

# A fuzzy time series forecasting model based on data differences

HongXu Wang & JianChun Guo
*Tourism Management College, Qiongzhou University, Sanya, China*

Hao Feng
*Science and Technology College, Qiongzhou University, Sanya, China*

HaiLong Jin
*Tourism Management College, Qiongzhou University, Sanya, China*

ABSTRACT: In 1993 Song and Chissom first proposed a fuzzy time series forecasting model, and applied to study 'the enrolment forecasting problem.' Over the past twenty years, despite the improved forecasting models that are constantly emerging, none have yet been given a fuzzy time series forecasting model to predict data for an unknown year. This paper presented a fuzzy time series forecasting model based on data differences. In this model there is a forecasting function for data in an unknown year.

## 1 INTRODUCTION

'During last few decades, various approaches have been developed for time series forecasting, but the classical time series methods cannot deal with the forecasting problems in wich the values of time series are linguistic terms represented by fuzzy sets' [1]. To overcome this drawback, Song & Chissom presented the theory of fuzzy time series in 1993 [2]. 'A fuzzy time series model deals with both linguistic & numerical values.' [3] For the sake of convenience, we are calling such questions 'the enrolment forecasting problem.'

The basic feature of the problem is the values of the time series represented by the linguistic terms and fuzzy numerical values. This paper present a fuzzy time series forecasting model based on data difference: simply written as Fuzzy Time Series Forecasting Model Based on Data Difference (FTSFMBDD). There is a simulation forecasting function about the historical data, but the model also has a forecasting function for the data of an unknown year. Facing a prediction problem for the enrolment, the University of Alabama, used to illustrate the application process of FTSFMBDD.

## 2 THE PREPARATION WORK ON FTSFMBDD

Using the enrolments in the University of Alabama to introduce the use of process of the FTSFMBDD, this section is the preparation work on the FTSFMBDD. The enrolments $E_i$ of University of Alabama in 1971 ~1992 are shown in Table 1, as given by Song and Chissom [2].

From table 1 we can establish the enrolment universe $E = \{E_{1971} = 13055, E_{1972} = 13563, \ldots, E_{1992} = 18876\}$. Using the data difference formula $D_j = E_j - E_{j-1}$ to compute the data difference in year to year enrolments, and to obtain each year's Data Difference (*DD)* and to fill in the table 1. $D = \{D_{1972} = 508, D_{1973} = 304, \ldots, D_{1992} = -461\}$ is called the data difference universe and is also filled in Table 1. Because $D^{\min} = -955$, $D^{\max} = 1219$, we take a unit step

$$S_1 = \frac{1}{5}\max\{|D^{\min}|, |D^{\max}|\} = \frac{1291}{5} = 258.2,$$

obtaining the step universe

$$S = \{S_{-5} = -1291, S_{-4} = -1032.8, S_{-3} = -774.6, S_{-2} = -516.4, S_{-1} = -258.2, S_0 = 0.00, S_1 = 258.2, S_2 = 516.4, S_3 = 774.6, S_4 = 1032.8, S_5 = 1291, S_6 = 1549.2, S_7 = 1807.4\}.$$

To add $S_6 = 1549.2$ and $S_7 = 1807.4$ in order to calculate the forecasting value of an unknown year. Modified inverse fuzzy numbers Jilani, Burney and Ardil [12] proposed the idea; we presented the forecasting formula of FTSFMBDD:

$$L_i = \frac{1.005}{\dfrac{0.005}{D_{i-1}} + \dfrac{1}{D_{i-1} + S_j}}, \tag{1}$$

$$F_i = E_{i-1} + L_i.$$

where i = 1973, 1974, ...; j = −5, −4, −3, −2, −1, 0, 1, 2, 3, 4, 5, 6, 7.

Table 1. The original data, data difference [2].

| Year | Enrollments $E_i$ | Data difference $D_i$ | Year | Enrolments $E_i$ | Data difference $D_i$ |
|---|---|---|---|---|---|
| 1971 | 13055 | | 1982 | 15433 | $D_{1982} = -955$ |
| 1972 | 13563 | $D_{1972} = 508$ | 1983 | 15497 | $D_{1983} = 64$ |
| 1973 | 13867 | $D_{1973} = 304$ | 1984 | 15145 | $D_{1984} = -352$ |
| 1974 | 14696 | $D_{1974} = 829$ | 1985 | 15163 | $D_{1985} = 18$ |
| 1975 | 15460 | $D_{1975} = 764$ | 1986 | 15984 | $D_{1986} = 821$ |
| 1976 | 15311 | $D_{1976} = -149$ | 1987 | 16859 | $D_{1987} = 875$ |
| 1977 | 15603 | $D_{1977} = 292$ | 1988 | 18150 | $D_{1988} = 1291$ |
| 1978 | 15861 | $D_{1978} = 258$ | 1989 | 18970 | $D_{1989} = 820$ |
| 1979 | 16807 | $D_{1979} = 946$ | 1990 | 19328 | $D_{1990} = 358$ |
| 1980 | 16919 | $D_{1980} = 112$ | 1991 | 19337 | $D_{1991} = 9$ |
| 1981 | 16388 | $D_{1981} = -531$ | 1992 | 18876 | $D_{1992} = -461$ |

Table 2. Forecasting result of the FTSFMBDD.

| Year | $E_i$ | $DDD_i$ | $D_{i-1} + S_j$ | Forecasting $F_i$ | $E_i - F_i$ | $(E_i - F_i)^2$ | $\|E_i - F_i\|/E_i$ |
|---|---|---|---|---|---|---|---|
| 1971 | 13055 | | | | | | |
| 1972 | 13563 | 508 | | | | | |
| 1973 | 13867 | 304 | $D_{1972} + S_{-1}$ | 13813 | 54 | 2916 | 0.003894 |
| 1974 | 14696 | 829 | $D_{1973} + S_2$ | 14681 | 15 | 225 | 0.001021 |
| 1975 | 15460 | 764 | $D_{1974} + S_0$ | 15525 | −65 | 4225 | 0.004204 |
| 1976 | 15311 | −149 | $D_{1975} + S_{-4}$ | 15189 | 122 | 14884 | 0.007968 |
| 1977 | 15603 | 292 | $D_{1976} + S_1$ | 15685 | −82 | 6724 | 0.005255 |
| 1978 | 15861 | 258 | $D_{1977} + S_0$ | 15895 | −34 | 1156 | 0.002144 |
| 1979 | 16807 | 946 | $D_{1978} + S_3$ | 16878 | −71 | 5041 | 0.004224 |
| 1980 | 16919 | 112 | $D_{1979} + S_{-3}$ | 16839 | 60 | 3600 | 0.003546 |
| 1981 | 16388 | −531 | $D_{1980} + S_{-2}$ | 16505 | −117 | 13689 | 0.007139 |
| 1982 | 15433 | −955 | $D_{1981} + S_{-2}$ | 15349 | 84 | 7056 | 0.005443 |
| 1983 | 15497 | 64 | $D_{1982} + S_4$ | 15511 | −14 | 196 | 0.000903 |
| 1984 | 15145 | −352 | $D_{1983} + S_{-2}$ | 15026 | −119 | 14161 | 0.007857 |
| 1985 | 15163 | 18 | $D_{1984} + S_1$ | 15051 | 112 | 12544 | 0.007386 |
| 1986 | 15984 | 821 | $D_{1985} + S_4$ | 15980 | 4 | 16 | 0.000250 |
| 1987 | 16859 | 875 | $D_{1986} + S_0$ | 16805 | 54 | 2916 | 0.003203 |
| 1988 | 18150 | 1291 | $D_{1987} + S_2$ | 18246 | −96 | 9216 | 0.005289 |
| 1989 | 18970 | 820 | $D_{1988} + S_{-2}$ | 18926 | 44 | 1936 | 0.002319 |
| 1990 | 19328 | 358 | $D_{1989} + S_{-2}$ | 19275 | 53 | 2809 | 0.002742 |
| 1991 | 19337 | 9 | $D_{1990} + S_{-1}$ | 19428 | −91 | 4281 | 0.004706 |
| **1992** | **18876** | **−461** | $\mathbf{D_{1991}+S_{-1}}$ | **19046** | **−170** | **28900** | **0.009006** |
| AFER | | | | | | | 0.442495% |
| MSE | | | | | | 6825 | |
| EPER | | | | | | | 0.9006% |

## 3 FORECASTING ANALYSIS BASED ON HISTORICAL DATA

Using the enrolments of Alabama University during 1971∼1992 as historical data we can make simulation and forecasting analyses. Note that in the forecasting formula (1) in the parameter there are 13 values, so the application of the forecasting formula (1) in each of the years of simulation and forecasting the value should be 13. We give the following definitions:

**Definition1** In the forecasting values of a certain years the forecasting value of the closest historical data is called the prediction value.

According to definition 1 we calculated the prediction values of the enrolments in the University of Alabama during 1974∼1992 as filled in Table 2.

**Definition 2** The maximum prediction error rate of historical simulation forecasting values is called the Error Rate of Experience Prediction (EREP).

From Table 2 according to the definition 2, we obtained the EPER = 0.9006%.

## 4 COMPARISON WITH EXISTING METHODS

The forecast results of the proposed FTSFMBDD for the enrolments in the University of Alabama during 1973∼1992 based on historical data and the existing literature on [2], [3], [4], [7], [8], [10], [11], [13], [14], [15], [16] a comparison of model predictions is shown in Table 3. From Table 3 we see the forecast value AFER of the enrolments obtained by the FTSFMBDD is smaller.

Table 3. Comparison of different forecasting models [3], [16].

| Year | Enrollments | Song, Chissom [2] | Saxena, Sharma, Easo [3] | Hwang, Chen & Lee [4] | Chen [7] | Song, Chissom [8] | Chen [10] |
|---|---|---|---|---|---|---|---|
| 1971 | 13055 | – | – | – | – | – | – |
| 1972 | 13563 | – | 13486 | – | – | 14000 | 14000 |
| 1973 | 13867 | – | 13896 | – | – | 14000 | 14000 |
| 1974 | 14696 | – | 14698 | – | 14500 | 14000 | 14000 |
| 1975 | 15460 | 14700 | 15454 | – | 15500 | 15500 | 15500 |
| 1976 | 15311 | 14800 | 15595 | 16260 | 15500 | 16000 | 16000 |
| 1977 | 15603 | 15400 | 15600 | 15511 | 15500 | 16000 | 16000 |
| 1978 | 15861 | 15500 | 15844 | 16003 | 15500 | 16000 | 16000 |
| 1979 | 16807 | 15500 | 16811 | 16261 | 16500 | 16000 | 16000 |
| 1980 | 16919 | 16800 | 16916 | 17407 | 16500 | 16813 | 16833 |
| 1981 | 16388 | 16200 | 16425 | 17119 | 16500 | 16813 | 16833 |
| 1982 | 15433 | 16400 | 15657 | 16188 | 15500 | 16789 | 16833 |
| 1983 | 15497 | 16800 | 15480 | 14833 | 15500 | 16000 | 16000 |
| 1984 | 15145 | 16400 | 15214 | 15497 | 15500 | 16000 | 16000 |
| 1985 | 15163 | 15500 | 15184 | 14745 | 15500 | 16000 | 16000 |
| 1986 | 15984 | 15500 | 15995 | 15163 | 15500 | 16000 | 16000 |
| 1987 | 16859 | 15500 | 16861 | 16384 | 16500 | 16813 | 16000 |
| 1988 | 18150 | 16800 | 17965 | 17659 | 18500 | 19000 | 16833 |
| 1989 | 18970 | 19300 | 18964 | 19150 | 18500 | 19000 | 19000 |
| 1990 | 19328 | 17800 | 19329 | 19770 | 19500 | 19000 | 19000 |
| 1991 | 19337 | 19300 | 19378 | 19928 | 19500 | 19000 | 19000 |
| 1992 | 18876 | 19600 | 18984 | 15837 | 18500 | – | 19000 |
| AFER | | 3.11% | 0.34% | 2.44% | 1.52% | 4.38% | 3.11% |
| MSE | | 407507 | 9169 | 226611 | 86694 | 775687 | 321418 |

Table 4. Comparison of different forecasting models [3], [15], [16].

| Year | Enrollments | Huarng [11] | Jilani, Burney, Ardil [13] | Jilani, Burney, Ardil [14] | Jilani, Burney, Ardil [15] | Stevenson, Porter [16] | FTSFMBDD |
|---|---|---|---|---|---|---|---|
| 1971 | 13055 | – | 14464 | 13579 | 13579 | – | |
| 1972 | 13563 | 14000 | 14464 | 13798 | 13798 | 13410 | |
| 1973 | 13867 | 14000 | 14464 | 13798 | 13798 | 13932 | 0.003894 |
| 1974 | 14696 | 14000 | 14710 | 14452 | 14452 | 14664 | 0.001021 |
| 1975 | 15460 | 15500 | 15606 | 15373 | 15373 | 15423 | 0.004204 |
| 1976 | 15311 | 15500 | 15606 | 15373 | 15373 | 15847 | 0.007968 |
| 1977 | 15603 | 16000 | 15606 | 15623 | 15623 | 15580 | 0.005255 |
| 1978 | 15861 | 16000 | 15606 | 15883 | 15883 | 15877 | 0.002144 |
| 1979 | 16807 | 16000 | 16470 | 17079 | 17079 | 16773 | 0.004224 |
| 1980 | 16919 | 17500 | 16470 | 17079 | 17079 | 16897 | 0.003546 |
| 1981 | 16388 | 16000 | 16470 | 16497 | 16497 | 16341 | 0.007139 |
| 1982 | 15433 | 16000 | 15606 | 15737 | 15737 | 15671 | 0.005443 |
| 1983 | 15497 | 16000 | 15606 | 15737 | 15737 | 15507 | 0.000903 |
| 1984 | 15145 | 15500 | 15606 | 15024 | 15024 | 15200 | 0.007857 |
| 1985 | 15163 | 16000 | 15606 | 15024 | 15024 | 15218 | 0.007386 |
| 1986 | 15984 | 16000 | 15606 | 15883 | 15883 | 16035 | 0.000250 |
| 1987 | 16859 | 16000 | 16470 | 17079 | 17079 | 16903 | 0.003203 |
| 1988 | 18150 | 17500 | 18473 | 17991 | 17991 | 17953 | 0.005289 |
| 1989 | 18970 | 19000 | 18473 | 18802 | 18802 | 18879 | 0.002319 |
| 1990 | 19328 | 19000 | 19155 | 18994 | 18994 | 19303 | 0.002742 |
| 1991 | 19337 | 19500 | 19155 | 18994 | 18994 | 19432 | 0.004706 |
| 1992 | 18876 | 19000 | 18473 | 18916 | 18916 | 18966 | **0.009006** |
| AFER | | 1.529% | 2.38% | 1.02% | 1.02% | 0.57% | 0.442495% |
| MSE | | 86694 | 227194 | 41426 | 41426 | 21575 | 6825 |

Table 5. The forecasting values of the unknown 1993 year.

| Year | $D_{1992}+S_j$ | Forecasting $F^j_{1993}$ | $D_{1992}+S_j$ | Forecasting $F^j_{1993}$ | Year |
|---|---|---|---|---|---|
| $1993_{-5}$ | $D_{1992}+S_{-5}$ | 17148 | **$1993_2$** | **$D_{1992}+S_2$** | **18932** |
| $1993_{-4}$ | $D_{1992}+S_{-4}$ | 17399 | **$1993_3$** | **$D_{1992}+S_3$** | **19192** |
| $1993_{-3}$ | $D_{1992}+S_{-3}$ | 17651 | $1993_4$ | $D_{1992}+S_4$ | 19454 |
| $1993_{-2}$ | $D_{1992}+S_{-2}$ | 17904 | $1993_5$ | $D_{1992}+S_5$ | 19718 |
| $1993_{-1}$ | $D_{1992}+S_{-1}$ | 18159 | $1993_6$ | $D_{1992}+S_6$ | 19983 |
| $1993_0$ | $D_{1992}+S_0$ | 18415 | $1993_7$ | $D_{1992}+S_7$ | 20249 |
| $1993_1$ | $D_{1992}+S_1$ | 18673 | EPER | | 0.9006% |

## 5 ANALYSIS AND PREDICTION OF THE UNKNOWN 1993 YEAR

Application prediction formula (1), parameters with $j=-5, -4, -3, -2, -1, 0, 1, 2, 3, 4, 5, 6, 7$ by the enrolment of 1992 can be calculated to predict the number values of the 13 enrolments of 1993 as shown in Table 4.

**Analysis.** We observed from Table 2, that the enrolments during 1989–1992 are between 18876–19337.

The data does not change radically, and belongs to the 'stable' period of development.

If other unexpected events occurred and assuming there were none, in the forecast values for 1993 we seek after the forecast values among in 18876 to 19337. In Table 4 there are 2 values 18932 and 19192 that meet the conditions, we use boldface to indicate them. So the prediction value of the 1993 enrolment is 18932 or 19192.

If there is good information, select the forecast value as 19454, or even 19718.

If there is bad information, select the forecast value as 18673, or even 18415.

**Prediction error rate estimation.** The forecast error rate of the enrolments was not greater than the (empirical prediction error rate) EPER.

Application prediction error rate estimation, and the prediction error rate of the prediction value of the 1993 enrolments are not greater than the experience prediction error rate EPER = 0. 9006%. This is estimated by a form and is not binding.

## 6 CONCLUSIONS

FTSFMBDD is a short-term fuzzy time series forecasting model of 'the enrolment prediction problem'

The advantages of this model is that the prediction formula has the advantages of a simple structure and convenient calculations, while the average forecast error rate of the prediction value of the historical data is smaller. The FTSFMBDD not only has a simulation prediction function for the historical data, but it also has a prediction function for the data of an unknown year. The application scope of FTSFMBDD is not only for 'the enrolment forecasting problem'.

## REFERENCES


S M Chen and Hsu. A new method to forecasting enrollments using fuzzy time series. International Journal of Applied Science and Engineering, Vol.2, no. 3, pp. 234–244, 2004

Q Song, B S Chissom. Forecasting enrollments with fuzzy time series–part 2. Fuzzy Set and Systems, Vol. 62, pp. 1–8, 1994

P Saxena, K Sharma, S Easo. Forecasting enrollments based on fuzzy time series with higher forecast accuracy rate. Int. J. Computer Technology& Applications, Vol. 3, No.3, pp. 957–961, 2012

J R Hwang, S M Chen, C H Lee. Handling forecasting problems using fuzzy time series. Fuzzy Sets and Systems, Vol. 100, pp. 217–228, 1998

S Melike, K Y Degtiarev. Forecasting enrollments model based on first- order fuzzy time series. Proceeding of World Academy of Science, Engineering and Technology, Vol. 1, pp. 132–135, 2005

K Huang. Heuristic models of time series for forecasting. Fuzzy Sets and Systems, Vol. 123, pp. 369–386, 2001

S M Chen. Forecasting enrollments based on high-order fuzzy time series. Cybernetics and Systems: An International Journal, Vol. 33, pp. 1–16, 2002

Q Song, B S Chissom. Forecasting enrollments with fuzzy time series – part 1. Fuzzy Set and Systems, Vol. 54, pp. 1–9, 1993

Q Song, B S Chissom. Fuzzy time series and its models. Fuzzy Set and Systems, 1993, Vol. 54, pp. 269–277, 1993

S M Chen. Forecasting enrollments based on fuzzy time series. Fuzzy Sets and Systems, Vol. 81, pp. 311–319, 1996

K Huang. Heuristic models of fuzzy time series for forecasting, Fuzzy Sets and Systems, Vol. 123, pp. 369–386, 2002

T A Jilani, S M A Burney, C Ardil. Fuzzy metric approach for fuzzy time series forecasting based on frequency density based partitioning. Proceedings of World Academy of Science, Engineering and Technology, Vol. 34, pp. 1–6, 2007

T A Jilani, S M A Burney, C Ardil. Multivariate high order fuzzy time series forecasting for car road accidents. International Journal of Computational Intelligence, Vol. 4, No. 1, pp. 15–20, 2007

T A Jilani, S M A Burney, C Ardil. Fuzzy metricapproach for fuzzy time series forecasting based on frequency density based partitioning. Proceedings of World Academy of Science, Engineering and Technology, Vol. 34, pp. 333–338, 2007

T A Jilani, S M A Burney. M-factor high order fuzzy time series forecasting for road accident data [M]. In IEEE-IFSA 2007, World Congress, Cancun, Mexico, June 18–21, Forthcoming in Book series Advances in Soft Computing, Springer-Verlag, 2007

M Stevenson and J E Porter. Fuzzy time series forecasting using percentage change as the universe of discourse. Proceedings of World Academy of Science, Engineering and Technology, Vol. 55, pp. 154–157, 2009

*Frontiers in Computer Education – Wang (ed.)*
*© 2015 Taylor & Francis Group, London, ISBN 978-1-138-02797-8*

# The design and implementation of a campus network automatic certification in a router

Ting Zhao & XiHan Li
*School of Computer Science and Technology, Tianjin University, Tianjin, China*

Lei Zhang
*School of Computer Science and Technology, Tianjin University, Tianjin, China*
*Tianjin Key Laboratory of Cognitive Computing and Application, Tianjin, China*

Gang Zhang
*School of Computer Science and Technology, Tianjin University, Tianjin, China*

ABSTRACT: Against the tedious process of PPPoE campus network access authentication on the web page, the logic flow of the client used in the router to complete the automatic certification, and part of the kernel technology are given to develop the software available for routers using OpenWrt system. Network communication details are obtained and further simulated through Firebug tool, the graphical user configuration interface, based on the LuCi framework, and the bridging technology is used to solve the IPv6 connectivity issues in subnet devices. Automatic account login and certification can be completed by the router, and the concise interface enables users to timely view their account status, modify their settings and monitor traffic. Through compiling and debugging, the software has successfully achieved these functions.

## 1 INTRODUCTION

Campus networks are to provide teaching, research and integrated network information services [12] for teachers and students within a school. The traditional campus network originally accesses the Internet using a fixed IP, however, with the depletion of IPv4 addresses in worldwide, more campus networks launch the way of PPPoE to use IPv6 resources. Point-to-Point Protocol over Ethernet (PPPoE)) is a point to point protocol over Ethernet achieving authentication, encryption and compression functions. Compared with IPv4, IPv6 networks can achieve barrier-free access to Google, Wikipedia and other network services deployed IPv6 [6], which will greatly benefit the research study and work life for university teachers and students.

However, there are some inconveniences in the current way to access campus network.

### 1.1 *Complicated login authentication*

After using the PPPoE dial-up software to certify their own legitimate status, users still need to pass the web verification on the PC side. The restart of router requires users to re-authenticate on the PC side after power failure. In general, the campus laboratories and dormitories will be power off at night for safety, so it is a cumbersome process for users to repeat certification on the PC side every day.

### 1.2 *Subnet devices*

IPv6 uses stateless address allocation mechanism [9], subnet devices accessing to low-level commercial routers cannot go on IPv6 communication with the parent network.

### 1.3 *Subnet devices*

Part of campus networks make charges, based on the use of traffic, providing different monthly traffic subscription plans. If users do not promptly login the self-service system to query the traffic amount, causing the amount having been used exceeding the limit, the user will be required to pay a relatively expensive cost.

Owing to the drawbacks in using the campus network, this paper aims to develop a program which runs in router to replace the external network login authentication on the web page, login self-service system, account cancellation, and a series of manual operations. Meanwhile, the software needs to timely feedback the traffic usage to the user in an intuitive way and allow subnet devices to successfully access IPv6 resources.

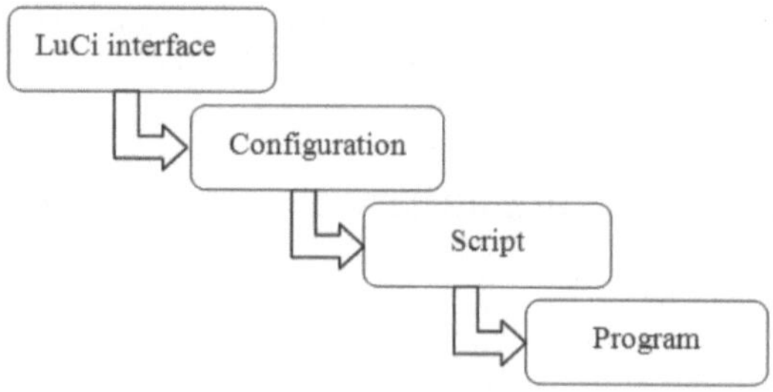


Figure 1. The execution flow of general software in OpenWrt system.

## 2 A DEVELOPMENT ENVIRONMENT

### 2.1 *OpenWrt*

This design uses one router installing the OpenWrt system to do development. OpenWrt is a Linux distribution for embedded devices and it provides a writable file system for adding packages, and so developers can directly develop required applications without building the entire effort required for firmware.

### 2.2 *LuCi*

LuCi [1] is one part of OpenWrt components, providing a free, simple, extensible, and easy maintenance Web user interface for embedded devices.

### 2.3 *OpenWrt Buildroot*

OpenWrt Buildroot [7] does cross-compile for the client software.

### 2.4 *Firebug*

Firebug [5] acts as a packet capture tool for analysing the Web authentication process.

In addition, the TP-LINK WR703N router that we used in this design was modified. To achieve the basic requirements to install and run the client program, the original 4MB flash memory of the router was expanded to 16MB flash memory and 64MB RAM.

## 3 MODULE DESIGN

### 3.1 *Program module in OpenWrt*

The execution flow of general software in OpenWrt system is shown in Figure 1. In accordance with the design objectives and the program invocation in OpenWrt, we divide this software into three parts: the network operating module, logic control module, and user interface.

### 3.2 *Network operating module*

This module is a pre-compiled Shell program, corresponding to the 'Program' function in Figure 1. The Shell program sets up four kinds of invocation, respectively corresponding to login to the self-service system, login to the Internet, maintain the login state and logout operations. The Shell program obtains parameters from corresponding invocation commands, and the formats of these commands are as follows.

- Login the self-service system: Tju_Login 1 Username Password
- Login the Internet: Tju_Login 2 Username Password
- Maintain the login state: Tju_Login 3
- Logout: Tju_Login 4 Username Password

However, the function of this client we design is not just the successive implementation of the above-mentioned four basic communication operations. In order to maintain online state and monitor the use of traffic, the client needs to constantly repeat communication operations with the authentication server. In addition, the client needs to make appropriate judgments taking into account the occurrence of various abnormal conditions, such as the wrong login information, flow depletion, and other issues. Therefore, we need a dedicated logic control process to manage the four basic communication actions in this module.

### 3.3 *Logic control module*

This module is a bash script, defining two sub functions,

Tju_init() and Tju_Login(). Between them, Tju_init() specifically achieve the function of logging in the self-service system, checking the remaining traffic, and disabling the account with excessively low remaining traffic. While the role of Tju_Login() is to take use of the available account checked by Tju_init() function to execute login operation, and then call the 'stay on the line' function of the program to maintain login state. When the remaining traffic of one account is insufficient, it will log off the account and switch to the next available account. The implementation process of Tju_Login() is shown in Figure 2.

When the router starts, by calling the Shell program, this module achieves the function of checking status, maintaining an online state and monitoring traffic for every account it reads.

### 3.4 *User interface*

This design uses a defined LuCi framework to build the user interface, and adopts the Module, View, Controller (MVC) three-layer architecture [10][4]: Module layer defines the configuration variables; it reads the value of variables from the Web and then writes the value into a configuration file, completing the operation of modifying the program configuration on the Web page. The view layer is mainly for defining the exclusive visual effect of the interface. The controller

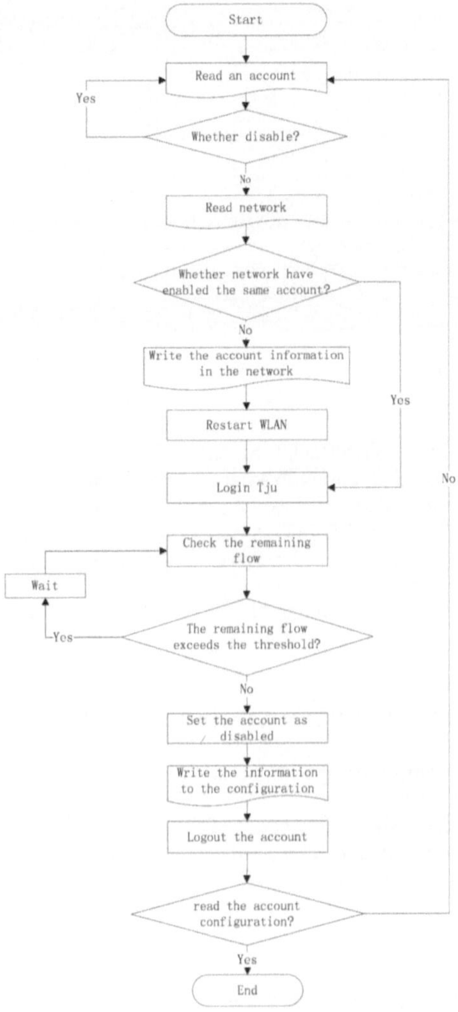


Figure 2. The control flow of Tju_Login().

layer defines the entrance of the module, equivalent to registering a dedicated configuration interface in LuCi. In this design, the file path of the controller layer is /usr/lib/lua/luci/controller/.

Users can set the username and password in the LuCi interface, and set the login and logout threshold of the remaining traffic. Besides, users can also add multiple different account information. When the remaining flow of one account is lower than the threshold, the program will logout this account and enable the next available one to log in.

## 4 PART OF THE CORE TECHNOLOGY

### 4.1 *HTTP communication*

This design uses the library function of LibCurl [8] to carry out HTTP communication with the authentication server [2], similar to the multi-layer structure

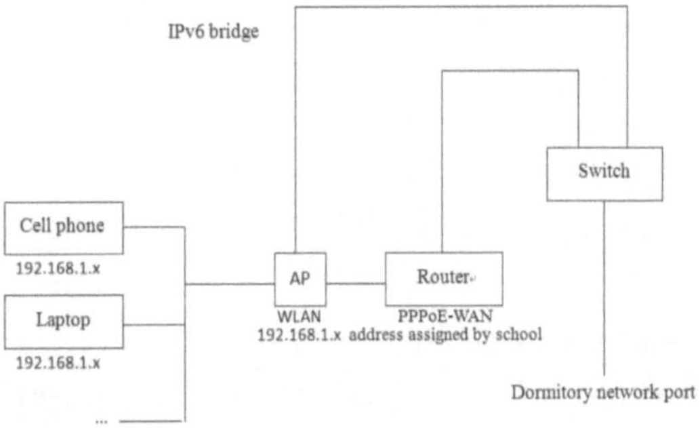


Figure 3. IPv6 bridge schematic diagram.

mode in network communication. LibCurl provides the configuration of the underlying protocol and a calling interface for developers, while shielding the concrete steps. The programs based on LibCurl mainly adopt the call back function to complete their transfer tasks. The formats of call back functions are as follows. Wherein, *contents* points to the received data, while *userp* is specified by the user.

**size_t** WriteBackToString(**void** * contents, **size_t** size, **size_t** nmemb, **char** * userp)

The following codes are used to implement the declaration and initialization of the *crul* handle. Among them, the *curl_easy_setopt* is to configure the *curl* handle.

CURL * curl;

CURLcode res;

curl = curl_easy_init();

curl_easy_setopt(curl, CURLOPT_URL, url); // Set *url*

curl_easy_setopt(curl, CURLOPT_POSTFIELDS, data);//Set *post*

curl_eas _setopt(curl, CURLOPT_COOKIEJAR , cookie);// Receive and save to a specified file *cookie*

curl_easy_setopt (curl, CURLOPT_WRITEFUNCTION, DoNothing);// Set the callback function and call it after the Libcurl receiving the data

curl_easy_setopt(curl, CURLOPT_WRITEDATA, NULL) ;

curl_easy_setopt(curl, CURLOPT_HEADERFUNCTION, WriteBackToString)// Call the function Once receiving the head data of libcurlhttp

### 4.2 *IPv6 bridge*

Currently, the IPv6 portion of the campus network uses the Neighbour Discovery Protocol [11], which is a stateless address configuration, for address assignment. For gateway and subnet devices, they can get available addresses by directly accessing the network using this way of address distribution. However, the router accessing the network can obtain a valid IPv6 address with a prefix length 64, but cannot continue subnet division based on this address, therefore cannot continue to assign IPv6 addresses to those devices with access to the router via LAN or WLAN.

```
config_load network
brctl addbr brbr
ifconfig brbr up
brctl addif brbr eth0
brctl addif brbr eth1
brctl addif brbr wlan0
brctl addif brbr wlan1
ebtables -t broute -A BROUTING -p ! ipv6 -j DROP
```

Figure 4. IPv6 bridge script.

In this case, to allow subnet devices to successfully access the IPv6 network, our design uses IPv6 bridging technology [3], directly bridging the subnet and the parent network through script commands, and masks the IPv4 packet transmission in the bridge. The function of this part depends on the *ebtables* package in OpenWrt, and the schematic diagram and the script of the IPv6 bridging are shown in Figures 3 and 4.

In the above script, eth0 and eth1 corresponds to the WAN and LAN wired network card of the router, while wlan0 and wlan1 to the wireless network card. Tests proved that as long as the WAN port and the LAN port own at least one network interface, the script can be executed normally and the IPv6 bridge will be successfully built. Besides, the last *ebtables* command is used to shield the IPv4 transmit packets in the bridge.

### 4.3 *Child functions*

Functions of some child functions in the original Shell codes are given in this section.

- int Login_Internet (char *id, char *pass): Call the Post function to execute the operation of logging in the Internet.
- int Post (char *url, char *data, char *cookie, char *writeback): A customized Post operation, the callback function has been set as *WriteBackToString*. This function will read the http response header, if the header contains the *set-cookie* field, then the content of the field will be stored in the file indicated by the string *cookie*. This function is the first step towards logging into a self-service system.
- size_t WriteBackProcess (void *contents, size_t size, size_t nmemb, char *user p): The specific callback function when logging in the self-service system. This function finds the remaining flow value from the self-service system, and stores the value in the string indicated by *userp*.
- double Get_Rid_Of_Comma (char *str): This function will format the string containing the value of the remaining flow, and then return the value in the form of a double-precision floating point.
- int Keep_Alive(): This function will call the GET function to maintain login state, and output and display the remaining flow value.

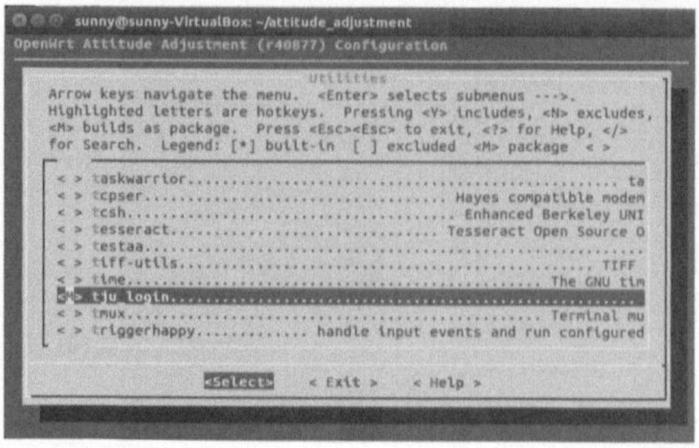


Figure 5. Menu config interface.

- size_t WriteBackToString (void *contents, size_t size, size_t nmemb, char *user p): A call back function called when executing the operation of HTTP POST or GET. This function writes the content returning from the server to the string pointed by *userp*.
- int Log_Out (char *id, char *pass): This function calls the Post function to perform logout operation.

## 5 PREFERENCES, SYMBOLS AND UNITS

All compiling operations are carried out in Ubuntu 14.04 LTS. In the OpenWrt Buildroot compiler environment, the source codes of the client were uploaded to the router, and then got installation and running after cross compiling.

### 5.1 *Compile*

The compilation process will first put the folder containing the source *tju_login* under the folder *package*, then enter the root directory of OpenWrt Buildroot, generating a corresponding configuration file. Related commands are as follows:

make defconfig
make menuconfig

Second, remove excess options and tick the *tju_login* in Utilities, as shown in Figure 5.

At last execute the compilation, and the following is the command: make package/tju_login/compile V= s CC= architecture-openwrt-linux-uclibc-gcc LD= architecture-openwrt-linux-uclibc-ld

### 5.2 *Instruction*

After compilation, the directory of the generated ipk installation package is /bin/ar71xx/packages/. In order to install the ipk package on the router, it should first be uploaded to the router device. The command of using SSH link to upload the ipk package to the *tmp* directory of router is as follows:

scp ./bin/ar71xx/packages/tju_login_1_ar71xx.ipk root@192.168.1.1:/tmp/tju_login_1_ar71xx.ipk

```
Try logging out profile WRONG...
username_error
Profile WRONG has a wrong password.
Try logging out profile DanDi...
logout_error
The flux left for HeShu is too low (2932.68).
Try logging out profile Sunny...
logout_ok
The profile Sunny has 16217.27 Metabytes to use.
All profile checked.
PPPOE profile switched to Sunny
Restarting WAN service...
Logging in Tju...
logged_in
Sunny has logged in.
Checking status of Sunny...
Flux left: 16217.27
Flux left: 16217.27
Flux left: 16217.27
Flux left: 16217.27
Flux left: 16217.27
```

Figure 6. The operating results of every account.

Then execute the *opkg* installation:

ssh root@192.168.1.1 opkg install /tmp/tju_login _1_ar71xx.ipk

### 5.3 *Debug*

Visit http://192.168.1.1, and login the router setting page, then set account information in the 'network/tju_login'. The information of three accounts in the configuration file is shown as follows.

Account 1 WRONG: set as the wrong login information.

Account 2 HeShu: login information is correct, but the remaining flow is lower than the threshold (set as 3000B).

Account 3 Sunny: login information is correct and the remaining traffic is higher than the threshold.

If submitting the debugger request in the user interface, we cannot observe the specific process of program execution. Therefore, we start the program directly by running one script. After writing the configuration file, we directly execute the following command to debug:

./etc/init.d/Tju_Login restart ≫ ./tmp/a.txt

Look at the operating results of every account in a.txt, and the corresponding results are shown in Figure 6.

See from the results, the program correctly filtered out the account "WRONG" with false login information and the account 'HeShu' which traffic was insufficient, while the account 'Sunny' showed normal access to external network. Besides it also updated the PPPoE login information and then performed the login operation.

### 5.4 *Detect boot from start*

In this section, we set the boot of the program with the starting up of the router to test its start situation. First, we need to add the program to the start up items of the system:

./etc/init.d/Tju_Login enable

Then execute the restart operation of the router:

reboot

After the restart is completed, we need:

- Detect the access to the Internet, visit http://cnbeta.com, and the result is normal.
- Detect the IPv6 access, visit http://ipv6.google.com, and the result shows normal.

## 6 CONCLUSION

This paper introduces the work done for the design and implementation of a campus network automatic authentication software running on the router side, mainly used to solve the drawbacks existing in the actual usage of a campus network. Through compiling, installation and testing, the software successfully implements the convenient IPv6 access to a campus network.

Innovation of this article:

- Change the originally cumbersome certification process on Web page into automatically authenticatation in the router for the campus network, avoiding the mannual operation on the PC side, which brings convenience to teachers and students in collage.
- Traffic monitoring. The software will automatically disconnect with the Internet when the remaining traffic value falls below a preset threshold, thus avoiding high network costs.
- Provide a LuCi interface based on the MVC three-tier architecture. Users can set account-related information, and timely view the usage state of the traffic in this interface.
- Provide barrier-free IPv6 accessment for subnet users by bridging technology, enabling them to smoothly access a wealth of educational resources at home and abroad.

## ACKNOWLEDGMENT

This work was supported in part by NSF China under grants No.61202379 and Doctoral Fund of Ministry of Education of China (No.20120032120041).

## REFERENCES

Arezki, R., A. Mokrane, et al. 2004. LUCI: A personalization documentary system based on the analysis of the history of the user's actions. 6th International Conference. 487–498.

Bing Zhou, Jingyuan Wang, Zixuan Zhou et al. 2012. Bandwidth estimation and rate adaptation in HTTP streaming.

2012 International Conference on Computing, Networking and Communications. 734–738.
Chung-Sheng Li, Yueh-Min Huang, Han-Chieh Chao. 2008. UPnP IPv4/IPv6 bridge for home networking environment. IEEE Transactions on Consumer Electronics. vol. 54: 1651–1655.
Geng Qiang. 2012. The Design and Building of Network Teaching Integrated Platform based on Model-View-Controller Framework and Knowledge Construction. Journal of Convergence Information Technology. vol. 7: 376–384. (in Chinese)
Hewitt, M. 2007. AJAX debugging with Firebug. Dr. Dobb's Journal. vol. 393: 22–26.
Li Dao-wang, Can-Wei Wang. 2011. Research Summary on Transition Strategy from IPv4 to Ipv6. Information Technology & Information. vol. 04: 62–67, 2011. (in Chinese)
Palazzi, C. E., M. Brunati, et al. 2010. An OpenWRT solution for future wireless homes. 2010 IEEE International Conference on Multimedia and Expo: 1701–1706.
Xiao, Hongyu Cui, Xiaojun Xu, et al. 2012. CNKI-based requirement-oriented system for searching paper. International Journal of Digital Content Technology and its Applications. vol. 6: 185–192.
Xiaoyu Song, Zeyang Qiu. 2005. Analysis of IPv6 Stateless Address Auto-configuration. Modern Electronics Technique. vol. 06: 81–82. (in Chinese)
Yinghui Wang, Yingjie-Wang, Yanjin Wang, HongbinWang. 2004. Study on Architecture of Software Interface Based on MVC. Application research of computers. vol. 21: 188–193, 2004. (in Chinese)
Zhi-gang Yang, Chang-he Zhang, Yue-fei Zhu. 2006. Research of IPv6 neighbor discovery protocol security mechanism. Computer Applications. vol. 26:38–941. (in Chinese)
Zhi-yun Xiong. 2001. Campus Net and University Information Resource Management. Journal of Hubei University. vol. 28:114–117. (in Chinese)

*Frontiers in Computer Education – Wang (ed.)*
*© 2015 Taylor & Francis Group, London, ISBN 978-1-138-02797-8*

# Research into a visual tracking approach based on a complicated environment

ZhenXia Liu, Yi Wang & LiWei Yang
*Information and Navigation College, Air Force Engineering University, Xi'an, PR China*

WeiHong Tong & Kai Shi
*Military Representative Office at Certain Company of PLA China, ChengDu, PR China*

ABSTRACT: Based on Dezert-Smarandache Theory (DSmT), the targets tracking platform which embedded location and colour cues into the PF was developed, and the basic framework and algorithm of fusing multi-source information were described. Results showed that the conflict strategy and DSmT combination model were available, and the proposed approach exhibited a significantly better performance for dealing with high conflict between evidences than a conventional PF. As a result, the approach was suitable for real-time visual target tracking, and it can easily be generalized to deal with larger number of visual targets and additional cues based on complicated environment.

## 1 INTRODUCTION

Visual targets' tracking is one of the research hotspots in the field of computer vision. In the last fifteen years we have witnessed a rapid development in the theory of Particle Filters (PF), and the corresponding algorithms of PF are widely applied in tracking fields [1,2]. Although lots of algorithms have been introduced to track moving targets in different cases, these approaches have mainly improved local performances by optimizing PF algorithms, and there still exist many key issues which need to be discussed further. Recently,

Dezert-Smarandache Theory (DSmT) by Dezert and Smarandache has to be viewed as a general flexible bottom-up approach for managing uncertainty and conflicts for a wide class of static or dynamic fusion problems where the information to combine is modeled as a finite set of belief functions provided by different independent sources of evidence [3]. The corresponding researches showed the conflicting focal elements were increased, as based on DSmT, and the computational effort was also increased in the course of reasoning. Based on this matter of fact, improved methods were presented to reduce the computational efforts. At present, some researches focused on visual target identification based on DSmT, while the related studies of visual target tracking based on DSmT are much fewer. The aim of this article is to present a novel approach to visual target tracking that will handle the targets of crosses and occlusions in order to attain an excellent information fusion [4–6].

## 2 TARGETS TRACKING

### 2.1 *Combining rule*

According to the above discussion, an improved combination rule of conflict redistribution is introduced by improving the disposal way for the conflict's focal elements. The basic idea of conflict redistribution rule based on DSmT is described as follows: (1) the conflict information is only produced by the conflicts between certain focal element information t (2) when the conflict information and uncertain information or conflict information between them are combined, the uncertainty of the combining rule is increased greatly. (3) when the conflict information and single focal element of producing the conflict are combined, the conflict information is redistributed to the single focal element according to the combination results.

In order to describe the total level of conflicts between evidences efficiently, an influence coefficient $R$ is introduced to address the conflict level. Based on the mass function, the conflict level can be corrected by $R$, and the corresponding expression of $R$ is given by [7, 8].

$$R = \left( \sum_{A \subseteq U} m_1(A) m_2(A) + \sum_{B_i \cap C_j = \varnothing} m_1(B_i) m_2(C_j) \right)^{-1} \left( \sum_{B_i \cap C_j = \varnothing} m_1(B_i) m_2(C_j) \right) \quad (1)$$

where A denotes the single focal element of identification framework, $m_1$, $m_2$ denotes the basic probability assignment function of two evidence sources, respectively.

According to Eq. (1), when the influence coefficient $R$ tends to 1, the conflict between evidences is strenuous, and the conflicts are partly transferred to a union of conflicting focal elements. When the influence coefficient $R$ tends to 0, the conflicts between evidences are very weak, and they are partly transferred to the conflicting focal elements. As a result, the corresponding combining rule of conflict redistribution between evidences is defined as follows.

$$m(A) = \sum_{\substack{B_i \cap C_j = A \\ B_i, C_j \subseteq \Theta}} m_1(B_i) m_2(C_j) + \quad (1-R)\sum \frac{r_1 m_1(A)}{r_1 m_1(A) + r_2 m_2(X)} m_1(A) m_2(X) \tag{2}$$

$$r_i = \frac{S_i}{\sum_{j=1}^{n} S_j} \max\left\{ \frac{\sum_{j=1}^{n} S_j}{S_j} \right\} \tag{3}$$

where r denotes the absolute reliability of evidence, and S denotes the degree of support based on the D-S evidence theory.

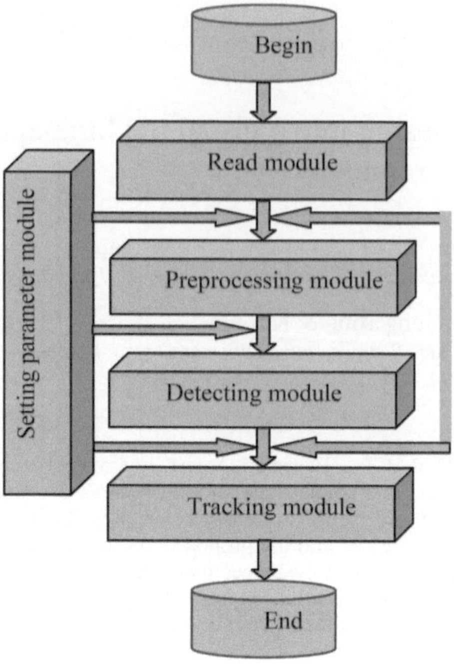

Figure 1. The basic framework of tracking process.

## 2.2 *Tracking model*

According to the above discussion, the following established objects tracking model of objects tracking by using colour and position cues, based on conflict redistribution DSmT. In order to describe conveniently, the tracking model of two objects was established as an example.

For two objects, $\Theta$ was defined as follows:

$$\Theta = \{\theta_1, \theta_2, \overline{\theta_1 \mathrm{Y}\, \theta_2}\} \tag{4}$$

Let define $\{q_{FA}(u)\}_{u=1}^{m}$ as the histogram of the scene from which subtract the histogram of objects 1 and 2.

$$q_{FA}(u) = \max\{q_{scene}(u) - q_1(u) - q_2(u), 0\} \tag{5}$$

The probability that $s_{t,j}^{(n)}$ belongs to the false alarm hypothesis was given by

$$p_{t,FA}^{(n)} = \frac{1}{\sqrt{2\pi}\,\sigma} e^{-\frac{\left(\sqrt{1-\sum_{u=1}^{m} h_{t,j}^{(n)}(u) q_{FA}(u)}\right)^2}{2\sigma^2}} \tag{6}$$

The mass functions of particle $n$ according to color can be evaluated as follows:

$$m_{t,2}^{(n)}(\overline{\theta_1 \mathrm{Y}\, \theta_2}) = \frac{p_{t,FA}^{(n)}}{p_{t,1}^{(n)} + p_{t,2}^{(n)} + p_{t,FA}^{(n)}},$$

$$m_{t,2}^{(n)}(\theta_j) = \frac{p_{t,j}^{(n)}}{p_{t,1}^{(n)} + p_{t,2}^{(n)} + p_{t,FA}^{(n)}}, j = 1,2 \tag{7}$$

The weight of $s_{t,j}^{(n)}$ particle within the posterior $p(X_t|Z_t)$ distribution is calculated using belief (or the plausibility) function for target $j$, is given by

$$\pi_{t,j}^{(n)} = Bel_t^{(n)}(\theta_j) = m_t^{(n)}(\theta_j) + m_t^{(n)}(\theta_1 \mathrm{I}\, \theta_2) \qquad j = 1,2 \tag{8}$$

Based on the above discussion, the generalization of the tracking scheme described in this section to $\tau$ targets can be carried out by defining a frame of discernment $\Theta = \{\theta_1, \ldots, \theta_\tau, \theta_1 \cup \cdots \cup \theta_\tau\}$, where $\theta_j$ are individual targets and $\overline{\theta_1 \cup \cdots \cup \theta_\tau}$ is the false alarm hypothesis. Hence, the same method can be adopted to establish a dynamic combination model of $\tau$ targets in the framework of the DSmT.

## 2.3 *Tracking realization*

Based on the above analysis of combination strategy and a dynamic model for targets tracking in targets crosses and occlusions, the corresponding algorithm of multi-source information fusion was realized in the framework of the DSmT. The algorithm by merging location and colour cues was embedded in the tracking frame of PF, and the whole tracking process included a read module of video sequences, a setting parameter module, a preprocessing module of video images, a detecting module of moving targets, and a tracking module of moving targets. Figure 1 shows the basic framework of the tracking process.

Figure 2. The tracking process of main frames and tracking result.

## 3 TRACKING EXAMPLES

A tracking example under occluding conditions was tested to validate the correctness of the tracking method by using improved combination rules based on the DSmT. The image scene came from a small region, and there were only two cross and closely spaced objects in the region. For the sake of clarity, the girl was denoted as object 1, and the boy was denoted as object 2. In order to analyse them conveniently, the two persons in the video scene kept uniform motion along the moving direction of cross and occlusion. The tracking sequence was divided into three phases: phase 1 was the pre-occlusion sequence, phase 2 corresponded to the occlusion sequence, and phase 3 was the post-occlusion sequence. Tracking in phase 2 was challenged due to the closeness of the object, which perturbs the measured cues and might have led to a false identification. In the process of objects' occlusion locally or wholly, there were high conflict problems including scale variations, cross and closely spaced objects. As a result, the tracking experiment was achieved based on the improved combination rule, and each object was tracked using 25 particles averagely. Figure 2 shows the tracking process of main frames and tracking result.

It could be seen from frame 145 to frame 175 that the tracking accuracy was maintained even when the target was blocked locally or wholly. Especially, the position cue lost gradually its ability to separate objects 1 and 2 as they converged to the intersection point in frame 145. However, the position cue remained a valid measurement because it was independent from the relative position of objects with respect to the camera (occluding or occluded). The colour cue was extremely sensitive to the occlusion.

Based on the above experiment's results, the corresponding error analysis of tracking objects in the direction of level and vertical was discussed. The value of object tracking minus the value of a real object was defined as the mean error. Figure 3 shows the tracking error for target 1 by comparing with the conventional PF approach.

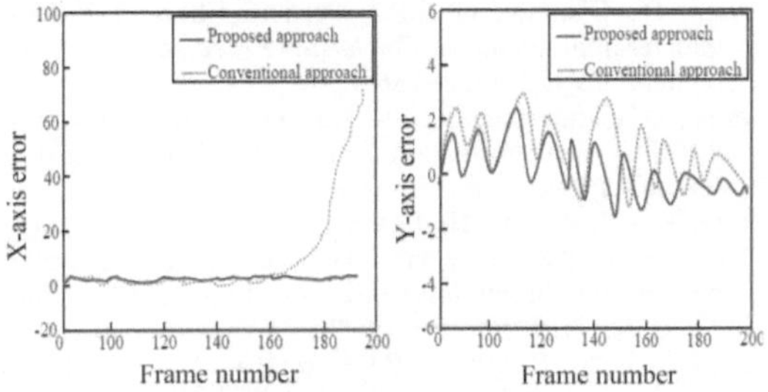


Figure 3. The mean error curves of estimated position for target 1.

It was very obvious that the introduced method had a small tracking error when the object was occluded locally or wholly, but the error change was very visible from frame 145 to frame 200 by using the conventional PF. The tracking mean error was increased greatly when high conflict evidences were handled by using the conventional approach. Especially, the tracking almost failed from frame 145 to frame 175, owing to the occluded objects being fully based on the conventional PF. Hence, the correctness of the introduced method was unsuspected according to the tracking results and error analysis.

## 4 CONCLUSIONS

Based on DSmT, this article mainly addressed visual target tracking in a complicated environment, and the introduced approach has been tested and evaluated. Experimental results have been demonstrated that this approach ameliorated the interference immunity for tracking targets. DSmT is a useful theory for dealing with uncertainty problems. Study demonstrated that DSmT fusing high conflict among evidences had lots of difficulties for analysing a conflict's focal element. An improved combination rule of conflict redistribution based on DSmT was introduced to settle high conflict among evidences. Simulation results shown that the basic belief assignment of main focal element A was increased efficiently by the suggested combination rule, and the value of focal element A could be converged stably by a fusing of 10 times.

## ACKNOWLEDGMENTS

The authors are grateful for the support provided by the Natural Science Foundation of Shaanxi Province, China (grant no. 2012JM8004), the Science and Technology Plan Project of Shaanxi Province, China (grant no. 2013K07-17).

## REFERENCES

Brun, O., Teulieve, V and Garcia, J.M. 2002. Parallel particle filtering. *Journal of Parallel and Distributed Computing*, 62: 1186–1202.

Crisan, D., Obanubi, O. 2012. Particle filters with random resampling times. *Stochastic Processes and their Applications* 122: 1332–1368.

Dezert, J. Smarandache F. 2004. Advances and applications of DSmT for information fusion. Rehoboth American Research Press.

Djuric, P.M., Kotecha, J.H., Zhang, J. 2003. Particle filtering: a review of the theory and how it can be used for solving problems in wireless communications. *IEEE Signal Processing Magazine*, 20: 19–38.

Han, Z.J., Jiao, J.B., Zhang, B.C., Ye, Q.X., Liu, J.Z. 2011. Visual object tracking via sample-based adaptive sparse representation. Pattern Recognition, 44: 2170–2183.

Kazuhiro. H. 2009. Adaptive weighting of local classifiers by particle filters for robust tracking. *Pattern Recognition*, 42: 619–628.

Tehamova, A., Dezert, J., Semerdjiev T. 2005. Multitarget tracking applications of Dezert-Smarandache theory. NATO Advanced Study Institute, Albena, Bulgaria, 66–69.

Yang, H.C., Chen, K., Yang, J.A. 2010. Fusion of communication interception information based on DSmT. *Computer Engineering and Applications*, 46: 129–132.

*Frontiers in Computer Education – Wang (ed.)*
*© 2015 Taylor & Francis Group, London, ISBN 978-1-138-02797-8*

# A friendship network based on random and triadic-closure in a fixed community

D.S. Zhao*, K. Zhao*, J. Yang & J.R. Sha
*College of Modern Educational Technology, Northwest University for Nationalities, Lanzhou, Gansu, China*

ABSTRACT: In this paper, we proposed a friendship network model in a fixed community based on such connection mechanisms as: random selection and triadic-closure. That means each node in the network can strike up friendships with strangers randomly or its two-hop neighbours (friends of its friends). We simulated the network structure with NetworkX and found it exhibits important network topological properties: short diameters, high clustering coefficients, and various shapes of in-degree distributions. We found that clustering coefficients and the shapes of in-degree distributions are both related to the ratio between triadic-closure probability and random probability. Accordingly, we gave a simple method to distinguish the effect of triadic-closure on the structure of friendship networks in the real world from a random connection.

## 1 BACKGROUND

In recent years, there has been considerable interest in complex networks arising in social life, largely because their contribution to human understanding of various social structures and processes, such as how people meet and make friends from strangers (Van et al. 2003), search information and obtain job opportunities (Granovetter 1973). The studies of social networks mainly focus on two related tracks: the static properties of network topological structures observed from the real world data, e.g., movie actors networks (Watts & Strogatz 1998), scientific collaboration networks (Newman 2001; Barabâsi et al. 2002), online friendship networks (Backstrom et al. 2012), the dynamical evolving processes measured from some network generation models, e.g., Erdös-Rényi model (ERDdS 1959), Barabási-Albert model (Albert et al. 1999; Bianconi & Barabási 2001), copying model (Kleinberg et al. 1999), forest fire model (Leskovec et al. 2007).

Social networks are composed of a lot of people or groups who are linked together according to a variety of relationships. The variety of connection mechanisms lead to complex network structures. In the previous studies, we have learned that social networks are very different from random and regular networks. They always have properties of small shortest paths (Backstrom et al. 2012), high clustering coefficients, scale-free degree distributions, and so on. Some of these features are also seen in the world wide web (Albert et al. 1999; Adamic & Huberman 2000; Broder et al. 2000), internet (Yook et al. 2002), and biological networks (Wagner & Fell 2001). How do these features emerge in social networks? There may be combinations of one or more mechanisms that are connecting people together. For example, homophily – the tendency of individuals to associate with those who are similar to themselves (Tarbush & Teytelboym 2013), mutual interests (Singer et al. 2009), the maximum-likelihood principle (Leskovec et al. 2008), the closure of short network cycles (Kossinets & Watts 2006), and drive people to interact with each other and social networks evolve over times. To observe these phenomena, a lot of network models have been proposed to reproduce statistical properties in the real world.

The most celebrated model is the Barabási-Albert (BA) model (Albert et al. 1999), in which when a new node joins the network, it creates a constant number of edges, where the destination node of each edge is chosen proportional to the destination's degree. The BA model reproduces the scale-free network with power law degree distribution, and reveals such a phenomenon 'richer get richer' in social networks. There are also some alternative models based on the preferential attachment mechanisms (Bianconi & Barabási 2001; Albert & Barabási 2000). David in his studies showed that 'winners don't take all', and gave an alternative model which combined preferential attachment and random connection (Pennock et al. 2002). Copying model (Kleinberg et al. 1999) and forest fire model (Leskovec et al. 2007) were based on attaching new nodes to the network by copying or burning through existing edges. These models reproduced densification and shrinking diameters in social networks. Jure, Lars, Ravi and Andrew (Leskovec et al. 2008) attributed the macroscopic network properties to the maximum-likelihood principle and proposed a model based on triadic-closure mechanisms. Jackson and Rogers

*These authors contributed equally to this work, and should be regarded as co-first authors.

(Jackson & Rogers 2007) additionally suggested that 'friends of a friend' are more likely to link in friendship networks. Based on these theories, Davidsen, Ebel and Bornholdt (Davidsen et al. 2002) modelled acquaintance networks based on local interactions. Vazquez (Vazquez 2003), Yuta, Ono, and Fujiwara (Yuta et al. 2007) respectively proposed CNN (connecting nearest neighbours) model and CNNR (connecting nearest neighbours with apparently random linkage) model both based on the basic assumption: the evolution of social connections is mainly determined by the creation of new relations between pairs of individuals.

The models mentioned above are all growth networks. In the whole process, new nodes join the networks continuously and link edges to existing nodes. Some models (Leskovec et al. 2008) also defined a lifetime for each node upon arrival. Though these models have properly defined individual node behaviours and reproduced some statistical properties such as power law node degree distributions and small diameters, there are still some social networks such as who-trusts-whom (Richardson 2002) which has a constant number of nodes. Singer H M, Singer I, and Herrmann (Singer et al. 2009) found that the mutual interest is the dominant factor for the friendships. According to this, they proposed a model to understand the structure of social networks in a fixed setting. Marta, Pedro and Hans (González et al. 2006) also proposed a model of mobile agents to construct social networks, based on a system of moving particles by keeping track of the collisions during their permanence in the system. In these two models, the numbers of nodes were constant. Deepayan, Yiping and Christos (Chakrabarti et al. 2004) gave a 'recursive matrix' (R-MAT) model which can generate weighted, directed, and bipartite graphs with a constant of nodes.

## 2 INTRODUCTION

In this paper, we try to propose a new model to construct a directed friendship network in a fixed community such as in a class of a university. This is an important scene in social network areas. The connection mechanism leading to individuals making friends with each other in the model combines random and triadic-closure rules. Triadic-closure is always called 'friends of a friend' in friendship network which also appears in previous papers (Jackson & Rogers 2007; Vazquez 2003; Yuta et al. 2007). It is really an important element driving the evolution of social networks.

There are three main ingredients in our model. First, the network is evolving in a fixed community. In the whole process, the number of nodes is a constant. No node is added in or moved out. And we do not model a lifetime to keep nodes from producing new edges. Second, the connection mechanisms in the network are based on both random and triadic-closure rules. That means each individual in the friendship network can strike up friendships with strangers who

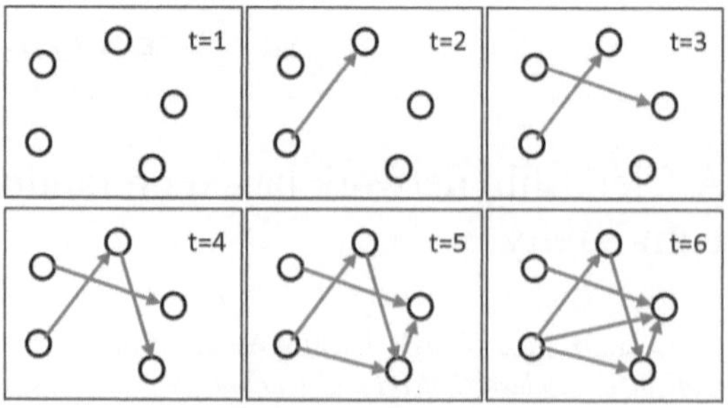


Figure 1. The friendship network evolving process: blue edges are linked randomly, red edges are linked according to triadic-closure rule.

are selected randomly, or selected from friends of his friends. Third, each edge in the network is directional. That means the friendships between each pair of individuals are not symmetrical. Figure 1 shows the connection mechanisms in the friendship network.

Our model is simple and easy to implement. It contains all the structure features listed above. In section 3, we show the simulations of the friendship network developing process based on our model and reproduce various properties such as diameters, clustering coefficients, in-degree distributions for different values of random probability and triadic-closure probability. In section 4, we give a brief summary of our works.

## 3 SIMULATION

Social networks always are analysed as graphs which are combined with a set of nodes and a set of edges. We define N nodes in the friendship network with zero initial edges. During each time interval, we select one node (i as identity) randomly who can interact and make friends with others. Node i chooses the destination node and produces a friendship link, based on a two basis mechanism: random selection and triadic-closure. If node i chooses the destination node randomly from the whole network, the probability of node i producing a new edge is p. If node i chooses the destination node from its two-hop neighbours (friends of its friends), the probability of node i producing a new edge is q. During the whole network developing process, these two mechanisms are effective simultaneously. Figure 1 also shows the evolving mechanisms in our model.

In each simulation, the number of nodes N is assigned a constant number 1000. The random probability p and triadic-closure probability q are assigned different values between 0 and 1. To smooth out measurement deviations of the friendship network properties, we simulated the network evolving process by a hundred times and averaged the measured values.

### 3.1 *The diameter*

Figure 2 shows the diameters of the friendship networks when average degrees rise to 10 and 50 (the

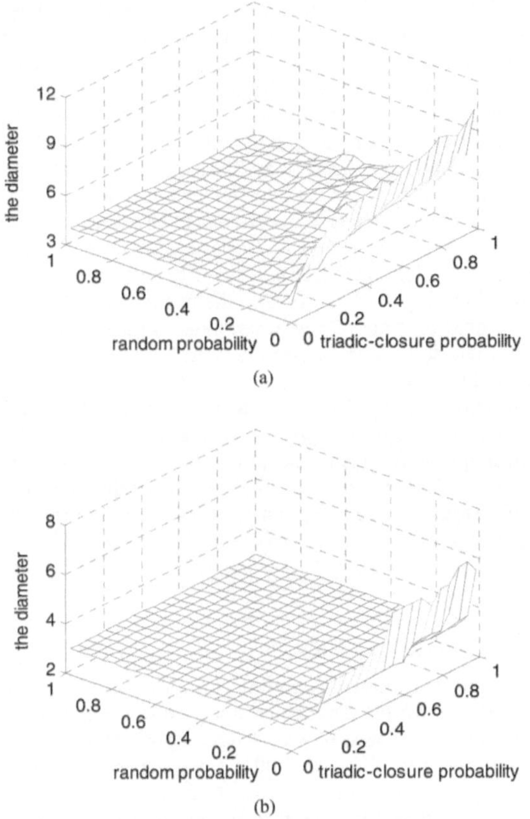


Figure 2. The diameters of the friendship networks for several different values of random probability and triadic-closure probability. The constant number of nodes N = 1000, (a) and (b) are measured while the average degree is 10 and 50.

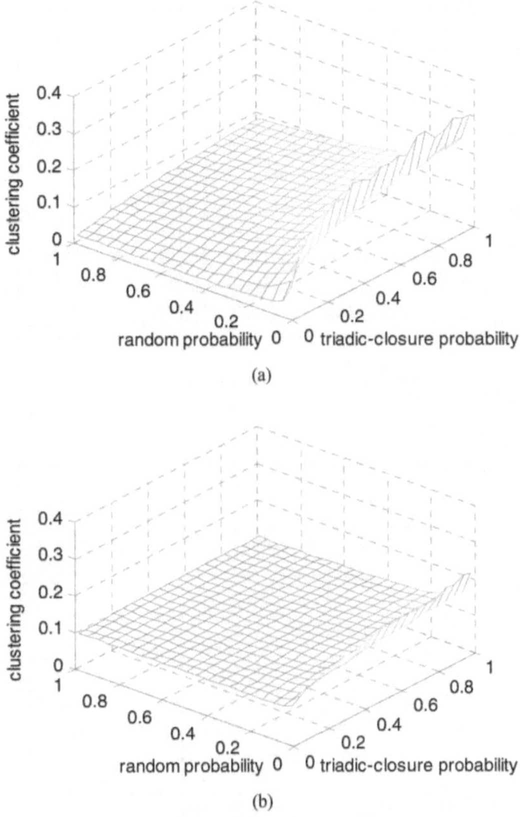


Figure 3. The clustering coefficients of the friendship networks for several different values of random probability and triadic-closure probability. The constant number of nodes N = 1000, (a) and (b) are measured while the average degree is 10 and 50.

total numbers of edges are 10000 and 50000). In Figure 2(a), when p is smaller than q, the diameters are a little larger. When p is larger than q, the diameters are stable and remain at a lower level. The same phenomenon also appears in Figure 2(b). The small diameters reveal the statistical phenomena of a small-world in social networks (Milgram 1967). In Figure 2(a), while random probability is larger than triadic-closure probability, there are at least 5747 edges which are produced randomly. But when random probability $p = 0.01$ and triadic-closure probability $q = 0.99$, the diameter is 10, there are only 1253 edges produced randomly. In Figure 2(b), the diameters are almost at the same level. Even though when $p = 0.01$, $q = 0.99$, there are 1895 edges produced randomly, and the diameter is only 6. The small diameters in Figure 2 show that random connection is an important element leading to a short distance between individuals in friendship networks. The results also match some studies in WS and NW small-world models (Watts & Strogatz 1998; Newman & Watts 1999). Both these two models generated small-world networks through reconnecting edges or adding new edges randomly in regular networks.

### 3.2 *Clustering coefficient*

Figure 3 shows the clustering coefficients in the friendship networks based on our model. The two figures both consist of two parts: $p > q$ and $p < q$. While $p > q$, the clustering coefficients are lower and the slope of the curved surface is gentler. As q rises, the clustering coefficients rise too. Especially while $p < q$, the clustering coefficients are much higher and the slope of the curved surface is steep. In Figure 3(a), the clustering coefficient is 0.0205 at coordinate ($p = 0.99$, $q = 0.01$) and 0.3052 at coordinate ($p = 0.01$, $q = 0.99$). In Figure 3(b), the clustering coefficient is 0.0980 at coordinate ($p = 0.99$, $q = 0.01$) and 0.2175 at coordinate ($p = 0.01$, $q = 0.99$). All the clustering coefficients are much higher than a random graph with the same nodes and edges. The results reveal that triadic-closure rule (connection based on friends of a friend) in the friendship network is an important element leading to high clustering coefficients.

In Figures 2 and 3, we can find that, at the lowest point of the curved surfaces, the friendship networks nearly approach to random networks which have properties of short diameters. At the top of the curved

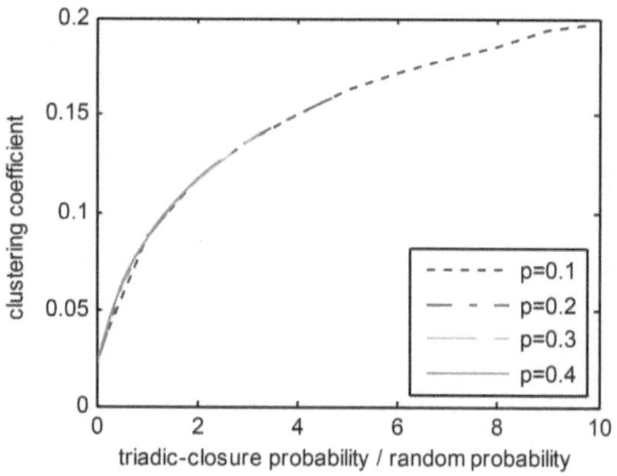


(a)

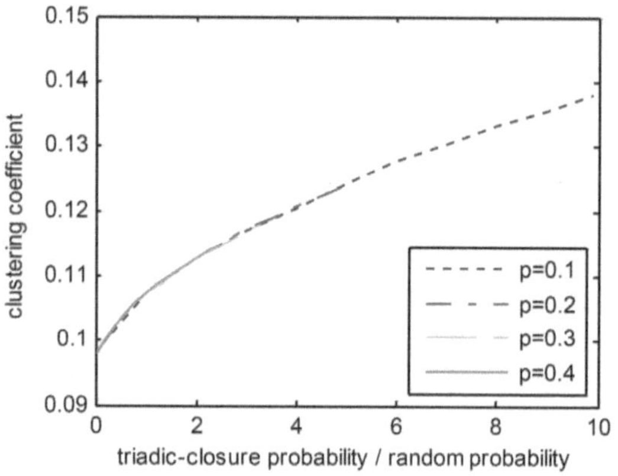


(b)

Figure 4. The clustering coefficients of the friendship networks for different values of q/p. The constant number of nodes N = 1000. (a) and (b) are measured while the average degree is 10 and 50.

surface, the friendship networks nearly approach to networks completely produced according to triadic-closure rules, which have properties of high clustering coefficients. The effect of triadic-closure probability q on clustering coefficients is closely related to the value of random probability p. As p increases, the effect of q on clustering coefficients reduces. In Figure 3(a), while p = q and average degree is 10, the clustering coefficients in the friendship networks are around 0.0870. In Figure 3(b), while p = q and average degree is 50, the clustering coefficients are around 0.1074. These results reveal such a phenomenon: while q/p is a constant, clustering coefficients nearly remain at a stable level. In simulations, we also find a truth: while p = q, no matter what values (0.01 or 0.99) assigned to them, the ratio of edges, produced randomly to edges and produced according to triadic-closure rules, appears to be nearly uniform. It is the main cause of the stable clustering coefficient.

Figure 4 additionally shows the relationship between a clustering coefficient and the value of $q/p$ while average degrees are 10 and 50. Both in Figure 4(a) and in Figure 4(b), we can observe that no matter what value $p$ is, the curves are nearly coincident. That means clustering coefficient of the friendship network, based on our model, is uncorrelated with the value of $p$. As long as $q/p$ remains constant, the clustering coefficient nearly stays the same. As the value of $q/p$ rises, the clustering coefficient increases. A clustering coefficient is proportional to the ratio of $q$ to $p$. It is easy to observe the relationship: the clustering coefficient increases non-linearly with the value of $q/p$. While $q/p$ is near to zero, the friendship networks are near to random networks, clustering coefficients are near to 0.02 in Figure 4(a) and 0.098 in Figure 4 (b). These values also match clustering coefficients in random networks which have the same numbers of nodes and edges. In Figure 4, we can get such a result: in a sparse friendship network, the value of clustering coefficients can indicate the effect of the intensity of triadic-closure rules that lead to the structures of networks.

### 3.3 *In-degree distribution*

Degree distribution is another important property in social networks. There are many studies that have revealed that social networks have a power law degree distributions (Newman 2001; Barabâsi et al. 2002). We give some in-degree distributions of the friendship networks for several different values of random probabilities p and triadic-closure probabilities q in Figure 5. In Figure 5(a), while p and q are very small (p = 0.01, q = 0.01), the in-degree distribution is Poisson. While p remains very small and q rises much higher (p = 0.01, q = 0.95), the in-degree distribution has such a shape which has a lot of small degree nodes and a heavy tail. While q is a little higher than p (p = 0.01, q = 0.1), the in-degree distribution has such a qualitative shape – unimodal body and heavy tail, which combines the two shape above (while q = 0.01 and q = 0.95). In Figure 5(b), p grows a little (p = 0.1), while q is less than or equal to p (q = 0.01 or q = 0.1), in-degree distributions are both Poisson. While q = 0.95, in-degree distribution also changes, it turns to the shape – unimodal body and heavy tail. In Figure 5(c), while p grows to 0.3, all in-degree distributions are Poisson, no matter how high q is. The larger q is, the lower the peak of Poisson is.

Figure 5 once again reveals that in-degree distributions in the friendship networks are not only related with triadic probability q or random probability p, but related to the value of q/p. As the value of q/p increases (q increases or p reduces), in-degree distributions turn from Poisson to heavy tail. It is similar to the phenomena of clustering coefficients in Figures 3 and 4. The friendship networks based on random and triadic-closure rules have various structure properties according to different values of p and q. Diameters are mainly related to p. The friendship networks exhibit short diameters once there are enough

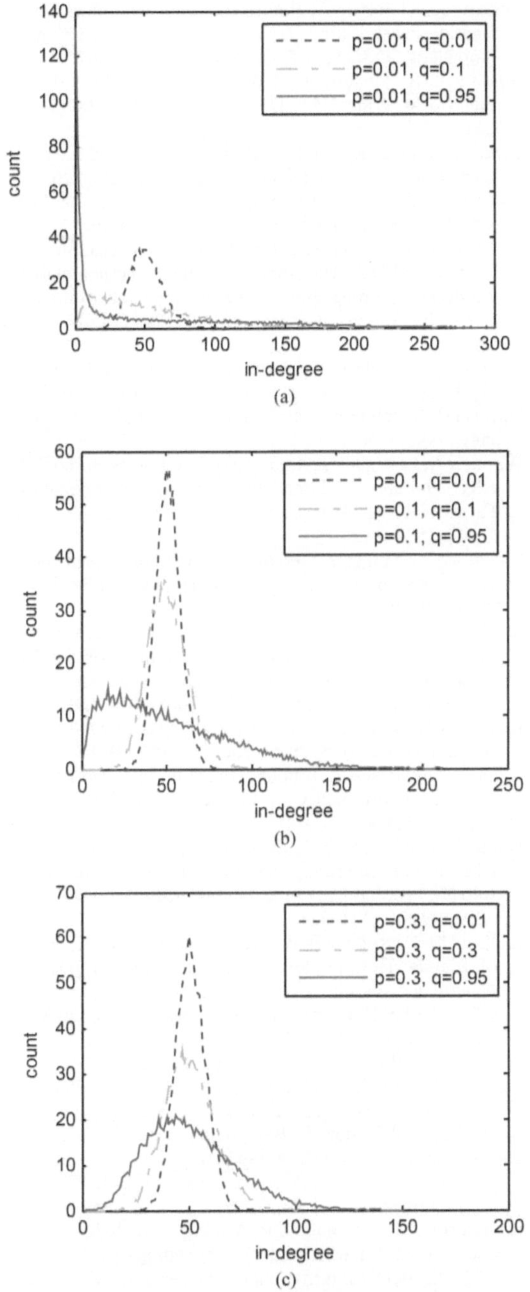


Figure 5. The in-degree distributions of the friendship network for several different values of random probabilities and triadic-closure probabilities. The constant number of nodes N = 1000, (a), (b) and (c) are all measured while the average degree is 50.

random connections. Clustering coefficients and in-degree distributions are mainly related to the value of q/p.

From all the results, we can get a method of how to judge which mechanism has more effect in real world friendship networks. The detailed steps of the method are as follows: first, we measure the numbers of nodes and edges in a friendship network and its clustering coefficient; second, we calculate the connection probability of the random network which has the same numbers of nodes and edges; third, we refer to the corresponding curves of clustering coefficients and the values of q/p which are shown in Figure 6. We then estimate the triadic-closure probability. Meanwhile, we can plot the figures of diameters and in-degree distributions as further references. Accordingly, we can make a judgment as to how much random connections or triadic-closure rules effect the structure of friendship networks in the real world.

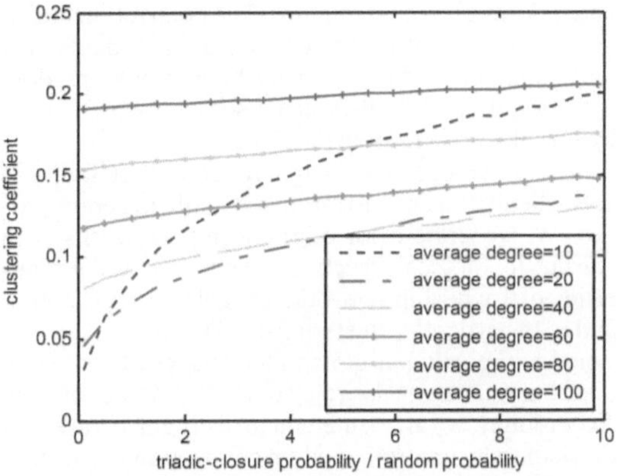


Figure 6. The clustering coefficients of the friendship networks for different average degrees. The constant number of nodes N = 1000.

In Figure 6, we can observe a phenomenon: while average degree is low, clustering coefficient varies apparently in a board range as long as the value of $q/p$ changes. As the average degree rises, the curves of clustering coefficients vary gently and increase on the whole. Each curve represents the relationship between the clustering coefficient and the ratio of triadic-closure probability to random probability in a friendship network with different average degrees. Thus, it is easy to judge the effect of triadic-closure rules or random connections in a sparse friendship network. The method is less effective when friendship connections are dense in a network. That is because clustering coefficients are almost at a stable level. It is difficult for us to distinguish the corresponding value of q/p.

## 4 CONCLUSIONS

Friendship is one of the most normal relationships between individuals in social life. It is also an important element constituting the complex social network. Though there are a lot of models proposed to explain how social networks form and evolve, they mainly focus on online friendship networks and growth networks. In this paper, we give a new model based on random and triadic-closure rules to generate friendship networks in a fixed community. The connection mechanism reveals that in the real world, individuals may meet someone and make friendships through random opportunities, while they can also make new friends by introductions from their friends

In simulations, we assign random probability p and triadic-closure probability q with different values. We generate a lot of friendship networks with various topological structures. We find that the properties of friendship networks are related with both p and q. In particular, clustering coefficients are proportional to the value of q/p and the shapes of in-degree distributions are also related to the value of q/p. Accordingly, we give a method to distinguish the effect of triadic-closure from random connection in a friendship network. The results give a specific explanation: how these various topological properties emerge in the friendship networks based on mixed connection mechanisms and which mechanism performs greater effects. There is no doubt that our work is useful for people to realize some special properties of social networks in the real world.

## ACKNOWLEDGMENT

This work is supported by 'the Fundamental Research Funds for the Central Universities' under No.zyz2011085, and Graduate Research Funds' under No.ycx14081, Z.D.S, Z.K, Y.J and S.J.R gratefully acknowledge the financial support from Northwest University for Nationalities.

## REFERENCES

Van Duijn M A J, Zeggelink E P H, Huisman M, et al. Evolution of sociology freshmen into a friendship network [J]. Journal of Mathematical Sociology, 2003, 27(2–3): 153–191.

Granovetter M S. The strength of weak ties [J]. American journal of sociology, 1973: 1360–1380.

Watts D J, Strogatz S H. Collective dynamics of 'small-world'networks [J]. Nature, 1998, 393(6684): 440–442.

Newman M E J. The structure of scientific collaboration networks [J]. Proceedings of the National Academy of Sciences, 2001, 98(2): 404–409.

Barabâsi A L, Jeong H, Néda Z, et al. Evolution of the social network of scientific collaborations[J]. Physica A: Statistical mechanics and its applications, 2002, 311(3): 590–614.

Backstrom L, Boldi P, Rosa M, et al. Four degrees of separation[C]//Proceedings of the 3rd Annual ACM Web Science Conference. ACM, 2012: 33–42.

ERDdS P, R&WI A. On random graphs I [J]. Publ. Math. Debrecen, 1959, 6: 290–297.

Albert R, Jeong H, Barabási A L. Internet: Diameter of the world-wide web A L. Bose-Einstein condensation in complex networks [J]. Physical Review Letters, 2001, 86(24): 5632.

Kleinberg J M, Kumar R, Raghavan P, et al. The web as a graph: Measurements, models, and methods [M]//Computing and combinatorics. Springer Berlin Heidelberg, 1999: 1–17.

Leskovec J, Kleinberg J, Faloutsos C. Graph evolution: Densification and shrinking diameters [J]. ACM Transactions on Knowledge Discovery from Data (TKDD), 2007, 1(1): 2.

Adamic L A, Huberman B A. Power-law distribution of the World Wide Web [J]. Science, 2000, 287(5461): 2115–2115.

Broder A, Kumar R, Maghoul F, et al. Graph structure in the web [J]. Computer networks, 2000, 33(1): 309–320.

Yook S H, Jeong H, Barabási A L. Modeling the Internet's large-scale topology [J]. Proceedings of the National Academy of Sciences, 2002, 99(21): 13382–13386.

Wagner A, Fell D A. The small world inside large metabolic networks [J]. Proceedings of the Royal Society of London. Series B: Biological Sciences, 2001, 268(1478): 1803–1810.

Montoya J M, Solé R V. Small world patterns in food webs [J]. Journal of theoretical biology, 2002, 214(3): 405–412.

Tarbush B, Teytelboym A. Friending: a model of online social networks [J]. 2013.

Wimmer A, Lewis K. Beyond and Below Racial Homophily: ERG Models of a Friendship Network Documented on Facebook1 [J]. American Journal of Sociology, 2010, 116(2): 583–642.

Singer H M, Singer I, Herrmann H J. Agent-based model for friendship in social networks [J]. Physical Review E, 2009, 80(2): 026113.

Leskovec J, Backstrom L, Kumar R, et al. Microscopic evolution of social networks [C]//Proceedings of the 14th ACM SIGKDD international conference on Knowledge discovery and data mining. ACM, 2008: 462–470.

Kossinets G, Watts D J. Empirical analysis of an evolving social network [J]. Science, 2006, 311(5757): 88–90.

Albert R, Barabási A L. Topology of evolving networks: local events and universality [J]. Physical review letters, 2000, 85(24): 5234.

Pennock D M, Flake G W, Lawrence S, et al. Winners don't take all: Characterizing the competition for links on the web [J]. Proceedings of the national academy of sciences, 2002, 99(8): 5207–5211.

Jackson M O, Rogers B W. Meeting strangers and friends of friends: How random are social networks? [J]. The American economic review, 2007: 890–915.

Davidsen J, Ebel H, Bornholdt S. Emergence of a small world from local interactions: Modeling acquaintance networks [J]. Physical Review Letters, 2002, 88(12): 128701.

Vázquez A. Growing network with local rules: Preferential attachment, clustering hierarchy, and degree correlations [J]. Physical Review E, 2003, 67(5): 056104.

Yuta K, Ono N, Fujiwara Y. A gap in the community-size distribution of a large-scale social networking site [J]. arXiv preprint physics/0701168, 2007.

Richardson M, Domingos P. Mining knowledge-sharing sites for viral marketing [C]//Proceedings of the eighth ACM SIGKDD international conference on Knowledge discovery and data mining. ACM, 2002: 61–70.

González M C, Lind P G, Herrmann H J. System of mobile agents to model social networks [J]. Physical review letters, 2006, 96(8): 088702.

Chakrabarti D, Zhan Y, Faloutsos C. R-MAT: A Recursive Model for Graph Mining [C]//SDM. 2004, 4: 442–446.

Milgram S. The small world problem [J]. Psychology today, 1967, 2(1): 60–67.

Newman M E J, Watts D J. Renormalization group analysis of the small-world network model [J]. Physics Letters A, 1999, 263(4): 341–346.

*Frontiers in Computer Education – Wang (ed.)*
*© 2015 Taylor & Francis Group, London, ISBN 978-1-138-02797-8*

# Exploring the application of cloud computing in university informatization

LiHua Zhang
*North China University of Water Resources and Electric Power, Zhengzhou, China*

ABSTRACT: Cloud computing is a current vital issue with the global IT community and universities are the cradles for cultivating talents for the country. In order to promote university informatization construction and improve the competitiveness of universities, it is the general trend to apply cloud computing technology to universities. This paper focuses on the application of cloud computing in university informatization construction and analyses the significance of the cloud computing to university development, combined with the characteristics of cloud computing.

*Keywords*: cloud computing; informatization; university

## 1 CLOUD COMPUTING

Cloud computing is considered both a technology and a kind of service model. From a technical point of view, it is another innovation after the information revolution. From the service model, it provides IaaS, PaaS, and SaaS, these kinds of service models that cover the entire IT ecosystem. Cloud computing as a new technology or service mode has different characteristics from those in former internet technology.

### 1.1 *Definition of cloud computing*

Cloud computing is a super computing model connected via the internet, including distributed computing, parallel computing, and grid computing technology, and it is a new service delivery and usage pattern, under the foundation of an open standard and service and the internet centre, to provide safe and fast data storage and computing services. The network for providing resources of data storage and computing services is known as cloud.[1] Cloud computing is a new shared infrastructure model, connecting a huge pool of resources together, as operators and customers, through the internet, providing various kinds of storage and computing resources for users. In the cloud computing environment, users using a variety of personal computer terminals, PDA, mobile phones, etc, put a large amount of information and processor resources together for collaborating work.[2] It is a large scale distributed computing mode, determined by the operators' economic scale, abstract, and a virtual and dynamic variable.

### 1.2 *Characteristics of cloud computing*

The characteristics of cloud computing are as follows.

#### 1.2.1 *The large scale power of data storage and computing*

"The cloud" has a considerable size, giving the user a calculating ability hitherto unknown. For example, Google cloud computing has about 1000000 servers Amazon, IBM, Yahoo, and Microsoft "cloud" have hundreds of thousands of servers; even though in general the common private enterprises' cloud computing also have hundreds to thousands of servers.[3] Thus, the competence of data storage and computing given by cloud computing is hitherto unknown.

#### 1.2.2 *Resource virtualization*

Virtualization refers to the underlying hardware comprehensive virtualization, including servers, storage and networking equipment, to build a shared resource pool under the basis of a demand assignment. Cloud computing supports users at any location using a variety of terminal and access application services.

#### 1.2.3 *High reliability*

"The cloud" guarantees high reliability service, using measures of more copies of data fault tolerance and computing and isomorphic interchangeable nodes, as well as being more secure than any local computer.

#### 1.2.4 *Universality*

Cloud computing does not specify particular applications, and one cloud supports different applications. The universality of cloud makes the utilization rate of resources more pronounced than any traditional system.

#### 1.2.5 *Dynamic scalability*

Cloud computing's scalability includes horizontal and vertical scalability. Horizontal scalability refers to combining multiple cloud servers to form larger cloud calculators for processing and storing internet data computing. Vertical scalability refers to the cloud node backwardly compatible and upwardly upgradable for meeting the ever-changing markets and different client demands.

#### 1.2.6 *An on-demand service*

"The cloud" is a huge pool of resources, providing enormous services. A user can buy it according to his needs, just as water, electricity, and gas is charged by the quantity consumed.

#### 1.2.7 *Very cheap cost*

As to the special fault tolerant measures of "the cloud", it can be formed by extremely low nodes." Enterprises costs less to the increasingly high data centre management by the automatically centralized 'the cloud' management.

## 2 THE SIGNIFICANCE OF CLOUD COMPUTING TO THE DEVELOPMENT OF UNIVERSITY INFORMATIZATION

'The education cloud' refers to cloud computing application in education. As a new service model, it is welcomed. Its service development is significant for promoting education informatization of our country in depth, improving the level of informatization in universities, and promoting the development of Chinese university education.

### 2.1 *The use of cloud computing for reducing the funds invested in software and hardware infrastructures in universities*

Cloud computing requires slow terminal equipment, making the memory, storage and computing capacity of a large number of distributed computers together into a virtual pool of resources, and through the network providing users computing services. With the escalation of the software version, the hardware requirements are also rising, so universities often have to update their hardware equipment, which will lead to funding problems and difficulties for them. In the cloud computing environment, most tasks will be handed over to the cloud, users need not worry about information security, software upgrades, hardware performance, storage and backup and other problems. As long as there is a network and a terminal, it can be used like water or electricity, at anytime, anywhere, on demand, accessed for educational resources and educational services.

### 2.2 *The application of cloud computing realizing the integration of university resources*

There are differences in educational developments in our country, for example, the development of northwest education is inferior to southeast educational development and the education fairness has always existed. The cloud computing platform establishes a virtual pool of resources crossing systems and regions, being unified, safe, and strong. From different universities, teachers and students can upload or download teaching resources in the resource pool, solving the problems of the slow updating speed of school resources and their low utilization rate.

### 2.3 *The cloud computing provides secure and dependable data storage centre for university*

In the Internet era of wide spread viruses, the security of data storage reliability becomes more and more important, especially in any university lacking professional information security. If it uses cloud computing services and storing data in the cloud, a professional team with the professional means for measures of safety control, the university need not worry about virus and hardware data loss caused by damage. In addition, the redundant storage and disaster recovery mechanism of cloud storage services are the reliable guarantees for data security.

## 3 THE APPLICATION OF CLOUD COMPUTING IN UNIVERSITY INFORMATIZATION

Cloud computing, a great IT technology innovation, leads us into a new information era – the cloud era. It provides a new mode for the construction and development of University informatization. It can help solve the informatization construction problems, such as when education information platforms cannot meet the needs of teachers and students, when education information and resources cannot be effectively shared, because of network transmission speed, lack of funds and data processing ability, etc. Using the cloud computing technology, integrating isolated islands of resources into a unified information platform, it provides better services for teaching, research, management, and the sharing of resources. We explore the application of cloud computing for university informatization in detail.

### 3.1 *Constructing large-scale sharing education resources library based on cloud computing*

Teaching resources are the key to cultivating talents, and the current university education teaching resources are unevenly distributed. These problems of low teaching resources sharing a low degree of duplication investment to education resources and teaching resources and high update costs, and a slow update speed to expand the differences of teaching quality. With cloud computing, domestic and foreign universities can build the sharing of teaching resources space, share high quality education resources, composed of a large number of university libraries, and then participating universities in cooperation will not only greatly reduce their costs, but also greatly meet teacher and student demands for resources. They access in cloud whenever and wherever possible and use any client, then, study or prepare for their works by using these high-quality educational resource libraries. So we need to use cloud computing, establish the open sharing of educational resources with flexible libraries. Cloud computing will integrate, to the greatest degree, the maximum range of higher education resources, narrow higher education resource gap, promote quality education resources with universal library sharing, reduce teaching costs, improve the quality of teaching, and improve the level of university education.

### 3.2 *Constructing a digital library based on cloud computing technology*

As cloud computing technology becoming more mature, the daring vanguards of science and technological progress, universities should timely apply the cloud computing to the digital library. This will bring new opportunities and challenges to university libraries, as a library will no longer be in a traditional service mode, but replaced by a virtual cloud management mode.

#### 3.2.1 *The development of service from classification to individuation*

In cloud age library services will tend to provide more personalized service. Through the establishment of the reader database, tracking the reader's personal hobbies and the characteristics of intellect, then analysing and researching their researching and studying orientation and characteristics, the cloud library will ultimately provide a personalized service.

#### 3.2.2 *Changing service mode from fixed to mobile*

Another development direction provides user information access in any place, at any time and with any different forms of information access, for example, mobile phones, PDAs and with notebook computer equipment, with the availability WiFi and cellular connection service coupling.

#### 3.2.3 *The integrating of all sorts of resources in society*

The construction of digital library resources in cloud computing should be presented in the macro national perspective and provide the convenience of the communication and integration of various information resources. Pushed by cloud computing, a digital network interconnection to all aspects of life can be constructed, and this can achieve a shared interconnection between the economic and the cultural resources of the whole society

#### 3.2.4 *Operating industrially*

With deepening economic reforms, the university digital library industrialization has been imperative, as the direction of the comprehensive development of knowledge, the economy and society, as the goal of promoting the development of the university digital library for the information industry, of strengthening the network systems and information construction, and establishing a new development approach to the university digital library.

### 3.3 *Constructing online office cloud in a university based on the cloud computing technology*

Online office cloud is a virtual desktop solving scheme based on cloud computing technology for any university. A modem-size cloud computer equipment can replace a PC and host, achieving the same operating system with similar graphical desktop functions as a PC, making the user feel as though he is using a PC. Software as a Service (SaaS) is a type of cloud computing service, providing software as an online service. Some commonly used software, such as office software and email, can adopt cloud computing services, using online office behind the cloud, and reduce the maintenance investment of software upgrade.

Online office cloud based on cloud computing technology effectively integrates the administrative resources in university, builds the platform of information exchanging, interacting and sharing among various departments, internal departments, leaders, administrative staff, teachers and students, serving many people online at the same time, changing their files, seeing others' editing, and greatly improving the cooperation efficiency, the public service, and the service quality and efficiency of university administration, policy transparency, and administrative efficiency.

### 3.4 *Constructing teaching and research cloud based on cloud computing technology*

At present, university research and the experimental environment becomes more and more complex, because of more and more information, which constricts factors of capital, resources, time and system load and others, resulting in a number of projects and research plans failing to be implemented. Cloud computing technology will help to build a university teaching and research cloud environment, making these projects and research projects be carried out smoothly. In this regard, IBM has cooperated with six worldwide universities, accelerating scientific research and project implementation for these university using IBM 'blue cloud' solutions. IBM's farsightedness in educational and cloud computing will give the global educational institutions of existing IT service mode positive, far-reaching significance.

### 3.5 *Constructing distance education cloud based on cloud computing technology*

Distance education cloud is a cloud computing migration technology in the field of distance education, the direction of future development in distance education, including the required computational resources of hardware and software in distance education. After virtualizing these computing resources, distance education institutions and students can rent these

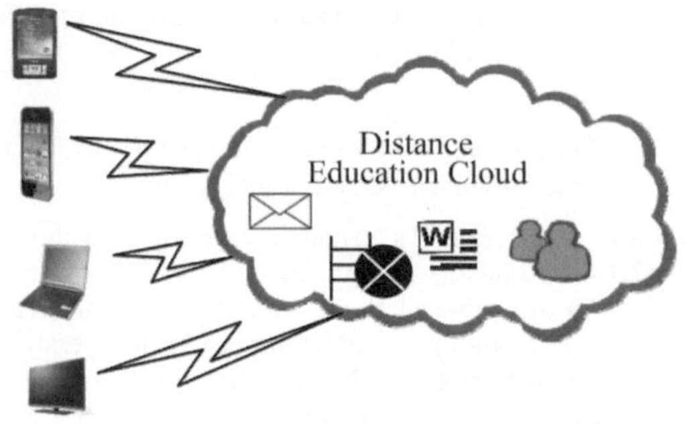


Figure 1. Students visiting distance education cloud.

computing resources. Distance education cloud infrastructure is consisting of the physical resource pool, a basic management layer, an application interface layer and an application layer of distance education. The core of the distance education cloud is the distance education application layer, mainly containing: 1. operating system procedures for distance education, teaching and management, including remote teaching platforms, homework systems, office automation systems, and educational administration management systems; 2. distance education application procedures, including e-mail, talk, video player, word and others; 3. long-distance education middleware, which is used by users as most convenient for developing their own desired distance education programs.

From the point of view of distance education institutions, its shape will be like the operation and service centres of Google, or Amazon, simply regarded as data centre + computing centre + interface or connector. By interface or connector, user can obtain all distance education cloud computing resources. Distance education institutions will store all the teaching resources in the cloud, by using them to do cloud business; also develop application procedures to provide service to the general user through a cloud computing service centre, with open API and distance education middleware. It will be the main trend of future distance education and cloud will cooperate between individuals or small and medium-sized institutions of distance education and a distance education cloud service centre.

From the perspective of students, distance education cloud is the distance education resource centre in Internet. Through various terminal configurations, such as television, mobile phone, PDAs, personal computers, etc, students can obtain a variety of distance education service. For example, through the mail to send and receive e-mail; through Talk to communicate with teachers and students, and through Word to write documents as shown in Figure 1.

### 3.6 *Establishing a home–university interaction platform based on cloud computing*

Family education is the basis of school education, the establishment of home–university interaction platform promoting communication between parents and teachers based on cloud computing, providing convenience for communication between home and university, and parents having more initiative, and improving parents' participation, which are mainly illustrated in the following three aspects.

#### 3.6.1 *Guiding parents to participate in school management*

A home–university interaction platform based on cloud computing is not a simple electronic bulletin board but an interactive gallery. Through this platform, parents can intuitively understand student's performance in the university, and propose their suggestions about whether the curriculum is reasonable, or whether the student management systems accord with the age and characteristics of students, etc. Let parents really participate in university management and give timely feedback for the university.

#### 3.6.2 *Promoting home–university exchanging educational concept and experience*

The most important feature of the home–university interaction platform, based on cloud computing, is the education concept of exchanges between home and university. At the same time, parents can share their educational experiences through the platform, and in the "help" section parents can express their problems meeting in the education process and receiving other parents or teachers' views on this issue to help parents avoid this predicament, and then form a good home–university relation.

#### 3.6.3 *Providing educational services*

Through the platform, university shows university education and life education to parents, provides one-stop education services by some successful cases of educational concepts and methods, and works together for students' healthy growth. Of course, there are more cloud service tools realizing education function, such as 'Good-looking Book (a photograph blog recording life and learning by photograph), Google forum and so on , which fully indicates that cloud computing application will become more prominent in education.

## 4 THE MAIN PROBLEMS NEEDED TO BE WATCHED

At present, cloud computing and its application develop swiftly and fiercely, which are still innovations. There are many urgent improvements in cognition and practice, which need a clear understanding in advance, planning ahead, and drawing on advantages and avoiding disadvantages to play its role fully.

### 4.1 *No cloud computing standards unified*

At present, international and domestic industries have not issued any relevant standards leading to no standards and unhealthy developments, unscaled and no coordinated industrial development. It has to be said that this is not only a pity but also an opportunity and a challenge lost.

### 4.2 *Core innovation technology needed*

At present, there are fierce competitors in the field of cloud computing, therefore, we must vigorously develop the 'Nuclear High-based' projects (core electronic components, high-tech general chips and basis software products) proposed by 'Development and Plan Outline of National Long-term Science and Technology (2006–2020)', vigorously promote core cloud computing technology and independent innovation products and continuously improve Chinese core competitiveness in the field of cloud computing.

### 4.3 *Cloud safety concerned mostly*

At present, the security of cloud computing is still the common issue of concern for industry and other users, especially since 2007 when the SaaS service provider — Salesforce.com – was attacked, resulting in a large number of users' privacy data being leaked lost. In 2010 two employees of Google intruded into Google Voice, Gtalk and other accounts, causing privacy data leakage; due to the lack of data encryption and dispersed isolated storage mechanism, database information of 6000000 users in the CSDN website was stolen and opened by a hacker[4]. Therefore, information construction based on cloud computing urgently needs to innovate security management mechanisms, improve security defence technology and products.

### 4.4 *Issues of environmental protection*

It should pay more attention to environmental protection issues because of the intensive, large-scale cloud computing system. Faced with an increasingly nervous energy situation, environmental protection consciousness should be set up; scientific and reasonable techniques and means should be adapted; and environmental cloud computing digital resources should be constructed with accompanying energy-saving and emission reduction initiatives.

## 5 CONCLUSION

The rapid development of cloud computing is changing the pattern of Chinese university informatization development. There are no revolutionary changes about cloud competing, which is more like a development method transferring and extending university informatization computing platform and service to a better cloud computing platform. Cloud computing, a kind of development trend, will show great vitality in the near future, injecting new impetus to university education informatization, and contributing to the leapfrog development of IT industry.

## REFERENCES

[1] Liu Peng. Cloud Computing [Yun jisuan] [M]. Beijing: Electronics and Industry Press [Dianzi gongye chubanshe], 2010: 2–3.

[2] Cao Taifeng & Li Duanming. The Application of Cloud Computing in University Informatization Construction [Yunjisuan zai gaoxiao xinxihua jianshezhongde yingyong yanjiu] [J]. Modern Intelligence [Xiandai qingbao], 2010, 30 (8): 167–170.

[3] Li Jiahou. Moving towards Educational Technology 'Cloud' Service [Zouxiang jiaoyu jishu 'yun'fuwu] (J). Distance Education Journal [Yuancheng jiaoyu zazhi], 2008, (3): 17–18.

[4] Zhuang Xiuli. Great Discovery in Education: Small Community Manua [Jiaoyu dafaxian:xiaoshequ shouce], [EB/OL].

*Frontiers in Computer Education – Wang (ed.)*
*© 2015 Taylor & Francis Group, London, ISBN 978-1-138-02797-8*

# Research on internationalization and localization of user interface

ZhengXu Zhao, He Bai, LiLong Han & YunXiu Meng
*Shijiazhuang Tiedao University, Shijiazhuang, Hebei, China*

ABSTRACT: Multi-language software is one of the main goals of software internationalization since there are various languages and writing types around the world. This paper introduces the concept and the main features of software internationalization and localization via comparing and summarizing some technologies of multilingual user interface in application software.

*Keywords*: Multi-language; Internationalization; Localization; User interface

## 1 INTRODUCTIONS

With the deepening of the international cooperation, exchanges between domestic and international corporations become more and more frequent. In order to make the software enter into the international market faster and easier, the idea of software internationalization has been gradually put on the agenda. It has contributed to the development of multi-language, and people are paying more and more attention to the development of multi-language, which has become the mainstream of the application of international software. There are many methods to achieve software localization. Most of software support user to choose multi-language interface, which can greatly improve user satisfaction, reduce the cost of the software, and make the products more available to the international market[1]. The core of software internationalization lies in the information processing of different languages. To put it simply, the mechanism of software internationalization can process the text information from all over the world. Thus, the software is not subject to any restrictions of any types of languages. Users only need to conduct simple set up or modification and they can see the user interface being displayed in their preferred language.

The internationalization of operating system, windows system, and programming library are crucial in the design of international software, among which the internationalization of GUI (short for Graphic User Interface) is the key. The internationalization of user interface not only needs to correctly show the words in any languages, but more importantly it also needs to cater for demands and habits of different users. Therefore, GUI system should possess the abilities to adjust to different styles of user interface.

## 2 SOFTWARE INTERNATIONALIZATION/ LOCALIZATION

Except the difference in language, countries often have their own set of numbers, dates, currency and so on. To meet the requirements of users from different religions and cultures, the software needs to be further modified so that it can display and process the native language, meanwhile supporting the different representations. Since this kind of modification requires the changes of a great number of source codes, it can cost huge amounts of labor materials and financial resources, and it will also result in problems in the maintenance and upgrade of the software. Therefore, based on internationalization, people have come up with new method of software localization.

### 2.1 *Concepts*

Internationalization (I18n for short, where 18 stands for the number of letters between the first I and last n), is the process of designing a software application which can potentially be adapted to various languages and regions without any engineering changes. This way it can be used among a great number of countries and regions around the globe. Furthermore, software internationalization is a method for development without making any assumptions about the software system, including encoding style, users' writing habit, data display and religions, etc.

Localization (L10n for short, where 10 stands for the number of letters between the first L and last n), is the process of adapting internationalized software for a specific region or language by adding localized components and translated text. Specifically, it manages to makes the software comply with the language and writing habit of the local cultural.

Features required in the international software are as follows. (1) By incorporating local information of different countries and regions, the software can function smoothly in any part of the world. (2) A certain text messages, such as labels in the interface or prompts, are kept in the external resource files rather than being coded in the program. (3) It can support new languages without the need for recompilation. (4) The software allows the users to choose their preferred format for display, including those with the local characteristics in terms of numbers, dates, currencies, etc. (5) It can be localized very easily.

Compared with the traditional method of software localization, the new method based on internationalization has many advantages[2]. (1) It originates from one source program, which means it can be easily maintained and modified. (2) It can largely cut back long-term development fees, and reduce huge investments in human and material resources which are necessary in the traditional method. (3) It enables the company to promote the sales of the software simultaneously in the global market.

### 2.2 *The basic techniques*

#### 2.2.1 *The expression and processing of characters*

In the development stage of international software, the character encoding is a problem that cannot avoided. The differences in languages lead to different standards in character encoding, which presents as a challenge for the development of international software. Therefore, the crucial question lies in the encoding of different characters. The idea of using one uniformed coding system to process all the characters seems to be a convenient and efficient method[3]. It is the discovery of Unicode that makes the multi-language become possible. Unicode has the code formats of all the characters existed in the world and has 1.1 million of code points for processing. Thus, Unicode is viewed as a brilliant solution for processing multi-text and multi-language.

#### 2.2.2 *Information processing relating to locale*

The localization technique needs to extract part of information from the source code so that it can process different local expressions and grammar rules. Each region has its own locale information. It reflects the characteristic of localization and influences the outcome of the program. Moreover, it uses message catalogue to sort out hints and diagnostic information. Message catalogue can be a file or a storage area, which can be simply translated to the corresponding language during localization, while storing hints and the diagnostic information independent from the program.

#### 2.2.3 *Separate binary resource file*

In the development process of software, it is necessary to put the localized contents in a single localized resource file in order to separate the localized source code[4]. In the development of internationalized software based on Windows, the localized resource file should be complied into a single DLL file, which is called single binary resource file.

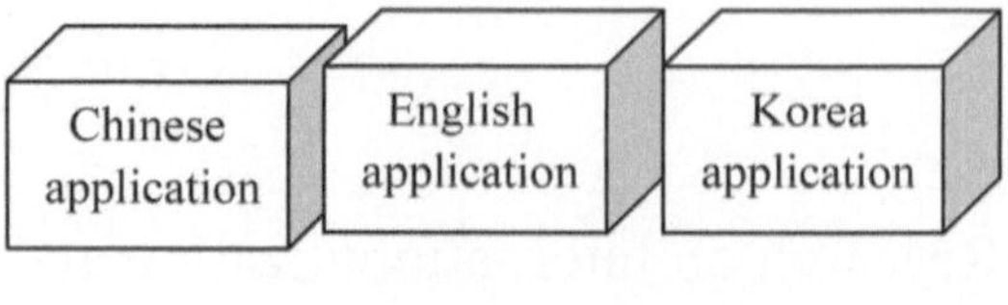


Figure 1. Multiple independent applications.

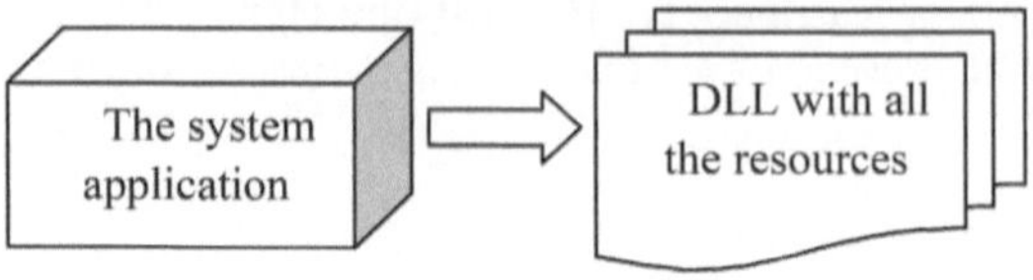


Figure 2. Integration resource of DLL.

Short for Dynamic Link Library, DLL is a library file which contains data and source code that can be used by multiple programs. Since DLL does not include executable code, it contributes to the sharing of data and resource. The application of DLL solves the problem of multi-language, reduces the difficulty of localization, and improves the efficiency of the development of international software.

## 3 TECHNIQUES SUPPORTING MULTIPLE LANGUAGES

The following techniques are currently available for the realization of multiple languages in international application software.

### 3.1 *Multiple independent applications*

Each language corresponds to one application. It is the easiest method that can thoroughly implement the software internationalization. The downside of this technology is that every language needs a series of source code which may result in huge wastes in disk space.

### 3.2 *Integration resource of DLL*

The application is compiled independently. The language-related resources should be compiled to DLL, and all the resources are stored in the same DLL. When the application needs maintenance and upgrade, one can just upgrade the program without recompiling the DLL. However, recompilation of the whole resource item is required when a new language is added into the application.

### 3.3 *Single language resource of DLL*

This is an improved technique of the previous ne. The resource files of different languages are compiled and stored separately in every single DLL[5]. When the program is operating, users can load the corresponding resource DLL based on their actual demand. When adding a new language, it is no longer necessary to

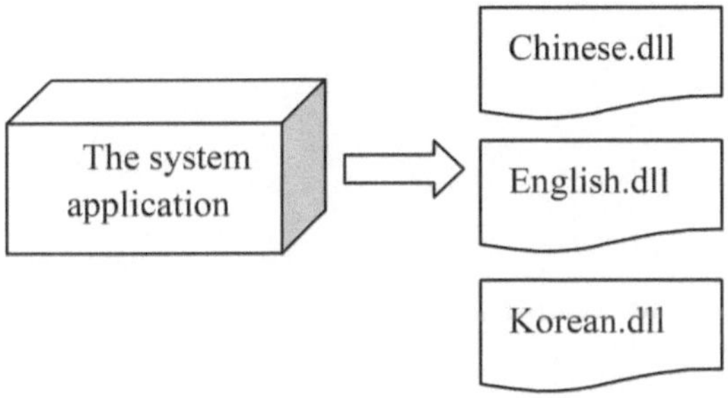


Figure 3. Single resource of DLL.

Table 1. Comparison of four technologies.

| Technologies | Advantages | Disadvantages |
|---|---|---|
| Multiple Independent Applications | It implements easily and thoroughly. | It needs high maintenance and development costs. |
| Integration Resource of DLL | It separates resources from source codes. It is maintained and upgraded without recompiling the DLL file. | It needs recompilation of the resources when adding a new language. There are unnecessary data in the DLL files, causing wasted disk space. |
| Single Resource of DLL | Every language has a separate DLL file. There is no need for recompilation of the whole DLL file. | It requires more maintenance. |
| The External Language Packs | It is easy to use and to be modified. | The length of words may hinder the normal function of user interface. |

revise the entire language resources, which saves labor and improves work efficiency thus, making up for the deficiencies existed in the above technique.

### 3.4 *The external language packages*

The key of this technology is to write all the words, which need to be displayed in the text files (*.ini or *.lng being the most common). When transferring languages, the system uses ID to read the associated contents, and then replaces the current messages. It is easy to use and to be modified, but for those large projects using this technology to transfer external resource files requires adding large amount of the code from the source text, which may lead to increased system expenses.

A comparison of the four technologies regarding to multi-language user interface is listed in Table 1.

The first two technologies are less commonly used at present. The implementation of the last may have the possibility of hindering the normal function of the user interface. At the moment, the third technology presents as a good solution to the international problem. According to the different settings of users, encapsulating resource by DLL can solve the problem of user interface and satisfy the needs of the users.

## 4 SUMMARIES

From the above analysis, we can see that single resource of DLL may be a better solution to the implementation of multiple languages than other technologies. It is also the preferred method to the internationalization of applications. The design approach of single resource of DLL can update the language resource promptly. When consumers use the software, they can choose to display the interface in the language that they need.

With the continuous development of international exchanges, the internationalization of application has become the main trend of the present software development. This paper analyzes the mechanisms of internationalization and localization, and the ways of solving the problems existed in the user interface of multi-language. Aforementioned technologies have their advantages and disadvantages. Thus further research is needed to enlarge and improve the technology of user interface of multi-language.

## REFERENCES

[1] Li Jiang, Jian Wu, Yufang Sun. 2005. Research on Internationalization of Graphical User Interface System [J]. Computer Science 32(12):128–134.
[2] Lina Chen. 2005. Research and implementation of Tibetan Office based on the Internationalization/Localization frame in OpenOffice.org [D]. Beijing: China Agricultural University.
[3] Jie Yuan. 2012. Research on Multi-language Supported Software base on Perl Script [D]. Hebei: Shijiazhuang Tiedao University.
[4] Weihua Zhao, Zhanwei Liu, Zhengxu Zhao. 2011. A community Cloud: Archive Retrieval in Multiple Language Services [M]. The Sixth International Conference on Internet and Web Applications and Services 237–240.
[5] Xuechun Wang, Zongxin Jin, Yajuan Zhang. 2010. Research on Multi-language User Interface Support under. NET [J]. Equipment Manufacturing Technology 3:71–72.

*Frontiers in Computer Education – Wang (ed.)*
*© 2015 Taylor & Francis Group, London, ISBN 978-1-138-02797-8*

# A study into a semi-supervised dimensionality reduction algorithm based on feature information integration

M.M. Qi
*School of Yuanpei, Shaoxing University, Shaoxing, China*

X. Liu
*The Third Research Institute of Ministry of Public Security, Shanghai, China*

B. Zhang
*College of Information, Mechanical and Electrical Engineering, Shanghai Normal University, Shanghai, China*

C. Luo, S.H. Yuan & D.D. Lv
*Department of Computer Science and Technology, Tongji University, Shanghai, China*

ABSTRACT: From the angle of the feature level information integration, through linear weighting integration method, this paper proposes a semi-supervised dimensionality reduction algorithm integrated with the sparsity reconstruction feature information and pair constraint feature information and applies the genetic algorithm to perform automatic optimization selection to the linear weighting balance parameters. Experiments show that, when comparing the algorithm with the previous method, it has good dimensionality reduction classification performance.

## 1 INTRODUCTION

The sparsity representation's sparsity reconstruction feature information is a kind of unsupervised information. The Sparsity Preserving Projection (SPP) put forward by Qiao, etc. keeps the geometrical structure relationship in the sparsity reconstruction during the dimensionality reduction process (Qiao, L.S., Chen, S.C. & Tan, X.Y. 2010). However, when the supervision information is insufficient, SPP performance will be easily impacted by the integrated structure of the sample data. Therefore, Gu, etc. introduced the category of marking supervision information through regular methods (Gu, N. N., Fan, M. Y., Qiao H., et al. 2012) and Gui, etc. increased the constraint conditions to introduce the largest category marking in edge criteria supervision (Gui, J., Sun, Z.N., Jia, W., et al. 2012). However, the algorithm of supervision information is a kind of sample based on category marking. As for the supervision of information, pair constraint includes the similar positive constraint set and different negative constraint set to ensure that the dimensionality reduction performance of the pair constraint feature information–oriented Pair Constraint Guidance Feature Projection (PCFP) will be easily impacted by pair constraint set compositions (Tang, W. & Zhong, S. 2006). It is noteworthy that, when comparing with the category marking supervision information, the pair constraint supervision information contains stronger classifications, which can be obtained by sample comparisons with even automatic operations of the machine.

This paper proposes a semi-supervised dimensionality reduction algorithm, integrated with the sparsity reconstruction feature information and pair constraint feature information. The algorithm applies the linear weighting mode to integrate the sparsity reconstruction feature information and pair constraint feature information as well as applies the genetic algorithm to complete the automatic optimization setting of the linear weighting balance parameters of two different types of information integration. Experiments show that, when comparing the algorithm with the previous method, it has good dimensionality reduction classification performance.

## 2 DETERMINATION OF SDRSRPC ALGORITHM TARGET FUNCTION

The paper, based on SPP and PCFP target function, proposes to apply the linear weighting mode to integrate the sparsity reconstruction feature information with the pair constraint feature information as well as complete the linear weighting integration under corresponding constraint conditions of these two types of information. Semi-supervised

Dimensionality Reduction fusing Sparsity Reconstruction and Pairwise Constrint (SDRSRPC) target function acquisition process is as follows:

(1) Firstly, based on SPP and PCFP target function, we list the sparsity reconstruction feature information, pair constraint feature information, and corresponding projection constraint conditions. Supposing that the sample $X=[x_1,\ldots,x_n]\in R^{d\times n}$, positive constraint set $ML=\{(x_i,x_j)|x_i$ and $x_j$ are the same type$\}$, the negative constraint set $CL=\{(x_i,x_j)|x_i$ and $x_j$ are not the same type$\}$, SPP target function description is as follows (Qiao, L.S., Chen, S.C. & Tan, X.Y. 2010):

$$\max_w w^{\mathrm{T}}X(S+S^T-S^TS)X^{\mathrm{T}}w \quad s.t.\ w^{\mathrm{T}}XX^{\mathrm{T}}w=1 \tag{1}$$

where, $S$ represents $X$'s sparsity reconstruction weight value matrix, $X(I-S-S^T+S^TS)X^T$ represents the sample $X$'s sparsity reconstruction feature information, $w$ represents the projection matrix $W=[w_1,w_2,\ldots,w_d]$'s column vector, and the constraint conditions $w^TXX^Tw=1$ is to avoid degenerate solution.

The PCFP target function's description is as follows[4]:

$$\max_w\left(\sum_{(x_i,x_j)\in CL}\|(w^Tx_i\text{-}w^Tx_j)\|^2-\sum_{(x_i,x_j)\in ML}\|(w^Tx_i\text{-}w^Tx_j)\|^2\right) \quad s.t.\ w^Tw=1 \tag{2}$$

where, $\sum_{(x_i,x_j)\in CL}\|(w^{\mathrm{T}}x_i\text{-}w^{\mathrm{T}}x_j)\|^2-\sum_{(x_i,x_j)\in ML}\|(w^{\mathrm{T}}x_i\text{-}w^{\mathrm{T}}x_j)\|^2$ represents the pair constraint feature information, $w$ represents the projection matrix $W=[w_1,w_2,\ldots,w_d]$ the column vector, and the constraint condition $sw^Tw=1$ is to avoid a general solution.

(2) According to formula (1) and formula (2), these two types of feature information are subject to the linear weighting information integration; in addition, the linear weighting method is also applied to integrate corresponding constraint conditions $sw^TXX^Tw=1$ and $w^Tw=1$ of two types of feature information. The SDRSRPC target function proposed by the paper is:

$$\max_w\left(w^{\mathrm{T}}\left[\beta XS_\alpha X^{\mathrm{T}}+(1-\beta)P_\alpha\right]w\right) \quad s.t.\ w^{\mathrm{T}}\left(\beta XX^{\mathrm{T}}+(1-\beta)I\right)w=1 \tag{3}$$

where, $w$ represents the projected vector, all different projected vectors are used to form the projection matrix $W=[w_1,w_2,\ldots,w_d]$ according to the corresponding feature value in a reversed order.

$$S_\alpha=S+S^T-S^TS \tag{4}$$

$$P_\alpha=\sum_{(x_i,x_j)\in CL}\|(x_i-x_j)\|^2-\sum_{(x_i,x_j)\in ML}\|(x_i\text{-}x_j)\|^2 \tag{5}$$

The weighting parameter $\beta$ represents the contribution degree of SPP in SDRSRPC, ranging between [0,1]. $1-\beta$ represents the contribution degree of PCFP in SDRSRPC. If $\beta=0$, SDRSRPC will be reduced to PCFP; if $\beta=1$, SDRSRPC will be reduced to SPP.

(3) The target function (3) can be converted to the minimization problem:

$$\min_w\left(-w^{\mathrm{T}}\left[\beta XS_\alpha X^{\mathrm{T}}+(1-\beta)P_\alpha\right]w\right) \quad s.t.\ w^{\mathrm{T}}\left(\beta XX^{\mathrm{T}}+(1-\beta)I\right)w=1 \tag{6}$$

By Lagrange multiplier method, the formula (6) can be converted to the minimum value optimization function:

$$L(w)=-w^{\mathrm{T}}\left[\beta X\mathrm{S}_\alpha X^{\mathrm{T}}+(1-\beta)P_\alpha\right]w+\lambda\left[w^{\mathrm{T}}\left(\beta XX^{\mathrm{T}}+(1-\beta)I\right)w\text{-}1\right] \tag{7}$$

Derive $w$ and make the derivative be 0. After compilation, the following can be obtained

$$\left[\beta X\mathrm{S}_\alpha X^{\mathrm{T}}+(1-\beta)P_\alpha\right]w=\lambda\left(\beta XX^{\mathrm{T}}+(1-\beta)I\right)w \tag{8}$$

According to the corresponding feature value $\lambda$, select the corresponding feature vector of the feature value to compose the projection matrix $W=[w_1,w_2,\ldots,w_d]$.

As for the linear weighting balance parameters, different $\beta$ values determine the specific gravity between the sparsity reconstruction feature information and pair constraint feature information. The following experiment applies the Genetic Algorithm (GA) to perform the dimensionality reduction performance analysis and the automatic selection of the linear weighting balance parameter, $\beta$ value.

## 3 EXPERIMENTAL ANALYSIS

Facial data are typical high-dimensional data; in the experiment, real face database with different features are selected as the experimental data for experiments and for analysis of the corresponding experimental result.

### 3.1 *Experimental data*

AR contains more than 4000 face images. Each face group has 26 face photographs. Every group of face pictures are taken during two periods, with an interval of three weeks. Each period has 13 face images with different expressions, illuminations, and cover. In the experiments, the AR face image has been adjusted to $30\times 30$ pixels. Figure 1 shows AR's one group of face samples.

### 3.2 *Experiment device*

In order to verify the SDRSRPC algorithm integration and dimensionality, the reduction performance

Figure 1. AR's one group of face samples.

Table 1. Specific parameter settings of each algorithm.

| Algorithm | Parameter Settings |
|---|---|
| SPP | None |
| PCFP | None |
| SDRSRPC ($p$,$c$) | $pc=0.6, pm=0.1$ |
| RLPP | $k=7, t=1$ |
| SSDR | $\alpha=1, \beta=20$ |
| NPSSDR | $\alpha=0.1, \kappa=5$ |
| DSPP | $\lambda=1$ |

proposed by the paper, in addition to selecting SPP and PCFP algorithms, also selects DSPP (Gu, N. N., Fan, M. Y., Qiao H., et al. 2012), SSDR (Zhang, D.Q., Zhou, Z.H. & Chen, S.C. 2007), RLPP (Yu, G., Peng H., Wei J., et al. 2010), NPSSDR (Wei, J. & Peng, H. 2008), semi-supervised sparsity dimensionality reduction algorithms for comparison. Table 1 shows the specific settings of each algorithm.

Where, $p$ represents the population size, $c$ represents the length of the chromosome, $pc$ represents the crossover probability and $pm$ represents the mutation probability. Usually, the crossover probability $pc$ and the mutation probability $pm$ are set to 0.6 and 0.1, respectively. For convenient expression, SDRSRPC ($p$,$c$) represents SDRSRPC with parameters $p$ and $c$.

In addition, the experiment takes the recent neighbouring classification algorithm as the classification algorithm. Each group randomly selects $L$ face images as the training samples, the remaining ones are treated as test samples. Take all marking samples to build the pair constraint set. In order to verify the performance of the algorithm, the retained feature dimensionality shall undergo increment and calculation of the corresponding accuracy of identification in accordance with the step length $D$. For each test, randomly select samples from the marking information to build the positive constraint set and the constraint set. All experiments will be carried out for twenty times and the average value will be made as the experimental result.

### 3.3 *Experimental result and analysis*

To accurately describe SDRSRPC's performance, the paper performs analysis from the dimensionality reduction performance and the linear weighting balance parameter optimization performance.

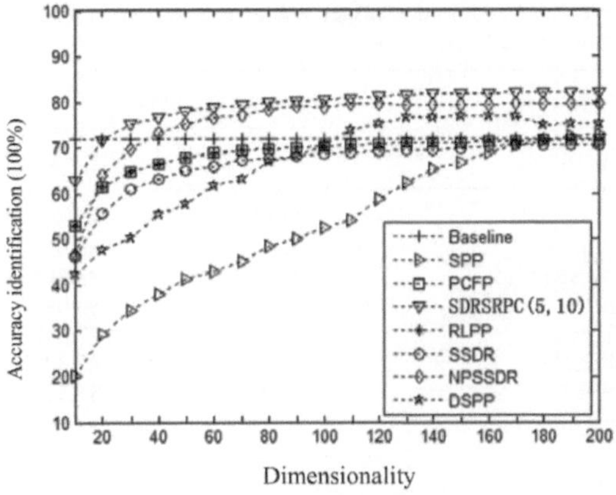


(a) $L = 5$ and $D = 10$

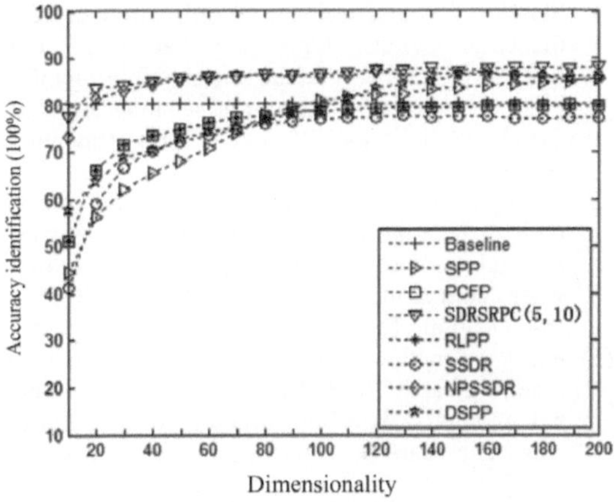


(b) $L = 10$ and $D = 10$

Figure 2. Accuracy identification (%) under different $L$ and $D$ of AR face database vs. dimensionality.

#### 3.3.1 *Dimensionality reduction performance analysis*

We select SDRSRPC (5,10) as the algorithm with parameters. The feature dimensionality increment is done according to the step length $D$. Figure 2 gives each group a training sample number $L$ and an accuracy identification under the feature dimensionality increment step length $D$.

In addition, to verify the best dimensionality reduction classification performance of SDRSRPC, the paper also lists the maximum accuracy identification of different algorithms under different training sample numbers $L$. Table 2 gives the corresponding experimental results.

Where, the boldface represents the maximum accuracy identification under a different training sample number $L$.

Table 2. Maximum accuracy identification (%) under different $L$ of AR.

| Algorithm | Sample Number 5 | Sample Number 10 |
|---|---|---|
| SPP | 72.55 | 85.53 |
| PCFP | 71.71 | 80.04 |
| SDRSRPC (5,10) | **82.15** | **88.39** |
| RLPP | 71.73 | 80.02 |
| SSDR | 70.71 | 77.97 |
| NPSSDR | 79.73 | 87.34 |
| DSPP | 77.01 | 85.53 |

According to the above experimental results, the following conclusions can be drawn:

(1) SDRSRPC's discriminant classification performance is obviously better than that of SPP and PCFP, which shows that the linear weighting integration method is effective, and it can ensure that SDRSRPC inherits SPP and PCFP's respective features.
(2) As the sparsity representation has good robust performance and sparsity reconstruction has good internal geometric feature description ability, when comparing with SSDR, SDRSRPC has significant performance advantages.
(3) RLPP strengthens the local manifold structure retaining capability but RLPP ignores the internal geometry manifold structure of the initial data. The sparsity reconstruction in SDRSRPC has good internal geometry information. Therefore, the SDRSRPC dimensionality reduction classification performance is better than RLPP.
(4) The NPSSDR keeps local structure features through regular methods. But the method for the NPSSDR to keep local structure features is still based on approximate nonlinearity. Therefore, in the face image with obvious nonlinear structure features, the NPSSDR's performance is inferior to that of SDRSRPC.
(5) As a semi-supervised sparsity dimensionality reduction algorithm, DSPP also inherits the sparsity representation features, but the performance of DSPP is inferior to SDRSRPC, the main cause is that DSPP's supervision information is based on category marking, and its classification performance is inferior to the pair constraint supervision information.

### 3.3.2 *Optimization performance analysis of the linear weighting balance parameter β*

To verify adaptive setting effectiveness of the balance parameter $\beta$ with GA, the paper selects some specific retention feature dimensionality and increases the step length of $\beta$ in SDRSRPC (5, 10) to 1 from 0.05 and calculates the corresponding accuracy of identification. In addition, it also, according to different dimensionalities, calculates the optimization weighting balance

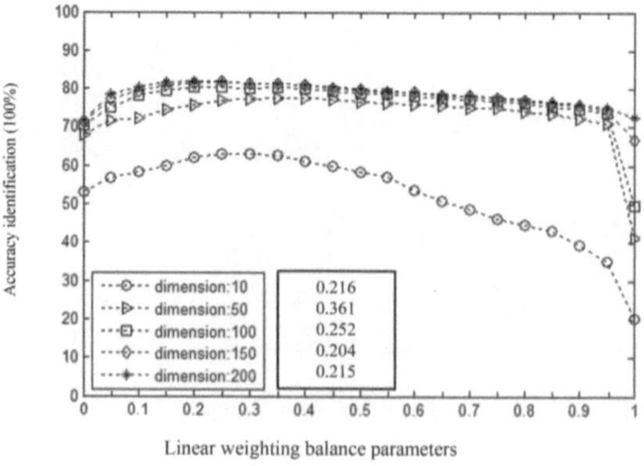


(a) $L$=5

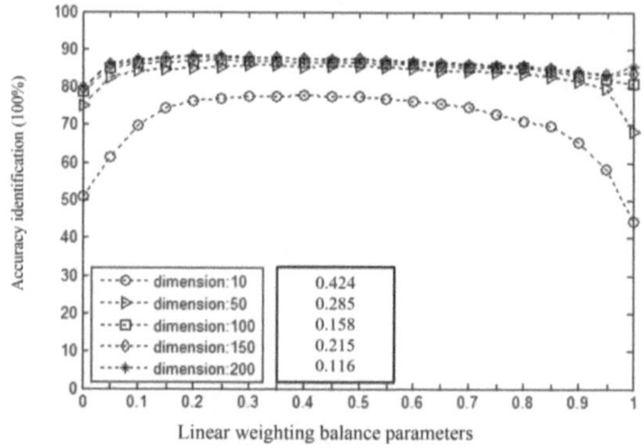


(b) $L$=10

Figure 3. Accuracy identification (%) under different L of AR face database vs. linear weighting balance parameter.

parameter value, based on GA in SDRSRPC (5, 10). Figure 3 shows the specific experimental result.

The analysis on above experimental results shows that:

(1) When the retained feature dimensionality is higher than a certain threshold value, the linear weighting balance parameter $\beta$'s impact on the accuracy of identification of SDRSRPC (5, 10) is almost the same, which shows that the selection of the linear weighting balance parameter, $\beta$ based on GA, has high stability.
(2) $\beta$ is in the [0,1] section, which increases with the step length of 0.05 and calculates the corresponding accuracy identification of SDRSRPC (5,10). The calculation method of the linear weighting balance parameter $\beta$ needs 21 times and the optimization value will be selected. However, SDRSRPC (5,10) only calculates 3 algebra individuals for 15 times. Accordingly, if applying GA to calculate the linear weighting balance of the parameter $\beta$ value, the searching time will be shorter.
(3) The linear weighting balance parameter $\beta$, calculated by SDRSRPC (5,10), and the optimization

linear weighting balance parameter $\beta$, obtained by [0,1] section traversal, are almost the same, which shows that the automatic setting method of the linear weighting balance parameter $\beta$ by the genetic algorithm perfect form (3) is effective.

## 4 CONCLUSIONS

The paper proposes a semi-supervised dimensionality reduction algorithm, integrated with the sparsity reconstruction feature information and pair constraint feature information. The algorithm applies the linear weighting method to integrate with the sparsity reconstruction feature information and pair constraint feature information, as well as applying the genetic algorithm to complete the automatic optimization setting of linear weighting balance parameters of two different types of information integration. The algorithm effectively solves the insufficiently supervised information and the failure of the automatic parameter setting in the existing sparsity representation of the semi-supervised dimensionality reduction algorithm. The experimental result shows that, when compared with the recent relevant semi-supervised algorithm, SDRSRPC has a better dimensionality reduction classification performance.

## ACKNOWLEDGMENTS

This work was supported by the National Natural Science Foundation of China (71171148, 61403238), the National Key Technology R&D Program (2012BAD35B01, 2012BAH13F04), the National High-Tech Research and Development plan of China (2012AA062203), and Key Lab of Information Network Security, Ministry of Public Security (C14602).

## REFERENCES

[1] Qiao, L.S., Chen, S.C. & Tan, X.Y. 2010. Sparsity preserving projections with applications to face recognition. *Pattern Recognition* 43(1):331–341.

[2] Gu, N. N., Fan, M. Y., Qiao H., et al. 2012. Discriminative Sparsity Preserving Projections for Semi-Supervised Dimensionality Reduction. *IEEE Signal Processing Letters* 19(7): 391–394.

[3] Gui, J., Sun, Z.N., Jia, W., et al.2012. Discriminant sparse Neighborhood preserving embedding for face recognition. *Pattern Recognition* 45(8):2884–2893.

[4] Tang, W. & Zhong, S. 2006. Pairwise constraints-guided dimensionality reduction. *Proceedings of SDM Workshop on Feature Selection for Data Mining:* 295–310.

[5] Zhang, D.Q., Zhou, Z.H. & Chen, S.C.2007. Semi-Supervised dimensionality reduction. *Proceedings of the 2007 SIAM International Conference on Data Mining:* 629–634.

[6] Yu, G., Peng H., Wei J., et al.2010. Robust Locality Preserving Projections with Pairwise Constraints. *Journal of Computational Information Systems* 6(5):1631–1636.

[7] Wei, J. & Peng, H.2008. Neighbourhood preserving based semi-supervised dimensionality reduction. *Electronics Letters* 44(20):1190–1192.

*Frontiers in Computer Education – Wang (ed.)*
*© 2015 Taylor & Francis Group, London, ISBN 978-1-138-02797-8*

# LustreSim, a simulator for the Lustre file system

Xiayu He & Guangming Liu
*Computer College, National University of Defense Technology, Changsha, China*

ABSTRACT: During practical applications, Lustre file system have shown their inadaptability for the specific application and design demands of the next generation of super computer storage systems. In order to simulate and analyse the I/O performances of Lustre file systems in different configurations and applications facing different I/O characteristics, we designed and realized in this paper, for the Lustre file system, a simulator, LustreSim. The experiment shows that LustreSim can effectively simulate a Lustre file system and obtain reliable results.

## 1 INTRODUCTION

On the TOP500 list published in Jun 2014, the TH-2 super computer, developed independently in our country, is the fastest computer in the world due to its 33.86 PFLOPS (Peta Floating Point Operations Per Second) operation performance, while the earliest ENIAC operation speed is only 5000 times/s. With the rapid improvement in computing power, the performance improvement of the storage system cannot meet the demand of the rapidly-growing computing power so that mismatching of computing performance with storage performance has limited the overall performance of high-performance computers.

As a branch of computer science, currently, the high-performance computer is standing on a rapid development stage and has a very important position in scientific research and technological service. Taking the National Super Computer Center in Tianjin for example, the application range of the super computer TH-1A used in the centre covered oil exploration, biological medicine, cartoon rendering, and design simulation, etc., in multiple fields besides scientific calculations. It has been widely used for administrative informationization, smart city and intelligent traffic etc. The TH-1A super computer has been in service for nearly four years in the Center in Tianjin. Through a period of operation, it shows that the system has a reasonable design, good integrated performance, and a stable and reliable system, and has successfully finished many large-scale, complicated calculation tasks. The TH-1A is provided with a Lustre file system as its storage system i.e. the system structure of calculation separated from storage. In the practical application for such as I/O, huge and intensive application for oil exploration data processing etc, the Lustre system for this structure shows some inadaptability, which has limited full play in the overall performance of this computer.

Owing to the inadaptability of the storage system for the specific application and research demands in a new storage hierarchy for the next generation of high-performance computers, as shown in the above TH-1A operation, it is necessary to simulate the storage system needed before the scheme is determined. Simulating the storage system is an efficient path to evaluate the storage system's performance during its design and to obtain a storage system design. At the same time, during the research for a new computer system structure, especially for a new storage system structure, the storage system simulator becomes the best tool to explore and research the new storage system structure because its performance cannot be verified on the existing system. This paper presented the storage system simulator LustreSim based on the Lustre storage system.

## 2 DESIGN AND REALIZATION OF LUSTRESIM

### 2.1 *Overview of Lustre file system*

Lustre is an object-oriented file system and is widely used for large-scale high-performance computing cluster system, provided with the super computer which its computing subsystem has separated from its storage subsystem. Figure 1 shows the system's structural diagram of Lustre. It consists of three components: Metadata Servers (MDS), Object Storage Servers (OSS) and Client. MDS mainly provides metadata service for the namespace administration of the overall file system to save and maintain the file distribution layout attribution, which contains data object index information. MDShas is a Metadata Target Device (MDT), used to save metadata objects such as file names, directory structures and access rights etc. OSS is used to lead out the block device and provide file I/O service. Each OSS is used to manage one or more Object Storage Target (OST) devices.

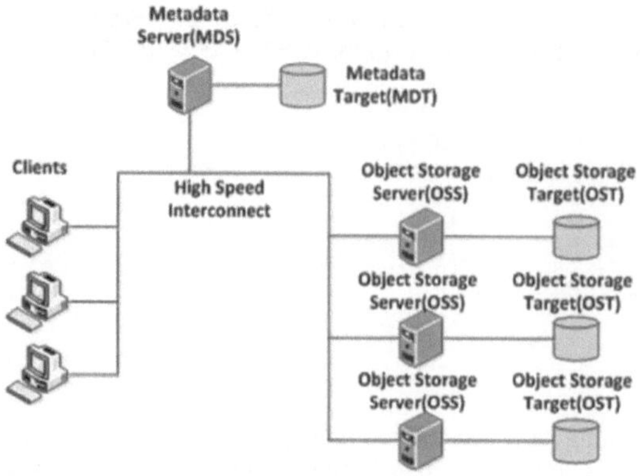


Figure 1. Structure of Lustre file system.

OST is used to save file data objects. The file data can be stripped to multiple objects located on OSTs to provide rapid concurrent read/write performance. The client realized interface of POSIX file system, such as open(), read() and write() etc, can access file service through a shared high-performance network. Generally, for the high-performance computing field, the client, MDS and OSS are connected mutually through a high speed network, such as InifiniBand. Lustre is provided with an Out of Band system to separate the control path from the data path. The client can directly set up data path to rapidly obtain data with OSTs after obtaining file metadata.

The operation process of the high-performance computer is as follows: when submitting parallel computation, the operator shall submit the operation description script. In the operation description script, the operator must appoint the relationship between the applied computing node and the data to be transmitted, and which files shall be transmitted to which node's local storage. The job dispatcher of the high-performance computer system can start data transmission when allocating computing resources for the parallel computation, and start computation after the data transmission is finished. During computation, the parallel program can guide its processes to write back the final result into the backend storage.

Figure 2 shows the I/O process of Lustre at the storage end. When the client accesses one file, he shall firstly send the file access request to MDS through MDC. MDS will inquire about the inode information of the file requested on MDT and return the layout information back to client. The data distribution information of the file is stored in the inode. Each object contains a part of the data of the file, which is stored on OSTs. After receiving the stripped information, Client can set up the data path for an I/O operation with OSCs and OSTs.

### 2.2 *Selection of drive mode*

Currently, the method to simulate the system structure mainly includes trace-driven and execution-driven drives.

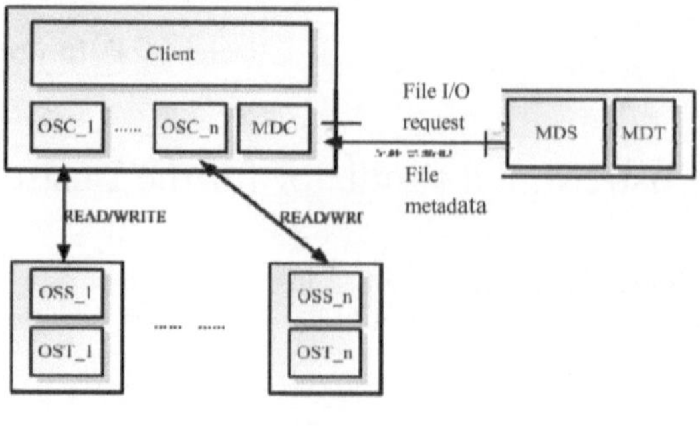


Figure 2. File access process in Lustre.

Trace-driven simulating is to use the trace of a process to drive the simulation. Firstly, the trace is obtained through the executed program in the real system. This real system must have a similar structure to the target device. After that, the trace is used as the input to run the simulator. Trace-driven simulating is roughly divided into three stages: recording traces from the program, trace selection, and targeted machine simulation. Trace-driven simulating has longer appearing time and is widely used in simulator research in the storage field.

Execution-driven simulating is used to combine program execution with target system simulation. Generally, it is a competitive process of multiple processes for the processor: Firstly, when the process is switched on, the process needs another processor to continue its work. Secondly, if the process switching among the processors does not occur, the current processor will execute the basic block in the program and promote the system clock. The repeated execution of the above two steps will promote the overall simulating process until each process is ended or the simulation has finished. Currently, execution-driven simulation is more used for the development of parallel compilers and an analysis of the application's performance.

Owing to the congenital advantage of the trace-driven selection on the storage system simulation, LustreSim uses trace-driven to set up the simulator, and uses a simplified set of I/O traces from the actual system for an input so that all the clients can competitively use storage nodes in the Lustre file system to obtain the results of the execution.

### 2.3 *Design of LustreSim system structure*

LustreSim, corresponding to the Lustre file system, consists of three modules: Client module, Metadata Server Module (MDS) and file data server module OSS. Client module is responsible for the generation of I/O requests, the promotion of a simulation clock, and interaction with the MDS and OSS buffer zones. MDS is mainly used to provide metadata service and return OST layout information. OSS is used to manage OST and provide I/O services. OST can provide different transmission bandwidths according to different RAID configurations.

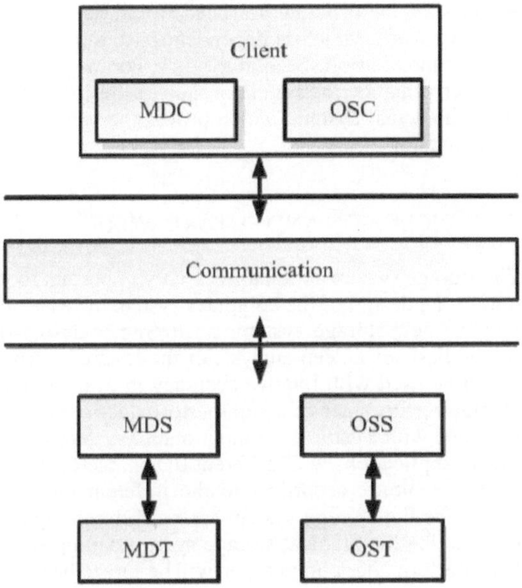


Figure 3. LustreSim system structure.

In order to study the I/O performance of the Lustre file system better, and discover the system's bottlenecks, LustreSim should have the following features:

1. For each I/O request, the time which it takes to pass through each of the system's components and the delay in response can be logged.
2. The input of the simulator can be obtained not only from the real trace, but also from a random generator according to any given I/O characteristics.
3. It can realize configurability of the scale and performance on client number, OST number, and stripped strategy, and a MDS's processing capacity, etc.

### 2.4 *Specific realization of LustreSim*

According to the module division of LustreSim, we use object-oriented methods to realize LustreSim.

During actual realization, if thousands of clients are configured with multithreads to realize parallel generation requested by the I/O may cause large time and space overheads, these overheads will greatly increase the actual operation time when the simulator executing the simulation process. Therefore, we want to use a single thread with multilevel lists to realize concurrent generations of multiple client I/O requests. Client can take each I/O request as one Event. Cid in Event shows the Client generating the request. The Event with the same Cid is saved in the same line in the multilevel linked list, constituting one request list, from the actual client (see Figure 4). Each line of the multilevel list is saved with all the I/O requests from one actual client.

At the beginning of the simulation, the first event on each line of the multilevel list is sent to an MDS buffer. When the client thread receives a completion message of the event each time, it can send the next event on the

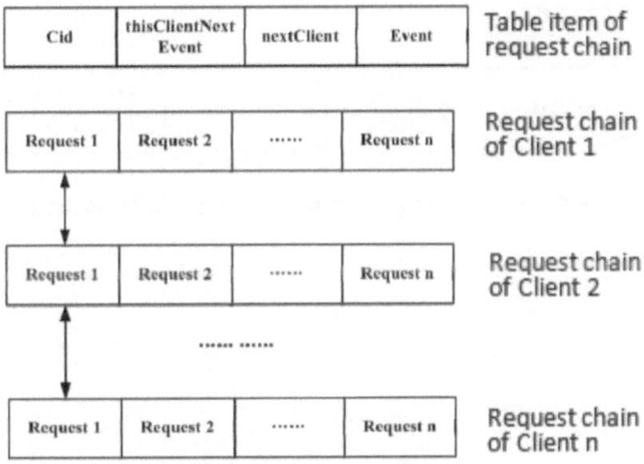


Figure 4. Multilevel list of client requests.

list to the MDS buffer and release the corresponding item on the list where the event is located. When all the request chains are empty, the simulator ends.

MDS

Generally, the actual Lustre file system is provided with two MDSs to provide service and data backup (active/standby). Therefore, during realization, the processing module for MDS metadata request only needs one thread. When the MDS processing module is empty in the MDS buffer, we took out the first event from the queue in the buffer and inquired about the metadata information of the event, then we filled its layout information to finish the metadata inquiring operation. During the actual operation of Lustre, after the client received the metadata information, it will start to set up the path for data transmission with the corresponding OSTs. The main time overhead affecting this process is the delayed time for the request from the queuing at the OSTs end. Therefore, in the model, the process can be simplified as MDS directly sends I/O request to the corresponding OSTs after the metadata information has been obtained.

The Metadata of Lustre is saved in the MDT in the form of an inode. It is possible to obtain the file layout information by asking the inode of the corresponding file through FileName. In our realization, metadata information of the file is saved in a HashTable. The FileID is used as the key, and the other parts are saved in HashTable, the file layout information. At the beginning, HashTable is empty. When a request appears, if HashTable is missed, we fill the table and return the metadata information according to stripe count and current unoccupied OST (feedback by OSS).

OSS

Corresponding to Client realization, it is unnecessary to allocate one thread for each OST that the realization is rather complicated. By referring to realization of the multilevel list, as well, a multilevel list can be designed for OST. Each line of the multilevel list saves all the requests that refer to its corresponding OST. The I/O bandwidth of OST is related to the configuration of RAID array. I/O bandwidth provided by RAID array can be obtained by testing the mathematical modelling or the actual system.

**HashTable**

| FileID | OSTNum Object1 | …… | OSTNum ObjectN |
|---|---|---|---|

Figure 5. Organizational form of metadata in HashTable.

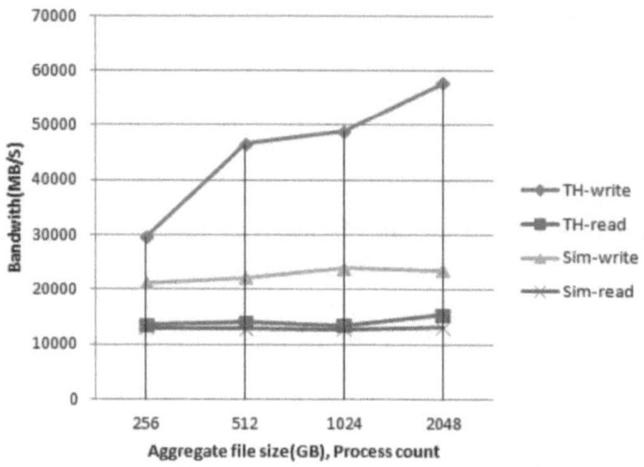


Figure 6. Testing data comparison of TH-1A system with the simulator.

## 3 THE LUSTRESIM TEST

In order to verify the validity of LustreSim, we performed a series of experiments based on the simulator and compared it with the testing data of the real TH-1A system. TH-1A uses Lustre as its I/O subsystem, and is provided with total four Lustre file systems. We use the largest file system for testing. The parameters for test case include 64 OSTs, 512, 1024 and 2048 cores (12 cores per node) for IOR FPP (File Per Processor). The I/O request block size is 1M. The aggregation file size is 512G, 1024G and 2048G respectively. The simulator is provided with same system conditions. Figure 6 shows the comparison of the performance testing data between them both. The result shows that the actual system of write performance is obviously better than the simulator, which is because the TH-1A is provided with a Seagate enterprise-level SATA II hard disk, which has a large disk buffer. For the write operation, the actual system would be taken as finished when written to the disk buffer but the simulator does not realize the buffer hierarchy. However, if the data block is increased to or higher than the disk buffer size, the disk's write speed would be close to the read speed, which is dependent on the magnetic density and rotation speed of the disk. For the parallel read performance, the simulator shows little difference from the actual system, which proves the validity of the simulator.

## 4 CONCLUSION AND FUTURE WORK

The storage system simulator is a very important tool during the design of the computer system or research into the new storage system. LustreSim is designed on the basis of a deep analysis of the Lustre system. It can be used with relative accuracy to evaluate the storage performance of a high-performance computer provided with Lustre as an input/output system when facing applications with different I/O characteristics.

In the future, according to the different requirements for the storage system facing different applications, the hierarchical storage system equipped by high-performance computation will be sure to become the vital research issue. Therefore, the future objective direction for LustreSim should be to develop an intermediate storage level based on LustreSim, which can simulate different storage mediums (such as SSD) through different performance parameters and configurations. In this way we can evaluate the impacts brought by the intermediate storage. Then we can provide a reference for the next generation of high-performance computer storage subsystems.

## REFERENCES

A white paper from cluster file systems, Inc. July 2007 Version 2

National Supercomputer Center in Tianjin Website. http://www.nscc-tj.gov.cn/ areas/areas_1.asp

Ping Zhu, Jiantao Zhu, et al. Key techniques of high performance computing storage. Research and development of the computer, 2011.

Sun Microsystems, Inc. Solving the HPC I/O Bottleneck: Sun™ Lustre™ Storage System [EB/OL]. April 2009.

Sorceforge Organization. IOR Benchmark [EB/OL]. https://asc.llnl.gov/sequoia/benchm_arks/IOR_summary_v1.0.pdf.

Top500 Supercomputer Sites P D. http://top500.org/lists/2014/06/

Zhibin Yu, Hai Jin, Nanhai Zou. Computer architecture software simulation technology. Journal of Software, 2008, 19(4).

*Frontiers in Computer Education – Wang (ed.)*
*© 2015 Taylor & Francis Group, London, ISBN 978-1-138-02797-8*

# A strong group quantization model in Go based on the stone influence function

Yuhan Zhou, Yuqi Zhang, Jiali Luo & Jianxiao Xie
*Department of Electronic Information Engineering, Nanchang Hangkong University, Jiangxi Province, China*

ABSTRACT: Quantization of strong groups in the Go game is critical for both professional players and Go programs. In this paper, we define a stone influence function named Strength, based on which a strong group quantization model named Filtration is established, and the model parameter is further optimized by multi-level species compete-die out algorithm. Through this approach various strong group quantization models based on different levels can be obtained. The results of the test show that compared with the original model, Filtration's accuracy and speed of operation are promoted by 51.7% and 48.4% respectively, which is precise enough to do positional analysis in the real time Go programs. Moreover, this model can be applied in researches on computer games, artificial intelligence and game software.

*Keywords*: Strong Group Quantization, Stone Influence Function, Go

## 1 INTRODUCTION

Go, known as the art of imaginative and logical thinking, is now regarded as one of the most difficult challenges for artificial intelligence[1]. The exhaustive research applied in the chess-playing program can hardly work here due to the vast search space in the Go game[2]. Presently, the strongest Go-playing program boasts a level of amateur 6 dan, which a human player can reach when learning for more than three years[3]. Hence, the heuristic search algorithm designed by referring to the thinking method of a human player is a workable approach to improve the level of a computer program[4].

Go researchers generally consider that the calculation of strong group value plays an important role in computer Go, the accuracy of which directly influences the final evaluation results especially during static evaluation. However, research on the calculation of strong group values makes slow progress because of the great difficulty in quantizing strong group values. Abe Yoshiteru presented a method, based on the stone quantity of a strong group, to quantize a strong group value in 1990s[5]. However, it had an extremely low accuracy which made it impossible to be applied in the computer Go programs. Whereas some famous Go programs such as MoGo[6], Crazy Stone[7], and GNUGO[8], etc., even define the quality of strong groups. The algorithm used to quantize a strong group value made its debut in a winning probability program – Winnable, which firstly adopted a quantification method to calculate strong group value and has gained good operational results[9]. Nevertheless, Fence, the strong group value quantization model in Winnable calculates with a relatively low speed and accuracy, because of its direct adoption of the stone influenced function of GNUGO, and the model was limited to only one certain Go level by the fixed parameters.

This paper presents a new method of strong group influence value quantification as follows: take the influence of the strong group as the diffusion of control probability in the two-dimensional space, therefore the strong group's value is calculated as the sum of the control probability. Meanwhile, the Multi-Level Species Compete Die out (MLSCD) algorithm is used to optimize the model's parameter and adjust the judgments according to different Go levels. Based on the levels of the players, the model finally optimizes the strong group quantification model parameter of professional and 1 dan level, which proves the accuracy and advance of the model. With the application of the method, the program can calculate the strong group value in the start game and middle game, so that the situation can be more accurately evaluated.

## 2 STRONG GROUP QUANTIZATION ALGORITHM

### 2.1 *Idea of quantization*

A chain of stones that has great influence in a broad area is referred as 'strong group'. Researchers of computer Go consider that a computable model should be defined so as to quantize and evaluate the influence of

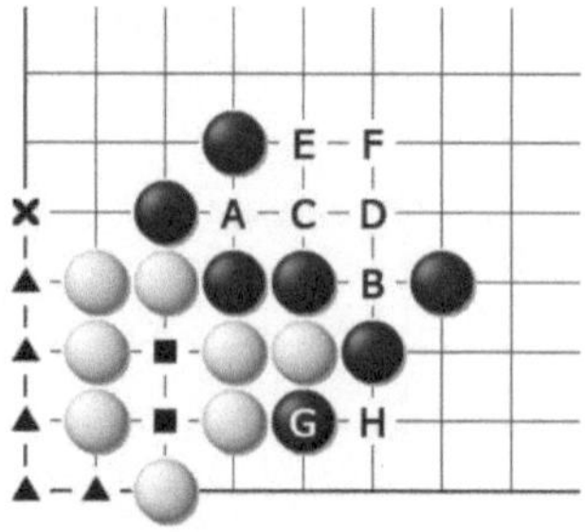


Figure 1. An example of territory and strong group.

the stones[10]. The traditional method of quantization assesses the group's value as the sum of the points it controls, which works well in quantizing territories. As Figure 1 shows, the white's territory marked triangle and square which cannot be invaded holds the value of 9 points considering two captured stones in the square marked area.

However, due to the potential influence of the strong group, the method of calculating the controlled points can hardly work here. As Figure 1 shows, though points C, D, E, F are greatly influenced by black, none will be captured if white occupies any of them, so they cannot be counted as black's territory. As a result, according to the traditional method, the strong group's value cannot be defined except by the two points marked A and B. On the contrary, however, the black group holds a value of 20 points[11].

Therefore, a new method of quantizing strong group's value is presented as follows: as strong groups have an influence over the vast space around them and the influence value of every unoccupied point varies from shapes and distances, we use probability to represent potential influence. Suppose black has a potential influence of $x\%$ over one certain point, which means the point has a $x\%$ probability to be black's one point, ie black has a value of $x\%$ point on this point.

Strong group's value can be taken as the diffusion of control probability in the two-dimensional space. We only need to obtain the strong group's value by adding up the influence probability of the unoccupied points around the strong group instead of calculating the points totally controlled by the strong group which is actually impossible to count.

### 2.2 *Stone's influence function*

When assessing the value of strong groups, proper influence function is required to quantize the influence of every move. The earliest influence function was proposed by Dr. Zobrist in his dissertation[12], and then various influence functions were successively established by Dr. Ryder[13] and Professor K. Chen[14]. However, these functions can merely be applied in confirming territories, dividing blocks and analysing the strength of groups.

To promote the accuracy and speed of the strong group's value quantization model, a stone's influence function needs to be established to meet the special requirements of the strong group value quantization.

According to the characteristic of the strong group and the results of the tests, it is concluded that stone's influence function should meet the following requirements:

(1) The influence value should be increased when the point gets closer to the stone.
(2) The degree of decay should get deeper when the point departs from the stone.
(3) The maximum influence distance of five generates the best effect according to the professional players' experience.
(4) The structure of the function should be concise and all procedures that may increase the calculation time should be reduced or revised in order to promote the speed.

According to the requirements above, the influence function Strength is established under the following rules:

(1) Given $(0, 0)$ is the coordinates of the influence source and the strength is $S(0, 0)$, the influence of the points on either x-axis and y-axis in the first quadrant would be:

$$\begin{cases} S(i,0) = \dfrac{S(i-1,0)}{1.5+0.5i} & (i \ge 1) \\ S(0,j) = \dfrac{S(0,j-1)}{1.5+0.5j} & (j \ge 1) \end{cases} \tag{1}$$

(2) The influence of the points on the angular bisector of the first quadrant would be:

$$S(i,j) = \frac{\sqrt{S(i,0)^2 + S(0,j)^2}}{2*3^{i-1}} \quad (i = j,\ i \ge 1,\ j \ge 1) \tag{2}$$

(3) The influence of the other points in the first quadrant would be:

$$S(i,j) = \frac{S(k,k)}{i+j} \quad (k = \min(i,j),\ i \ne j,\ i \ge 1,\ j \ge 1) \tag{3}$$

(4) Take y-axis as the axis of symmetry, the points in the second quadrant would be:

$$S(i,j) = S(-i,j) \quad (i \le 0,\ j \ge 0) \tag{4}$$

(5) Take x-axis as the axis of symmetry, the points in the third and the fourth quadrant would be:

$$S(i,j) = S(i,-j) \quad (j \le 0) \tag{5}$$

According to the rules above, the results would be as follows: Given $S(0,0) = 210$, the influence zone formed by the independent influence source is shown in Figure 2, in which the numbers are actually floating point numbers rounded off for convenience.

### 2.3 *Quantization algorithm*

Based on the stones' influence function above, this paper establishes a new strong group value quantization model named Filtration and works out

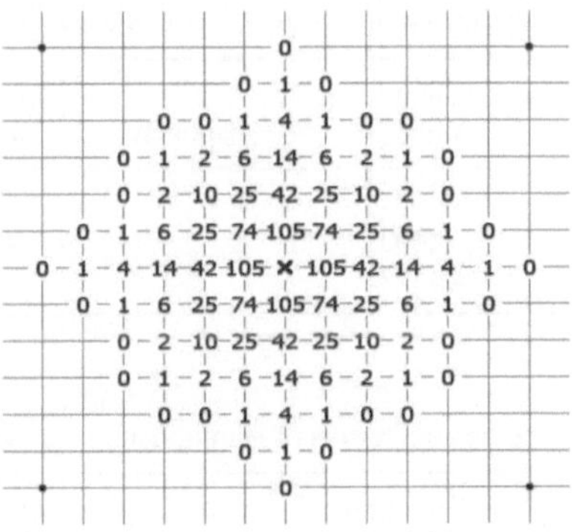


Figure 2. Strength's influence zone.

$I_{tol}$ which represents the sum of group influence value as follows:

(1) The black stones' influence value is a positive number while the white stones' is a negative number.
(2) Calculate the number of effective stones $n$: The stone with a number of liberty equal to or more than 3 is defined as an effective stone, to calculate its number. As Figure 1 shows, the black stone marked G has only 2 liberties, so it is not an effective stone. The black strong group's effective stone number $n = 6$.
(3) Calculate the threshold value: $V_{t1}$: $V_{t1} = \alpha * n$ and $V_{t2}$: $V_{t2} = \beta * n$
(4) Calculate normalize parameter $N$: $N = \gamma * n$
(5) Empty point adjust: If there are at least one black stone and one white stone beside an empty point, the empty point's influence value is 0.
(6) Calculate the influence value $I_i^*$ of the strong group on another point.
(7) Correct lines and corners: the correction method is

$$I_i^{**} = \begin{cases} \lambda * I_i^*, & \text{when } i \text{ on the No.1 line} \\ \mu * I_i^*, & \text{when } i \text{ on the No.2 line} \end{cases}$$

(8) Threshold value adjust: The larger the influence value, the more probability the empty point would be one point. However, no matter how large the influence value is, it is, at most, one point. For this reason, when $I_i^{**} \geq V_{t1}$, define $I_i = 1$. The points of which the influence value is the lesser, although it has an influence value, could not be conditioned by black in fact. Therefore, when $I_i^{**} \leq V_{t2}$, defines $I_i = 0$; other points, when $V_{t2} < I_i^{**} < V_{t1}$, define $I_i = \frac{I_i^{**}}{N}$.
(9) Calculate the sum of all points' influence value: $I_{tol} = \sum_{i \in U} I_i$ is the value of the strong group.

The strong group is closely related with the player's Go level, so it is highly subjective. Therefore, when setting the parameters, it is necessary to optimize the model parameters according to different Go levels.

Optimization means to find the best or quasi-optimal solution amongst lots of possible solutions to a problem. To optimize the parameters in the strong group quantization model means to form a combination of parameters, based on different Go levels, in order to assess the value of a strong group more accurately, so that the weight of the attacking or defensive moves can be adjusted and thus the best move can finally be determined upon.

## 3 OPTIMIZATION

### 3.1 *Genetic algorithm*

A Genetic Algorithm (GA) is adopted to optimize the parameters. GA is a computation module which simulates Darwinian's genetic selection and natural elimination in the process of creature evolution. The method is proved effective in resolving the optimization problem through the simulation of natural evolution. It was first proposed by Prof. J. Holland of Michigan University in 1975 and has now been widely used in the domain of artificial intelligence, machine learning, etc.[15].

### 3.2 *Fitness function and coding scheme*

The model is simplified as: $y = f(X, \psi)$. in which $X$ is an argument and $\psi$ represents the 5 parameters to be optimized. Least square method is usually adopted to define the object function of parameter optimization. Given the empirical data $\{(x_i, y_i) | i = 1, 2, \ldots, n\}$, in order to optimize parameters, the parameter combination $\psi$ should be obtained so as to minimize the function: $S(\psi) = \sum_{i=1}^{n} [y_i - f(x_i, \psi)]^2$.

There are 90 Joseki in[16], which represent one get territory and the other get strong group. First of all, a professional player is invited to judge the value of a strong group. Professional players present the highest Go levels, so the model of professional level could verify the correction and compare with the model of Winnable. Meanwhile, the worldwide top Go program is 2 kyu or so, and it is assumed that the Go program level equivalent of 1 dan will come out in two years. Therefore, we have prospectively optimized the parameters of a strong group quantization model based on the 1 dan level. Then 60 joseki are selected at random, used for optimizing parameters. Finally, a contrast test is made between the optimized professional-level model and the model in Winnable, in which 30 joseki are selected.

It can be inferred that the strong group quantization of optimized model can be represented by (6),

$$\min \sum_{i=1}^{n} [y_i - f(x_i, \psi)]^2 \tag{6}$$

In which $\psi$ represents the five parameters which need preparative optimization. After that, it is defined that the objective function of GA is (7).

$$F(\psi) = C - \frac{1}{n^2} \sum_{i=1}^{n} [y_i - f(x_i, \psi)]^2 \tag{7}$$

In the research of this paper, the constant $C = 600$.

The results of strong group quantization, calculated by the optimized model, are expected to be as close to the human players' judgments as possible, ie the larger $F(\psi)$ the better.

We adopt floating encoding in the coding scheme of this algorithm, for it is in correspondence with the variables and parameters of optimization problems, with high calculation accuracy and good convenience in realizing this algorithm. Since the winning probability model includes 5 parameters, the coding length of this algorithm $l = 5$. The value range of every gene in the coding string is determined by the Go characteristics of its corresponding parameters.

### 3.3 *Multi-level Species Compete-Die out algorithm model & optimized results*

To resolve the combinatorial optimization of control parameters and the genetic operator in GA, we introduced a multi-level species compete-die out algorithms model.

Species Compete-Die out (SCD) algorithm is a kind of genetic algorithm based on a niche evolutionary method and hierarchical fair competitions principles. It fully adopts the evolution phenomenon of the natural and the human society as well as the ideology 'survival of the fittest'. In terms of the initial fitness value, the niche evolution of initial species can be divided into groups. This SCD algorithm has been applied well in the parameter optimization of the growth models of cucumbers in greenhouses[17]. The Multi-Level Species Compete-Die out (MLSCD) algorithm model is based on the original SCD model, which has introduced the secondary GA; the primary and the secondary GA are embed with each other to realize parameter optimization of the winning probability model. Primary GA is applied to realize the parameter optimization of the winning probability model. SCD model is still used as the model frame for this algorithm, then the control parameters of the primary GA in the inner of species and between the species can be distinguished, and so can the genetic parameters; the execution strategy of the algorithm is determined by the secondary GA; the secondary GA is used for the realization of the combinatorial optimization of the control parameters and the genetic operators in the primary GA.

The MLSCD algorithms model can obtain better search results compared with the general GA model. The nested relational operation of the two-level GA model makes the iterative calculation increased by tens of times that lead to the rise in operational cost[18]. However, it is reasonably acceptable, for there is no real-time request.

The results of optimized parameters, shown in Tables 1 and 2, can be used for the program's instruction to calculate the value of a strong group.

The model of strong group quantization would finish after parameter optimization. Figures 3–5 are used as a reference for the convenience of description.

Table 1. Optimized results of filtration pro model.

| Parameter | Optimizing Value | Parameter | Optimizing Value |
|---|---|---|---|
| $\alpha$ | 28.6735 | $\lambda$ | 3.2298 |
| $\beta$ | 2.0031 | $\mu$ | 1.4508 |
| $\gamma$ | 28.5973 | | |

Table 2. Optimized results of filtration ama1 model.

| Parameter | Optimizing Value | Parameter | Optimizing Value |
|---|---|---|---|
| $\alpha$ | 28.6735 | $\lambda$ | 2.7904 |
| $\beta$ | 2.6843 | $\mu$ | 1.1263 |
| $\gamma$ | 30.1053 | | |

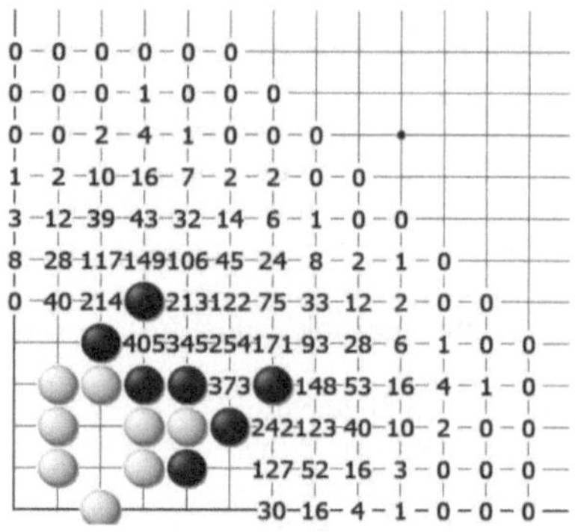

Figure 3. The initial influence value of a strong group.

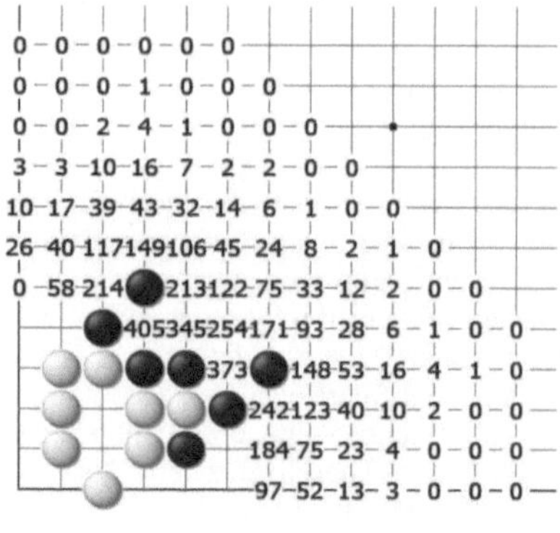

Figure 4. Influence value after correction of lines and corners.

Figure 3 shows the influence of the black's strong group in Figure 1 in the professional level. Figure 4 shows the influence after correction of the lines and corners. Figure 5 shows the normalized influence model. According to Figure 5, the sum of the strong group's influence value is 19.01 points, which is very close to 20 points judged by the professional player.

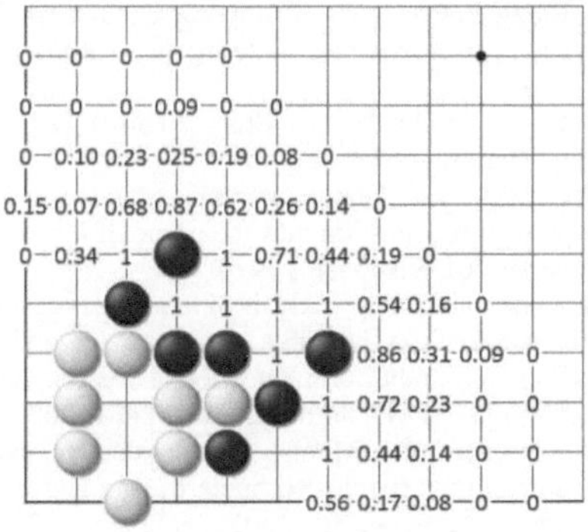


Figure 5. Ultimate influence value of a strong group.

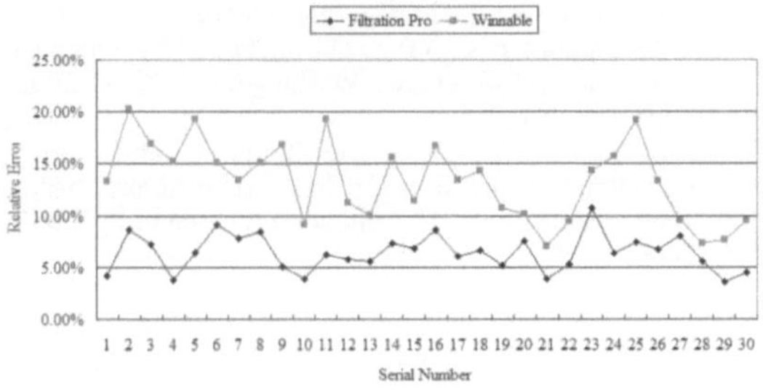


Figure 6. The comparison of results between winnable and filtration pro models.

## 4 EXPERIMENTAL RESULTS

We have used Intel Core 2 Duo CPU 3.0 GHz/4 GB RAM, and 30 joseki as experimental tests.

A professional player is invited to judge the value of the strong group, used as theory (*Va*). Then, the model is put into operation, and an experimental value (*Vb*) is obtained. According to (8) to (10), the parameters can be computed. The experimental results are shown in the Table 3 and Figure 6.

Relative error:

$$RE = \frac{|Vb_i - Va_i|}{Va_i} * 100\% \tag{8}$$

Average relative error:

$$ARE = \frac{\sum_{i=1}^{n} \frac{|Vb_i - Va_i|}{Va_i}}{n} * 100\% \tag{9}$$

Standard error:

$$\sigma = \sqrt{\frac{\sum_{i=1}^{n} |Vb_i - Va_i|^2}{n}} \tag{10}$$

It can be inferred from Figure 6 that the relative error of the filtration pro model is small compared with strength winnable model. We could calculate that compared with winnable, filtration's accuracy and speed of operation are promoted by 51.7% and 48.4% respectively.

Table 3. Performance of three models.

| Model | Number | Average Relative Error | Standard error | Time /s |
|---|---|---|---|---|
| Winnable | 30 | 13.37% | 2.0817 | 0.0186 |
| Filtration Pro | 30 | 6.46% | 1.1051 | 0.0096 |
| Filtration Amal | 30 | 10.83% | 1.8917 | 0.0097 |

It can be inferred from Table 3 that the relative error of filtration pro model is small compared with filtration amal model. The main reason is professional player's ability to make correct judgments; while 1 dan amateur makes unstable judgments, so *Va* can be inaccurate, and thus influences the final result. In order to offset the fluctuation, we can get the average value from the judgments of several 1 dan amateurs.

## 5 CONCLUSION

Unlike the territories, the strong groups' value can hardly be assessed by calculating every single point. Therefore, a quantization method is proposed with the introduction of probability and the establishment of the stone's influence function named Strength. Due to its high accuracy and speed, Filtration can be applied in the beginning game and middle game modules and assist the program to conduct positional analysis. When the level of the Go program varies, the parameters in the model can be adjusted correspondingly. Therefore, this model can be applied to all Go levels because of its good transportability and universality.

The introduction of probability theory makes it possible to substantiate the vague concepts and conduct a quantitive analysis, so that the dimensionality of the game is broadened. This approach can also be used to solve other computer game problems.

## ACKNOWLEDGMENT

We would like to thank professional 4P player Feng Pan, 2P player Ziji Zhao and 1d amateur Hong Xu for their assistance in parameter setting and Go level analysis.

This paper is supported by a grant from the Natural Science Fund Project in Science and Technology Department of Jiangxi Province (No: 2014BAB207002), the Science and Technology Research Project in Education Department of Jiangxi Province (No: GJJ14543) and the Doctor Start-up Fund (No: EA201104183).

## REFERENCES

[1] Xin W, Yinglong S, Hui Y, et al. The research of pattern symmetry problem in learning in computer

Go [C] // Control and Decision Conference. Chinese: CCDC-2012, 2012:3301–3306.
[2] Lei Yu, Chunni Dai, Xiaojun Zhang, Jingao Liu. Solving Winning Probability Problems in Go [C] // Proc of 2010 the 2nd International Conference on Advanced Computer Control. ShenYang: ICACC-2010, 2010: 25–28.
[3] Deng Chao, Wu Lin, Chen Lei, et al. Evaluation of local UCT algorithm in tsume go [J].Information Technology. 2013, (3):23–27.
[4] Yu Lei, Wei Jun, Tian Feng, Liu Jin-gao. Computer solution of static evaluation problems based on positional analysis and genetic algorithms [J]. Journal of Jilin University Engineering and Technology Edition, 2011, 41(6): 1694–1698.
[5] Abe Yoshiteru. Strong group calculation method [M]. Beijing: Chinese Olympic Press, 1990.
[6] Sylvain Gelly, David Silver. Combining online and offline knowledge in UCT [C] // In International Conference on Machine Learning. Corvallis: ICML-2007, 2007: 273–280
[7] Coulom R. Computing elo ratings of move patterns in the game of go [J]. Computer Games Workshop, 2007, 30(4):198–208.
[8] Coulom R. Efficient selectivity and backup operators in Monte-Carlo tree search [M]. Edmonton: Springer Berlin Heidelberg, 2007.
[9] Lei Yu, Xiaojun Zhang, Chunni Dai, Jingao Liu. Improved Winning Probability Model in Go Based on Strong Group Quantization and Multi-level Species Compete-Die Out Algorithms [C] // Proc of 2010 the 2nd International Conference on Future Computer and Communication. WuHan: ICFCC-2010, 2010:21–24.
[10] B. Bouzy, T. Cazenave. Computer Go: An AI-Oriented Survey [J]. Artificial Intelligence, 2001, 132(1): 39–103.
[11] Takagawa Shukaku. Go Story (Third) [Z]. Beijing: Weiqi Tiandi, 2007.
[12] Albert L. Zobrist. Feature extractions and representation for pattern recognition and the game of Go [D]. USA: University of Wisconsin, 1970.
[13] Jon Ryder. Heuristic analysis of large trees as generated in the game of Go [D]. USA: Stanford University, 1971.
[14] K. Chen. Computer Go: Knowledge, Search, and Move Decision [J]. ICGA Journal, 2001, 24(4):203–215.
[15] Dai Chun-ni, Yao Meng. New model for evolutionary algorithms: species compete-die out algorithms [J]. Computer Applications, 2005, 25(1): 224–225
[16] Yoshio Ishida. Complete volume on Joseki [M]. Sichuan: Shurong Press, 1988.
[17] Yao Meng, Dai Chun-ni, Pei Min, et al. The species compete-die out (SCD) algorithms model for evolutionary computation [C] // Proc of Genetic and Evolutionary Computation Conference. Washington: GECCO-2005, 2005:158–160
[18] Dai Chun-ni, Yao Meng, Xie Zhu-jie. Parameter optimization for growth model of greenhouse crop using genetic algorithms [J]. Applied Soft Computing, 2009, 9(1):12–19.

*Frontiers in Computer Education – Wang (ed.)*
*© 2015 Taylor & Francis Group, London, ISBN 978-1-138-02797-8*

# A study in watermarking relational databases based on genetic algorithms

Qingtao Wu
*School of Art & Design, Zhengzhou Institute of Aeronautical Industry Management, Zhengzhou, China*

Pengsong Duan
*Zhengzhou University Software Institute, Zhengzhou, China*

Jinfa Shi
*Zhengzhou Institute of Aeronautical Industry Management, Zhengzhou, China*

ABSTRACT: This paper studies a novel scheme of watermarking relational databases for copyright protection based on genetic algorithms. In this algorithm, a GA is used in a watermark signal processes, and associated novel watermark insertion algorithms and detection algorithms are proposed. Thus, the watermark signal in this method is expected to be more meaningful and has closer correlative relationship to the copyright holder. The experimental results verify that the proposed algorithm is feasible, effective, and robust.

## 1 INTRODUCTION

Relational database watermarking protects the intellectual property in today's internet-based application environments and in many content distribution applications. We present a mechanism for proof of ownership based on the secure embedding of a robust imperceptible optimization watermark in relational database. Based on the rational database watermark, researches in database watermarks are being conducted by people such as R. Agrawal and R. Sion. Including previous research into watermarking, they pay attention to arithmetic embedding and extracting.

A Genetic Algorithm (GA) is a search technique that is based on the principles of natural selection or survival of the fittest. Genetic algorithms are one of the best ways to solve a problem about which little is known objectively and functions directly in the search. Then, by using evolving operations such as crossover, mutation, and selection, the GA creates successive generations of solutions that evolve and inherit the positive characteristics of their parents and thus gradually approach optimal or near-optimal solutions. By using the objective function directly in the search, GA's can be effectively applied in nonconvex, highly nonlinear, and complex problems.

The main contributions to this work include: presenting an effective GA technique to generate watermark signal; and present a new technique to insert and detect watermarks using a mark bit position, together with a decision-making algorithm by pattern matching. In order to improve robustness, the scheme embeds a watermark into the database for several times and recovers the detected watermark by a majority voting mechanism.

## 2 ALGORITHMS FOR WATERMARKING RELATIONAL DATABASES

In this section, the watermark embedding and extraction algorithms are presented. We assume that some minor changes of some attributes values can be tolerated. And we will embed copyright information into these attributes. We consider the character copyright information as a sequence of 0 and 1, the marks of 0 and 1 are small errors in the relational data. All the marks of 0 and 1 represent integrated copyright information. The problem of watermarking relational databases can be modelled as follows.

Suppose relation R contains a primary key P and numerical attributes A0, A1,..., Av-1. Assume that it is acceptable to change one of $\xi$ Least Significant Bits (LSB). Copyright information which will convert as a sequence of 0 and 1 is to be embedded into relation R for the purpose of copyright protection. Table 1 summarizes the important parameters used in our algorithms.

Our technique aims to mark only numeric attributes. The data owner is responsible for deciding which attributes are suitable for marking. He also helps to decide two important parameters $\xi$ and $\gamma$, which describe the limits of modification to a database. We also suppose the adversary should not remove or modify the primary key attributes for the integrity and availability of the database. An attacker cannot guess

Table 1. Notations used in this paper.

| | Generation 0 | |
|---|---|---|
| 1 | String $x_i$ | fitness $f(x_i)$ |
| 1 | 1101101 | 109 |
| 2 | 1100101 | 101 |
| 3 | 1101110 | 110 |
| 4 | 1100111 | 103 |
| 5 | 1101101 | 109 |
| 6 | 1100001 | 97 |
| 7 | 1101001 | 105 |
| 9 | 1101100 | 108 |
| 10 | 1101110 | 110 |
| 11 | 1100111 | 103 |

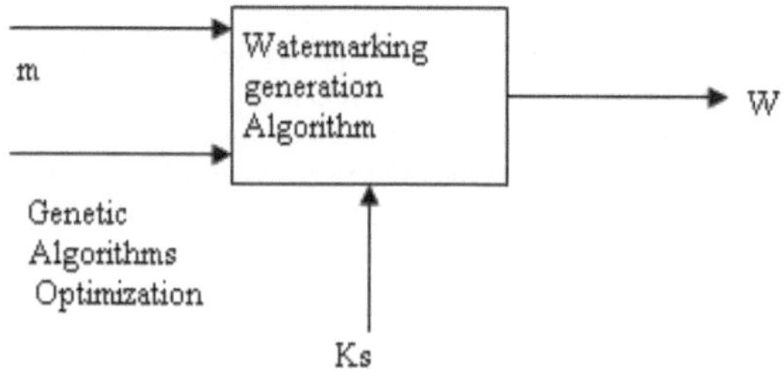


Figure 1. Watermark generation algorithm.

all of these for the secret key and other parameters are private to the owner.

## 2.1 *Watermark generation algorithm*

Figure 1 shows the watermark generation algorithm. It can be express as W = G (m, Ks). W included a secret key Ks only known to the copyright owner and identity information "m".

The detailed process of watermark generation algorithm by using GA is as follows:

Step 1. Encoding. Encoding of chromosomes is one of the problems, when you are starting to solve problem with GA. Encoding very much depends on the problem.Traditionally, solutions are represented in binary as strings of 0s and 1s.

The original copyright information, identity information and confidential information or other relevant information 'm', is converted into binary code, using function "dec2bin ()" to obtain binary code. In binary encoding every chromosome is a string of bits, 0 or 1. Table 2 shows the 0 generation.

Step 2: Selection, recombination, and mutation. We formed binary code in Step 1, using a binary string of symbols to express each individual.

(1) During each successive generation, a proportion of the existing population is selected to breed a new generation. Individual solutions are selected through a fitness-based process, where fitter solutions (as measured by a fitness function) are typically more likely to be selected. Certain selection methods rate

Table 2. The 0 generation.

| Symbol | Number of tuples in the relation |
|---|---|
| $\nu$ | Attribute numbers to be marked |
| $\gamma$ | Fraction of tuples marked |
| $\xi$ | Number of markable bits |
| $\omega$ | Number of tuples actually marked |
| ks | The secret key |
| T* | Optimal decoding threshold |
| Gi | Us ability constraints |

the fitness of each solution and preferentially select the best solutions. Other methods rate only a random sample of the population, as this process may be very time-consuming. Normally can be used 'selection ()' function to achieve the process of selection:

```
function[evol_gen,bes_ind,max_fitness]=
selection(old_gen, fitness) [min_fitn,expo(b)]=
min(fitn);
  [max_fitn, expo (a)]=max_ (fitn);
  Popusize=length (fitness);
  bes_ind=old_gen (expo (a), :);
  expo=[1:popusize];expo(expo(a))=0;
expo(expo(b))=0;
  expo=nonzeros (expo);
  evol_gen=old_gen (expo, :);
  evol_fitness=fitness (expo, :);
  evol_popusize=popusize- 2;
  posel=evol_fitness/sum (evol_fitness);
  poselcum=cusum (posel);
  r=rand (1, evol_popusize);
  selected=1+sum (poselcum*ones (1,evol_popusize)
<ones (evol_ popusize, 1)*r);
  evol_gen=evol_gen (selection, :);
```

(2) The next step is to generate a second generation population of solutions from those selected through genetic operators: crossover (recombination), and/or mutation. For each new solution to be produced, a pair of 'parent' solutions is selected for breeding from the pool selected previously. By producing a 'child' solution using the above methods of crossover and mutation, a new solution is created which typically shares many of the characteristics of its 'parents'. Normally 'recombination ()' and 'mutation ()'functions can be used to achieve the process of recombination:

```
  function [new_gen] = recombination (old_gen, pc)
  [nouse, match]=sort (rand (size (old_gen, 1), 1));
  match_gen=old_gen (match, :);
  pairs=size (match- gen, 1)/2;
  bit_n=size (match_gene, 2);
  string=rand (pairs, 1) <pc;
  crossp=randint (string, 1, [1, bit_n]);
  crossp=string.*crossp;
  for i=1: pairs
  new_gen ([2* i- 1 2* i], : )= [match_gen ([2* i- 1
2*i],
  1: crossp (i)) match_gen ([2* i 2* i - 1], crossp (i)
+1:bin_n)];
```

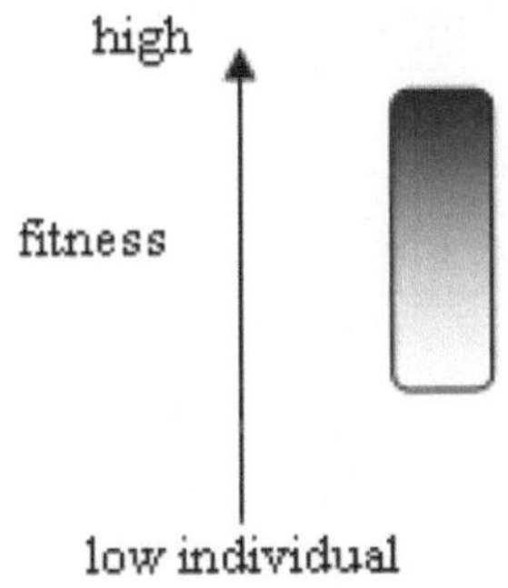


Figure 2. Watermark anagenesis sketch map.

```
function [new_gen] =mutation (old_gen, pm)
mpoints=find (rand (size (old_gen)) <pm);
new_gen=old_gen;
new_gen (mpoints) =1- old_gen (mpoints);
```

Step 3: To generate watermark W. After selection, crossover and mutation the watermark signal was created. Assuming bits of watermark signal is Wk∈ {0, 1}, k = 1, 2…, k, decided by capability and intensity of database. Figure 2 shows the process of watermark anagenesis sketch map.

### 2.2 *Embedding algorithm*

We use a one-way hash function result, decided by the primary key P and the secret key Ks, to choose where to mark and what to mark. Usually it has the form of h = H (M), where M is the seed. Besides, It bears the following characteristics: given M, it is easy to compute h; yet when given h, it is hard to compute M, such that H (M) = h; given M, it is hard to find another message M0 such that H (M) = H (M0). Several hash functions, such as MD5 and SHA, are good choices for this purpose.

Step 1. We transform a meaningful watermark W into a bit flow E (EMC, Encrypted Mark Code) of certain length.

Step 2. We decide which tuple (attribute and bit) should be marked by the hash value. The remainder of h module L (Ks = h mod L) decides which bit of the EMC is used at one time. Since the hash result is expected to be uniform distributed, we can divide the relation into groups of varied but similar sizes. Thus we have L (the number of bits of the watermark) groups. The i-th bit of the watermark will be inserted into the i-th group. The ascending order of I ranging from 0 to L-1 naturally preserves the sequence of the marks. This is helpful in the detection phase.

Step 3. We use id (id = Hash (Ks, P, Ks)) and L (the number of bits of the watermark) to calculate the remainder I for each tuple, and then collect tuples with same values of I into the same group. The i-th bit of the watermark will be inserted into the i-th group. The ascending order of I ranging from 0 to L-1 naturally preserves the sequence of the marks. This is helpful in the detection phase.

Algorithm: watermark embedding algorithm
Input: parameters Ks, L, $\gamma$, and $\nu$
Output: marked R

1) Calculate L-bit EMC E [L] = H (ks concatenate M)
2) For each tuple R do
3) t = H ((Ks concatenate r.P)
4) if (t mod $\gamma$ equals 0) then // mark this tuple
5) attribute_ index I = t mod $\nu$ // mark attribute Ai
6) bit_index j = t mod $\xi$ // mark j th bit
7) watermark_index Ks = t mod L //use the k-th bit of EMC to mark
8) mark_bit m=Ek XOR (Ks mod 2) // get the value of marked bit
9) set the j-th least significant bit of r.Ai to m
10) if (not within_usability(new_data)) // check the availability
11) rollback
12) return R

### 2.3 *Watermark extraction algorithm*

If the owner of the dataset suspects that some datasets are illegal copied or tampered from his relational database R, he or the third party can use watermark extraction algorithm to verify the ownership of the suspicious dataset. Now we introduce our watermark extraction.

Because of the identical distribution of the hash function when seeded by the same key, the selected tuples in each group is of the same order as in the insertion algorithm .For each marked bit, we count the numbers of its value to be zeroes or ones respectively, and then a majority voting mechanism is to decide the final value of this bit. The detected result is a binary sequence which includes the copyright information of the dataset. The copyright holder can make use of the watermark signal to prove the copyright.

Step 1. Majority-voting mechanism is used to find effective bits in databases with keys. So a watermark code will be obtained again as an opposite way of insetting.

Step 2. After values of index by inverse process of watermark generation.

Step 3. The copyright holder can make use of the meaning and features of the signal to prove the copyright.

Algorithm: watermark extraction algorithm
Input: parameters Ks, L, $\alpha$, $\xi$ and $\nu$ are also private to the owner
Output: a watermark DM [ ]

1 for s = 0 to L-1 do
2 DM[s] = " // initialize detected mark code
3 count[s] [0] = 0, count[s] [1] = 0
4 for each tuple r∈R do
5 t = H1(Ks concatenate r.P)
6 if (t mod $\gamma$ equals 0) then // select this tuple
7 I = select_attribute () // mark i-th attribute
8 j = t mod $\xi$ // select j-th bit
9 Ks = t mod L // mark the k-th bit of EMC

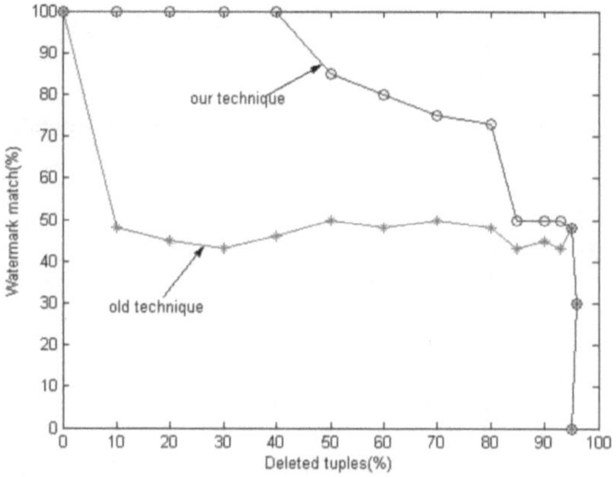


Figure 3. Compare our with old watermark technique.

```
m = (j-th LSB of r.Ai) XOR (k mod 2)
count [Ks] [m] = count [Ks] [m] +1
for s=0 to L-1 // get the watermark
if (count[s] [0]>=count[s] [1]) // majority voting
then DM[s] = 0 else DM[s] = 1 //the final bit
   value return DM [ ]
```

## 3 EXPERIMENTS AND ANALYSIS

We perform experiments on a computer running Windows XP Professional with 2.4 GHz CPU and 25 MB RAM. Watermark generation algorithm of Matlab has been given. Algorithms are implemented on Java Eclipse Platform Version 3.0 using JDBC to visit Microsoft Office Access 2003. We select before 50,000 tuples as a database. We choose MD5 as the one-way hash function and the significance level is 0.01.

The testing results of the various attacks on a relational database and the necessary analysis were given by the following.

### 3.1 *Deletion attack*

The attacker attempts to destroy the tuples from the watermarked data set. Furthermore, in this test we compare our implementation with the old watermark approach. Figure 3 shows the experimental results, where it clearly show that our watermarking technique is resilient to the random deletion attack. Using our technique the watermark was successfully extracted with 100% accuracy even when over 40% of the tuples were deleted.

### 3.2 *Inserted attack*

In this attack, the attacker takes a Bernoulli ($\kappa$) sample of the n tuples in relation R to form a subset R' $\subset$ R. He then takes a Bernoulli ($1 - \kappa$) sample U' from a collection U of n tuples from other sources and mixes the tuples in U' and R' to create his relation X of approximately the same size as R.

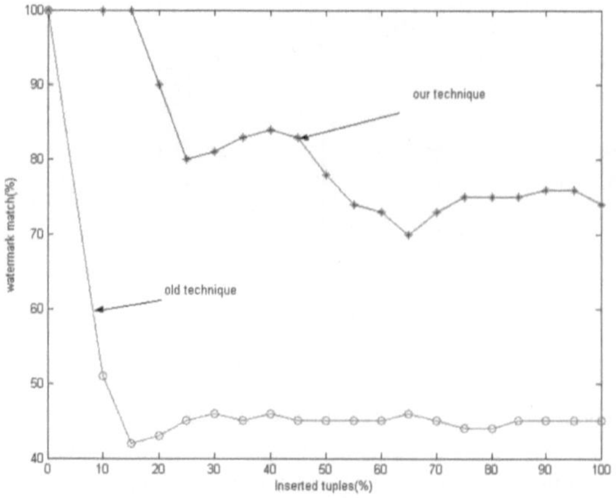


Figure 4. Compare our with old watermark technique.

The comparison show that our technique is resilient to selected attack. While on the other hand, the old technique deteriorates just after adding 10% of the data set size. Figure 4. shows the resilience of our watermarking technique to insertion attack, where the watermark was recovered with 100% accuracy even when up to 15% of the data set size tuples were inserted.

### 3.3 *Alteration attack*

In this form of attack, violators try to alter some attribute values of the relational database, which had been watermarked, and stochastically alter the attribute values. Suppose that an attacker magically knows the values of the $\nu$ and $\xi$ parameters, used by the database owner. Since the attacker does not know which bits have been marked, he randomly selects a Bernoulli ($\kappa$) sample from the n tuples. For each selected tuple, he flips all of the bits in all $\xi$ bit positions in all $\nu$ attributes. It is important to bear in mind that when the owner marks a tuple, he only perturbs a single bit, whereas Mallory perturbs $\xi \times \nu$ bits, thereby inflicting much more damage on each tuple that he marks.

To be noted, the robustness of our algorithm has close relation with marking frequency. The higher marking frequency we get, the more robust it is. Yet a compromise between robustness and imperceptibility must be reached in the implementation phase. Figure 5 shows the experimental results; they clearly show that our watermarking technique is resilient to a random modify attack.

## 4 CONCLUSIONS

Much research has done on watermarking relational databases. However, very few works focus on the optimization of the watermark signals of relational databases by using genetic algorithms. The results verify the effectiveness and usefulness of our approach. Furthermore, compared with existing proposals, it also

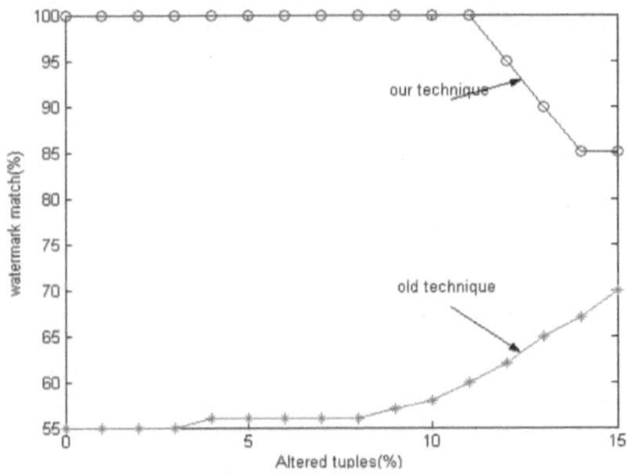


Figure 5. Compare our watermark with old technique.

implies the superiority of our approach over the previous ones. We have presented a resilient watermarking technique for relational data that embeds watermark bits in the data statistics. The algorithm proved to have immunity to popular attacks on relational databases and needed the smallest available bandwidth. In the future, we intend to conduct research on improving the robustness of the scheme by using GAs to watermark embedding algorithms.

Foundation item. Supported by the National Natural Science Foundation of China (No. 71371172), Aeronautical Science Foundation of China of China (No. 2012ZG55023) and Youth research foundation of Zhengzhou Institute of Aeronautical Industry Management (No. 29010014).

## REFERENCES


David, "Query-preserving Watermarking of Relational Databases and XML Documents". PODS 2003, San Diego, CA, 2003, pp. 191–201.

Goldberg, D.E. "Genetic Algorithms in Search, Optimization and Machine Learning", Addison Wesley, Reading, MA, 1989.

Hsu, C. T. and Wu, J. L. "Hidden Digital Watermarks in Images," *IEEE Transactions on Image Processing,* vol. 8, No. 1, pp. 1–2.

Huang, C.H and Wu, J. L. "A Watermark Optimization Technique Based on Genetic Algorithms," SPIE Electronic Imaging 2000, San Jose, 2000.

Ja-Ling Wu, Chun-Hung Lin, and Chun-Hsiang Huang, "An Efficient Genetic Algorithm for Small Search Range Problems and Its Applications", *Intelligent Multimedia Processing with Soft Computing,Springer-Verlag*, pp. 253–280.

Mohamed Shehab, Elisa Bertino, Arif Ghafoor, "Watermarking Relational Databases using Optimization Based Techniques", *IEEE Transactions on knowledge and data engineering,* 2006.

Rakesh, A., Peter, J. Haas and Jerry, K., "Watermarking Relational Data: Framework, Algorithms and Analysis", *VLDB Journal*, 2003, pp. 157–169.

Radu, S. Mikhail, and Sunil, A. "Rights Protection for Relational Data", *Proceedings of ACM SIGMOD*, 2003, pp. 98–109.

Tang, K. S., Man, K. F., Kwong, S. and He, Q., "Genetic algorithms and their applications", *IEEE Signal Processing Magazine,* 1996, pp. 22–37.

Wu, J. L., Lin, C. H. and Huang, C. H. "An Efficient Genetic Algorithm for Small Range Search Problem", *Intelligent Multimedia Processing with Soft Computing, Springer-Verlag*, 2005, pp. 253–280.

*Frontiers in Computer Education – Wang (ed.)*
*© 2015 Taylor & Francis Group, London, ISBN 978-1-138-02797-8*

# Performance prediction and reverse design of a glass composites pultrusion process based on RBF

Zhiping Zhang
*School of Humanities and Social Sciences, North China Institute of Science and Technology, Sanhe, Hebei, China*

Hanwu Liu, Shuai Luo & Zhiwen Guo
*School of Mechanical and Electrical Engineering, North China Institute of Science and Technology, Sanhe, Hebei, China*

ABSTRACT: Most glass composites products in engineering are manufactured by a pultrusion process. Because the pultrusion process is affected by many factors, it is not easy to control the performance of products. In order to obtain good performance and meet the requirements of glass composites products in specific parameters, artificial neural network technology was adopted to optimize and design the pultrusion process. The results showed that the curing degree of products increased gradually with the preheated mould temperature. During the reverse design of glass composites products, it was found that when the spread is 0.3, the measured temperature values of every stage based on curing degree were in good agreement with the forecasting values, which proved the correctness of the reverse design. The research method in this paper provides important theoretical basis and technical support for similar products.

*Keywords*: glass composites; RBF neural network; curing degree; mould temperature; performance forecast

## 1 INTRODUCTION

Glass reinforced plastic is commonly known as reinforced composite plastic, also known as Glass Fiber Reinforced Plastic (GFRP). It is made of synthetic resin as an adhesive and glass as a reinforcing material. Composite materials are a combination of different bases or reinforcing materials, such as organic polymer material, metal and inorganic non-metallic, in various processes. In addition to retaining the main features of component materials, composite materials can also obtain unique properties that component materials do not have through composite effect. With the development of the times, cement or non-metallic based composite materials have been gradually replaced by GFRP. Most cable protection tubes made of GFRP have better corrosion resistance and span with a smooth surface which is beneficial for cleaning.

Pultrusion is a typical process for manufacturing GFRP continuously and economically with many features, such as high production efficiency, good reproducibility, arbitrary length, raw material saving, energy saving. In pultrusion process of GFRP, temperature is one of the key factors affecting the forming process. When the curing degree meets the requirements, if the temperature is the optimum value, the energy consumption can be reduced and the costs can be saved with the competitiveness of the products increasing.

In this paper, the process plan of a GFRP pipe was firstly determined by analysing its composition, studying its properties and the pultrusion process. According to the material properties, the temperature of each stage was analysed and calculated. Then the network model of the temperature and curing degree was established, as based on the Radial Basis Function artificial neural network (RBF) by using MATLAB. After training, the network model could be used to predict the temperature of each stage and then get the optimum temperature, which could greatly cut down on the amount of time spent on experiments, reduce production costs, and improve the success rate of pultrusion forming.

## 2 TRAINING AND PREDICTION OF RBF NETWORK

The size of the GFRP pipe is: D = 22 mm, d = 6 mm, the total length of pipe in one production of 1000 mm, pultrusion rate of 20 m/h. This part is mainly installed in the crossing area of the communication cable and power line to prevent a short circuit caused by the break of power line which may cause the communication cable and wire rope charged and then to protect cables, switches, core boards and even the machine from being burned. Also this part can isolate magnetic field interference of the power line.

### 2.1 *Training algorithms and measurement indicators*

In general, there are two methods to train and test artificial neural network models. One is 'Leave-one-out', the other is 'dichotomy'. By comparison, 'Leave-one-out' can achieve a high forecasting accuracy and get the best, reliable forecasting accuracy, though it has a high workload and is time-consuming. In this paper, 'Leave-one-out' is used for training and testing RBF network model. Specifically, based on using GFRP temperature of each stage as input variables, the curing degree as output variables and then building a network model of three inputs and one output, the curing degree of GFRP can be predicted after repeated training. Afterwards, the temperature can be determined according to the curing degree through reverse design, which can reduce experimental costs with a high precision and high efficiency. 7 set of data are used to train the network model and by selecting different spread values, the optimal model can be achieved.

For the qualitative and quantitative measurement of the forecast performance of RBF, scatter plots and statistical indicators are selected. The three statistical indicators selected for quantitative measuring forecast performance are Mean Square Error (MSE), Mean Square Relative Error (MSRE) and Value of Fitness (VOF), their formulas are shown respectively:

$$MSE = \sqrt{\frac{\sum_{i=1}^{N}(X_{1i} - X_{2i})^2}{N}} \quad MSRE = \sqrt{\frac{\sum_{i=1}^{N}\left(\frac{X_{1i} - X_{2i}}{X_{2i}}\right)^2}{N}} \tag{1}$$

$$VOF = 1 + \frac{\sum_{i=1}^{N}(X_{1i} - \overline{X}_{1i})(X_{2i} - \overline{X}_{2i})}{\sqrt{\left(\sum_{i=1}^{N}X^2_{1i} - N\overline{X}^2_{1i}\right)\left(\sum_{i=1}^{N}X^2_{2i} - N\overline{X}^2_{2i}\right)}} \tag{2}$$

where: $X_{1i}$ is the measured value, $X_{2i}$ is the predicted value, $N$ is the sample size.

When the predicted value and the measured value get closer, which is when the forecast performance becomes better, MSE and MSRE get smaller and approach to 0, the VOF gets bigger and approaches to 2. When MSE and MSRE get bigger and farther from 0, the VOF get smaller and farther from 2, the forecast performance becomes worse.

### 2.2 *Training results*

#### 2.2.1 *Structure and training elements of RBF network*

The structure of an RBF Network model includes the number of layers in the model, the number of neurons in each layer, the type of the activation function and the spread value of the model, etc. Here the significance of spread value is explained. The model prediction accuracy is determined by the size of the spread value. A lower spread value means better forecasting accuracy but more time consumption, while higher spread value needs less time but means worse prediction accuracy. In order to ensure accuracy but also within a short time, it is necessary to adjust spread value constantly in the adjustment process so as to achieve the best combination of accuracy and time. The model training is selecting multiple sets of data to train the neural network, and selecting different spread values to determine the optimal model, which is ready for forecasting. After training the model, the structure of the network model has been adjusted and optimized. If a new variable is entered, the model can predict the result afterwards. As seen in table 1, 9 set of data have been selected to train, wherein the pultrusion speed is 20 m/h.

Table 1. Training and optimization of data (GFRP temperature and curing degree).

| Number | Temperature (°C) First stage | Second stage | Third stage | Curing degree (%) |
|---|---|---|---|---|
| 1 | 119 | 149 | 143 | 90.52 |
| 2 | 121 | 147 | 145 | 90.47 |
| 3 | 123 | 150 | 148 | 91.03 |
| 4 | 126 | 156 | 153 | 91.94 |
| 5 | 128 | 153 | 150 | 91.64 |
| 6 | 133 | 161 | 152 | 92.61 |
| 7 | 135 | 165 | 155 | 93.07 |
| 8 | 138 | 168 | 157 | 93.43 |
| 9 | 140 | 170 | 160 | 93.69 |

#### 2.2.2 *Influence of spread value*

7 set of data were randomly selected for training model, and group 1, 2, 3, 5, 6, 7, 9 were selected in this training.

Meanwhile the spread values were changed to observe the effect of different training. Figure 1 shows the training performance when the spread value is 0.4, 0.5, 0.6, 0.7, 0.8, and 1.

As can be seen from the Figure 1, when the spread value is 0.4, 0.5 or 0.6, the points on scatter plots distribute substantially along the 45° diagonal. However, the distribution of points along the 45° diagonal seems worse and worse with the spread values increasing. It can be seen that when the spread value reaches 1, the scatter distribution is substantially horizontal which indicates more training errors and a less effective training result. By synthesizing analysing the scatter plots with different spread values, the range of optimal spread value in the RBF training model can be qualitatively determined, and some of the changes between predicted and measured values in the graph can be easily read out. The training performance of the artificial neural network model can be directly qualitatively described through the scatter plot, but quantitative description must rely on statistical indicators.

#### 2.2.3 *Statistical indicators*

From Figure 1, it can be seen apparently that with the spread value increasing, the prediction error decreases

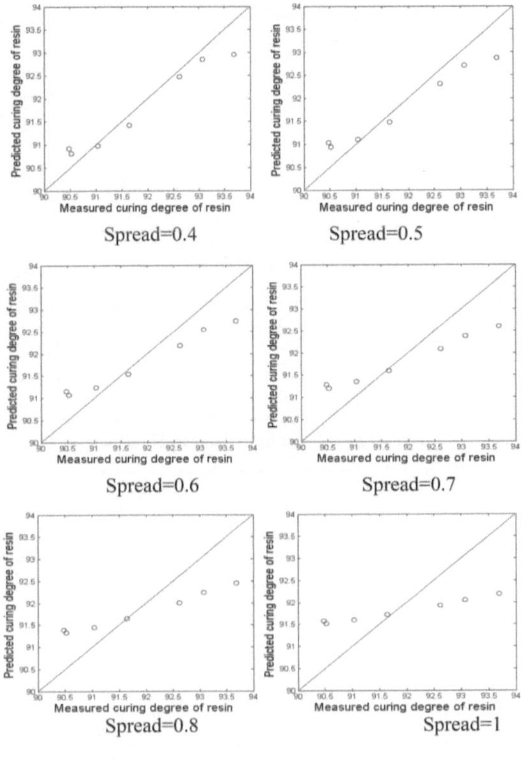


Figure 1. Scatter plots with different spread values.

Table 2. MSE, MSRE, and VOF.

| Spread value | MSE | MSRE | VOF |
|---|---|---|---|
| 0.4 | 0.3689 | 0.0040 | 1.9834 |
| 0.5 | 0.4515 | 0.0049 | 1.9877 |
| 0.6 | 0.5591 | 0.0061 | 1.9909 |
| 0.7 | 0.6728 | 0.0073 | 1.9925 |
| 0.8 | 0.7786 | 0.0084 | 1.9929 |
| 1 | 0.9475 | 0.0103 | 1.9918 |

firstly and then increases, which means that the prediction accuracy increases firstly and then decreases as the spread value increases. Different statistical indicators can be obtained by taking training results based on different spread values into the training program. Table 2 shows mean square error, mean square relative error, and the value of fitness.

By the table below, with the increasing spread values, mean square error and relative error are growing, indicating the greater the deviation of the predicted value and the measured value. When the spread value is of 0.4, the mean square error and relative error relative to the other groups are relatively small, and the fitted values relatively close to 2, it was decided to optimize the use of this spread is 0.4 forecast.

Table 3. The measured and predicted values.

| Number | 4 | 8 |
|---|---|---|
| Measured value | 91.94 | 93.43 |
| Predicted value | 91.7537 | 93.2159 |

## 3 FORECASTING AND IMPACT OF HEATING TEMPERATURE ON CURING DEGREE

After training and testing of the artificial neural network model, the structure has been optimized, with the best, most reliable prediction performance. The best spread value is 0.4, and on this condition the RBF model can be used for forecasting curing degree on the temperature of different stages. The samples were in two groups, numbered 4 and 8. Comparisons of measured and predicted results of temperature on curing degree are shown in Table 3.

Inputting measured and predicted values into program for calculation of statistical indicators, the results are as follows: $MSE = 0.2007$; $MSRE = 0.0022$; $VOF = 1.9987$. It can be clearly seen from statistical indicators that Mean Square Relative Error (MSRE) and Mean Square Error (MSE) are very close to zero, and the Value Of Fitness (VOF) is very close to 2. So the predicted results meet the ideal requirements, and this RBF model can be used in the pultrusion production of GFRP pipes for optimizing the forming process.

Figure 2 shows the influence of temperature on curing degree at different stages. As can be seen, with the temperature increasing, the curing degree also gradually increased with nonlinear relationship between the two. In actual production, the temperature is generally not too high considering the production efficiency and energy saving. Under the condition of obtaining appropriate performance, the temperature should not be too high.

## 4 REVERSE DESIGN OF PROCESS PARAMETERS BASED ON RBF NETWORK

According to the curing degree, reverse design is adopted to design and optimize the temperature of each stage for achieving the best predicted value. Similarly with predicting the curing degree, the artificial neural network model need training before using, called the 'learning phase of model'. The purpose is to adjust the structure of the artificial neural network model, reflect the intrinsic relationship between the input and output variables of training samples, and get the model that has the best prediction performance. Spread values 0.3, 0.4, 0.5, 0.6 were selected to train the neural network model. After training, it was found that when the spread value is 0.3, the mean square error and mean square relative error are relatively small, and the value of fitness is close to 2 with the best performance. Figure 3

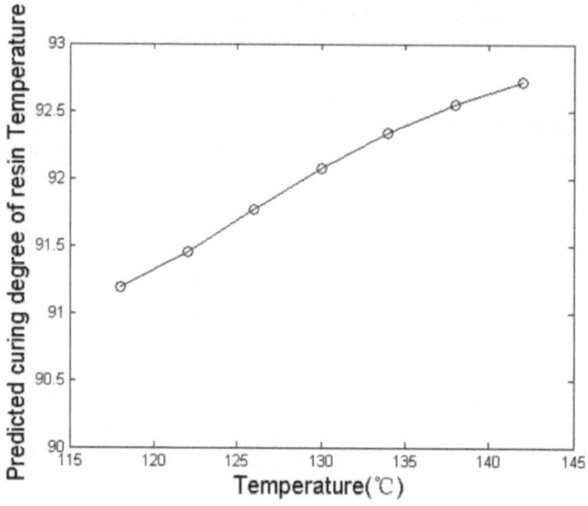


(a) The first stage

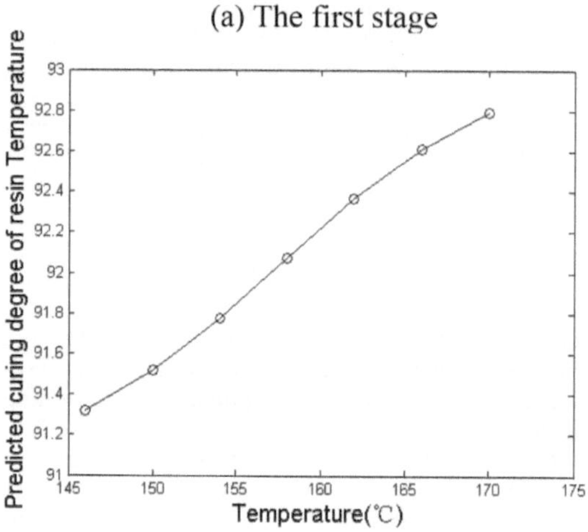


(b) The second stage

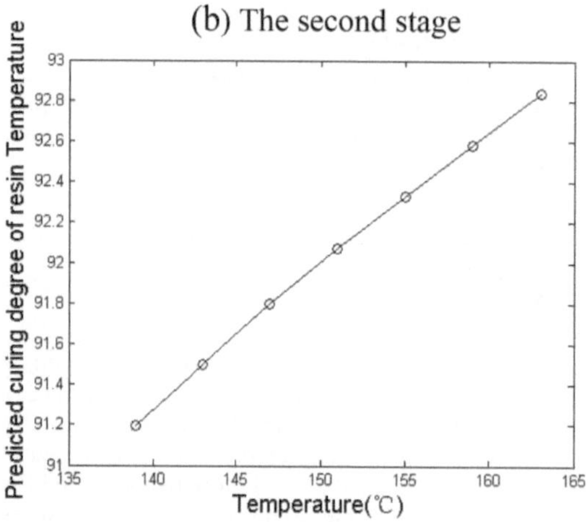


(c) The third stage

Figure 2. Influence of temperature on curing degree.

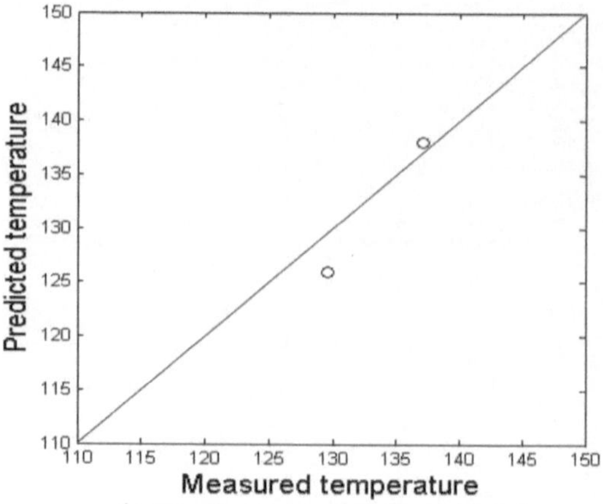


(a) The first stage when Spread is 0.3

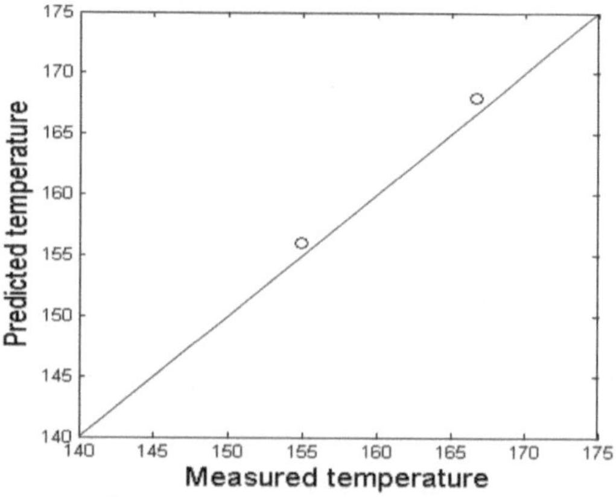


(b) The second stage when Spread is 0.3

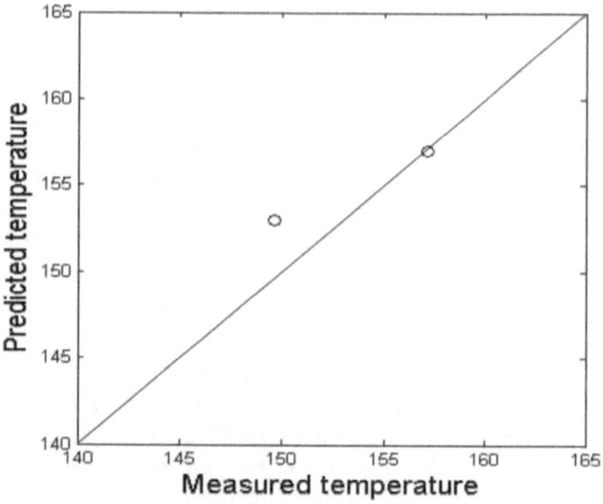


(c) The third stage when Spread is 0.3

Figure 3. Scatter plot of reverse design of curing degree on temperature.

shows reverse design of curing degree on temperature when Spread = 0.3.

As it can be seen above, when the spread value is 0.3, the forecast temperature data for each stage almost focuses on the 45 degree line, indicating the predicted temperature of each stage by curing degree has almost no error with the measured temperature. Due to the noise and quality problem of samples, the mean square error and mean square relative error do not equal zero, but it has basically reached the requirement. Meanwhile the value of fitness is close to 2, indicating that the reverse design of temperature for each stage is quite accurate.

## 5 CONCLUSIONS

According to the pultrusion curing degree of GFRP pipes, a neural network model was establish and after training it was used for predicting the relationship between temperatures and curing degrees at different stages of pultrusion. First, to meet relevant requirements, the spread value based on the training results should be modified. Then after training and testing, the sample could be predicted and the forecasting results were described by statistical indicators. The results show that:

1) By using an artificial neural network, not only the curing degree can be predicted according to the temperature of each stage, but also the temperature can be forecasted by curing degree with reverse design.
2) After forecasting, the influence of each input variable on the curing degree can be analysed to determine the optimal preheated temperature. Because of the small ranges of data, the calculation results do not change significantly, so the minimum spread value cannot be set as the optimal value during prediction. Thus the reverse design based on the curing degree is made to analyse the influence of the curing degree on the preheated temperature at different stages.
3) Due to the limited quantity and quality of samples, the prediction accuracy of RBF model based on these samples is surely affected. In order to analyse more accurately of the relationship between the curing degree and temperature in the pultrusion process of a GFRP pipe, statistical indicators are adopted in this paper to indicate that the prediction method is correct and feasible.

## REFERENCES


Eyassu W., Samuel K., Pang S. Optimization for buckling loads of grid stiffened composite panels. Composite Structures, 2003, (60): 159–169

Huaiqin Xie, Xingkai Chen. Unsteady temperature field, numerical simulation of curing degree and experiment of GFRP pultrusion process. GFRP/Composites, 2010, (1): 73–76

Jacob, A. Globalization of the pultrusion industry. Reinforced Plastics, 2006, 50 (5): 38–41

Jianhua Li, Sunil, C. Joshi etc. Curing Optimization for Pultruded Composite Sections. Composite Science and Technology, 2002, (62): 456–467

Lin He, Chuanzhen Huang. Artificial neural network and optimization method combination used in composite materials. Silicate Bulletin, 2004, (1): 85

Martin, J. Pultruded composites compete with traditional construction materials. Reinforced Plastics, 2006, 50 (5): 20–27

*Frontiers in Computer Education – Wang (ed.)*
*© 2015 Taylor & Francis Group, London, ISBN 978-1-138-02797-8*

# Strength analysis and fatigue life prediction on a steering knuckle

Hanwu Liu, Junhong Wang, Shengjun Fang & Shen Li
*School of Mechanical and Electrical Engineering, North China Institute of Science and Technology, Sanhe, Hebei, China*

ABSTRACT: In the process of operation, the loading condition of an automobile steering knuckle is very complex, which seriously affects its structure strength and fatigue life. In this paper, finite element analysis is applied to calculate and analyse the deformation and stress distribution in a steering knuckle under conditions of three kinds of limit load, which are the impulse force induced by rough ground, longitudinal impact caused by emergency braking, and the transverse impact generated when the car turned. Based on this, a prediction of the fatigue life on the more than 300000 times of cyclic loading is also carried out. The results show that under the three kinds of limit load conditions, the maximum stress value of the steering knuckle of 940 MPa, which is less than the yield strength of the steering knuckle materials, so the steering knuckle could not produce yield. Fatigue strength analysis shows that the fatigue life application factor is 0.9375, less than 1, which illustrates that fatigue damage will not occur. The study provides an important theoretical basis and technical support for the actual design and industrial production of automobile steering knuckles.

*Keywords*: steering knuckle; free modal; strength analysis; fatigue life prediction

## 1 INTRODUCTION

The steering knuckle is an important part of the vehicle axle. It is subjected to a tangential load, impact from the ground, and transfers the steering force from the steering knuckle, it is also necessary to leave enough space for brake to install Anti-lock Braking System sensor. At the same time, the steering knuckle is connected to the knuckle column, tie rod, shock absorber, brake caliper, etc, so its structure is very complex, and the requirements for strength, impact resistance, fatigue strength, and reliability are high. Therefore, research on steering knuckle life is necessary. In the motion process, vehicle parts are subjected to a variety of alternating loads. The alternating load is generally lower than the tensile strength limit, so the steering knuckle would not fail for strength reasons, but under the effect of the alternating load repeatedly, it may occur crack initiation and extension, and lead to a sudden rupture, known as fatigue destruction. It is extremely important to understand the influence of this damage form on the strength of the steering knuckle.

Most external loads which are withstood by rolling stock, are random loads which act repeatedly. Under the effects of these kind of random dynamic loads, many components will generate dynamic stresses and cause fatigue damage. As the fatigue damage accumulates, initial cracks will occur, extend, and finally cause fracture, and this is the main form of structural damage. Thus, the structural fatigue design of the rolling stock has gradually become one of the necessary processes in the early stages of new product development, while the prediction of the structure fatigue life through an effective calculating method is an important goal of structural design. This article will focus on strength analysis and fatigue life prediction on automobile steering knuckles, using the finite element simulation technology.

## 2 STRESS AND DEFORMATION ANALYSIS ON THE STEERING KNUCKLE

The bottom-up modeling method was used to create the model of the steering knuckle, and the basic modeling order of the steering knuckle is as follows: the hub, spherical area, the damping arm, the control arm, and the tire control point. The complete model is shown in Figure 1. The following work is then done, defining the type of element, setting the unit cell size, material attributes, and other parameters, using smart meshing technology to complete the knuckle model mesh, shown in Figure 2.

The load of an automobile steering knuckle generally consists of two parts, namely, the static load generated by the hanging sprung mass, and the dynamic load produced during the motion process. There are three kinds of stress limit working conditions, as shown in Table 1, Table 2, and Table 3.

Condition 1: The impact induced by the impulse force during the car running on rough road;

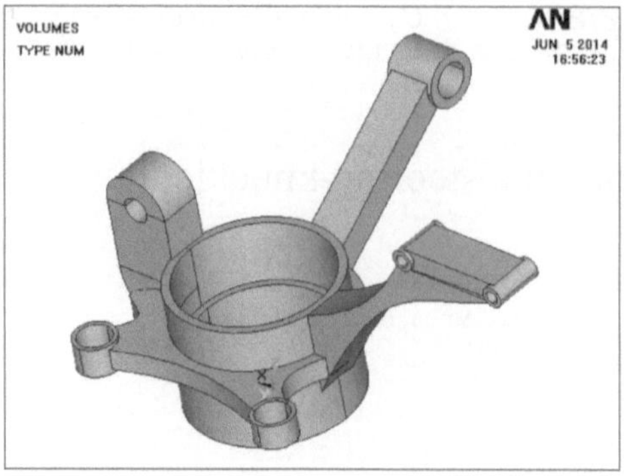


Figure 1. Mathematical model of steering knuckle.

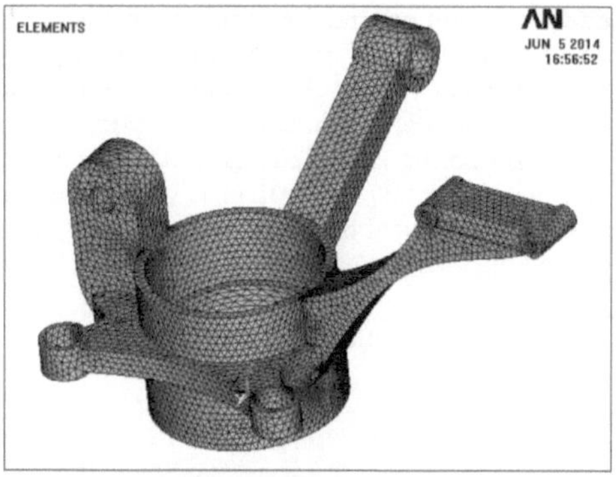


Figure 2. Finite element model of steering knuckle.

Table 1. Forces on condition 1 at different positions.

| Condition | Condition 1 | | |
|---|---|---|---|
| Position | control arm | damping arm | spherical pair |
| X-axis/N | 227 | −1123 | 1294 |
| Y-axis/N | 1109 | −6346 | 5429 |
| Z-axis/N | −193 | −19126 | −1041 |
| Force/N | 1148 | 20900 | 5800 |
| Pressure/MPa | 1.73 | 43.5 | 10 |

Table 2. Forces on condition 2 at different positions.

| Condition | Condition 2 | | |
|---|---|---|---|
| Position | control arm | damping arm | spherical pair |
| X-axis/N | 924 | −266 | −573 |
| Y-axis /N | −16061 | 3375 | −3023 |
| Z-axis /N | −1232 | −4627 | −230 |
| Force/N | 16000 | 5700 | 3060 |
| Pressure/MPa | 26.8 | 12 | 5 |

Table 3. Forces on condition 3 at different positions.

| Condition | Condition 3 | | |
|---|---|---|---|
| Position | control arm | damping arm | spherical pair |
| X-axis/N | −15497 | −4420 | −1049 |
| Y-axis/N | 11347 | −306 | −11008 |
| Z-axis/N | 908 | −4003 | −831 |
| Force/N | 18600 | 6000 | 11000 |
| Pressure/MPa | 30 | 12.4 | 18.5 |

Condition 2: Longitudinal impact from the ground and its own inertia force in an emergency stop;

Condition 3: Lateral impact in turning.

This paper analysed the force and deformation on the knuckle in the three different conditions, and the results are shown in Figure 3, Figure 4 and Figure 5.

It can be seen from Figure 3, in Condition 1, that the values of deformation and displacement are largest on the top of the damping arm of the steering knuckle, and can reach 22.7 mm, which shows that the design here is not very reasonable it is necessary to increase the thickness and width of the damping arm. The maximum stress point occurs in the middle of the damping arm, whose value is 940 MPa, which is less than material yield strength 1000 MPa, so the damping arm will not be damaged, but still needs reinforcement.

It can be seen from Figure 4, in Condition 2, that the values of deformation and displacement are largest on the top of the damping arm of the steering knuckle, and can reach 5.317 mm, which shows that the design here is reasonable. The maximum stress point occurs in the root of the damping arm, whose value is 520 MPa, which is far less than material yield strength 1000 MPa, so the damping arm will not yield.

It can be seen from Figure 5, in Condition 3, that the values of deformation and displacement are largest on the top of damping arm of the steering knuckle, can reach 2.029 mm, which shows that the design here is reasonable. The maximum stress point occurs in the top of damping arm, whose value is 706 MPa, which is less than material yield strength 1000 MPa, so the damping arm will not yield.

## 3 FATIGUE STRENGTH ANALYSIS ON THE STEERING KNUCKLE

In this paper, the prediction on the fatigue life of the steering knuckle is based on the elastic-plastic theory and Miner's linear cumulative damage rule. According to Miner's linear cumulative damage rule, it is assumed that the cumulative damage and the number of stress cycles have a linear relationship, and when the cumulative damage reaches 1, material damage occurs. It can be expressed by a mathematical expression as:

$$D = \sum_i \Delta D_i = \sum_i \frac{n_i}{N_i} = 1 \tag{1}$$

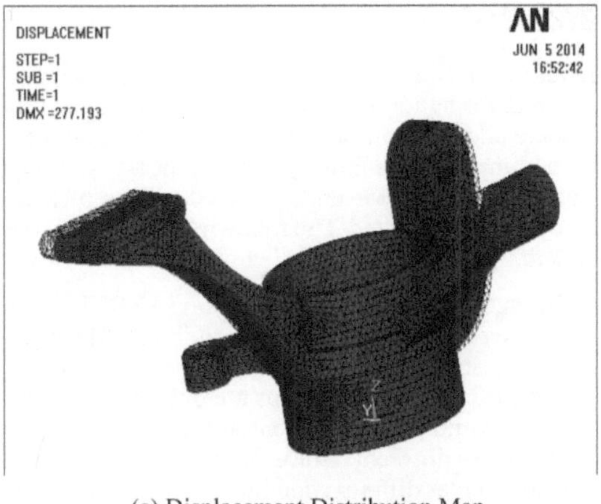


(a) Displacement Distribution Map

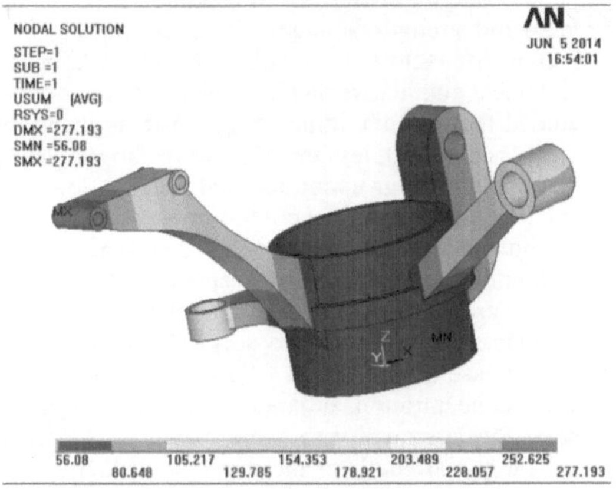


(b) Displacement Moir Map

(c) Stress Distribution Map

Figure 3. Finite element calculation results in condition 1.

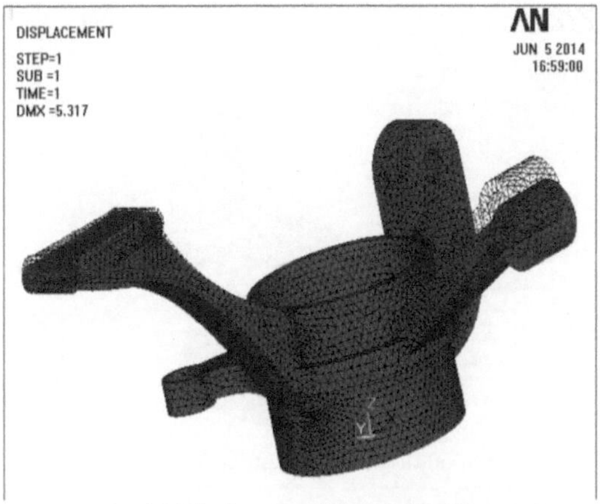


(a) Displacement Distribution Map

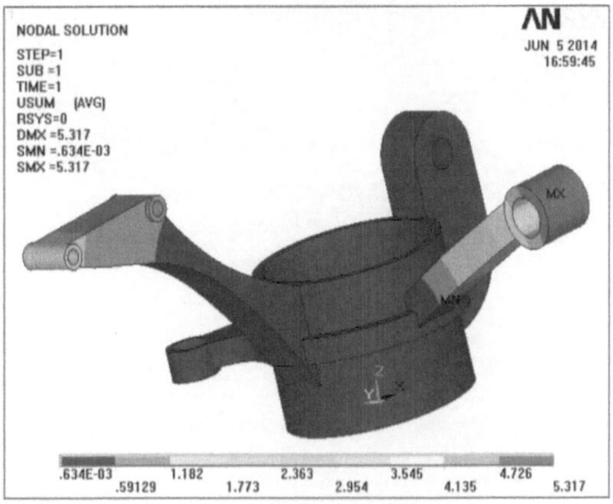


(b) Displacement Moir Map

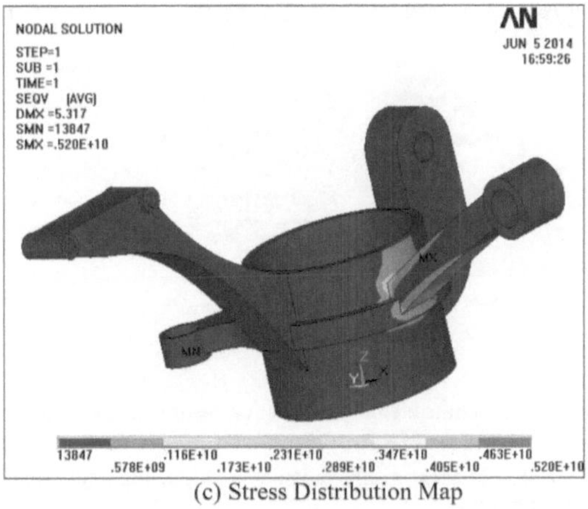


(c) Stress Distribution Map

Figure 4. Finite element calculation results in condition 2.

here, $\Delta D_i$ is the damage index under the i-th level constant amplitude cyclic stress whose amplitude is $S_i$, $n_i$ is the times of cycles on a corresponding level, and $N_i$ is fatigue life under $S_i$. The material of the steering knuckle is 40CrNiMoA.

The analysis results are shown in Table 4.

It can be seen from Table 4, that the values of the fatigue life application factor are all less than 1. According to the linear cumulative damage rule, only when the damage accumulation reaches 1, does material damage occur, thus, in the three conditions above, the steering knuckle which works under 300000 times of cyclic loading will not fail at all.

(a) Displacement Distribution Map

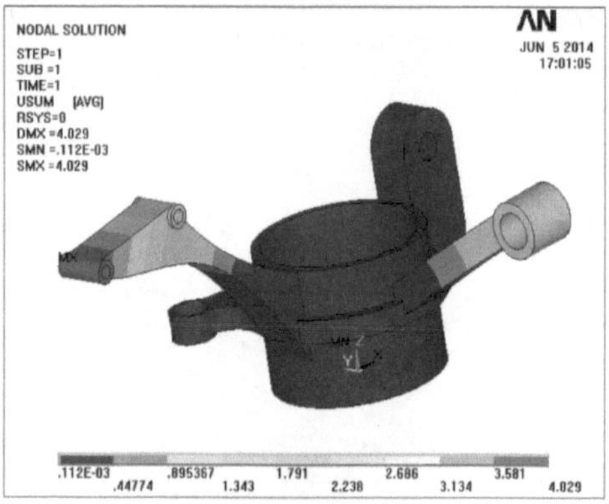


(b) Displacement Moir Map

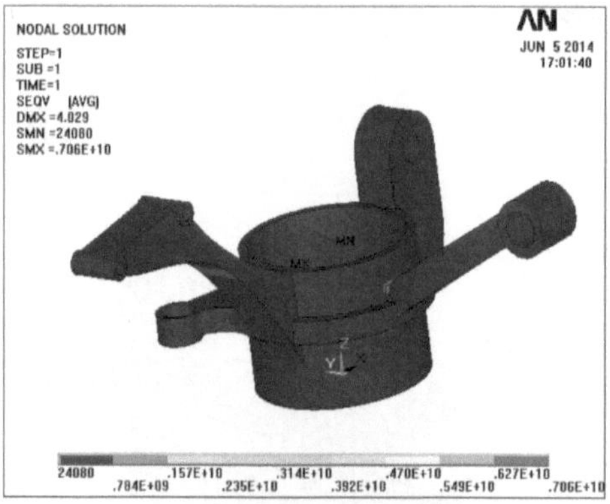


(c) Stress Distribution Map

Figure 5. Finite element calculation results in condition 3.

Table 4. Analysis results of knuckle fatigue life under three limit conditions.

| Load Condition | Node | Cycle times of loading | Fatigue longevity quotiety |
|---|---|---|---|
| Condition 1 | 37 | 375000 | 0.93750 |
| Condition 2 | 158 | 300000 | 0.75000 |
| Condition 3 | 37 | 375000 | 0.93750 |

## 4 SUMMARY

In this paper, finite element analysis is applied to analyse the deformation and stress distribution in a steering knuckle under conditions of three kinds of limit load. In combination with fatigue strength theory, a fatigue life application factor under 300 000 times of loading cycles is calculated. The following conclusions are gained:

1) Strength analysis is conducted using ANSYS. It can be seen that under different conditions, displacement distribution maps, displacement moir maps, and stress distribution maps are different, and also the maximum displacement and maximum stress values are different. However, the maximum stress values under different conditions are all within the yield strength, so the steering knuckle does not yield and strength damage will not occur.
2) Fatigue life factors are calculated based on ANSYS and linear cumulative damage theory. It can be concluded that the maximum fatigue life application factor is 0.93750, less than 1, that is, fatigue damage will not occur under 300 000 times of loading cycles in all of the three conditions.
3) In the process of simulation, a dangerous point is found, and it is the main cause of the initial crack, crack extension, and the flowing fatigue damage. The main way to solve this question is to: increase the transition fillet radius, relieve the stress concentration, and reduce the surface roughness. Surface quality to some extent also affects the fatigue strength of the parts, so reducing the surface roughness can reduce or avoid the generation of defects, thereby improving the fatigue life. Within a certain range, improving the parts surface hardness of the dangerous area on the parts surface can strengthen the anti-fatigue properties of parts. For this purpose, surface hardening methods can be used, such as carburizing, nitriding, shot peening etc.

## REFERENCES


Aijun Hu, Baozhan Lv. Development Tendency of Vehicle Steering System. Tractor & Farm Transporte. 2011, 12

Fei Li. Research on Durability Life Prediction of Passenger Car's Steering Knuckle. Changchun: Jilin University Ph.D. Thesis 2010

Wei Peng, Xiaoxiong Jin, Shuguang Zuo. Fatigue Life Prediction of a Car's Rear Axle Based on FEM Analysis. Automotive Engineering. 2004, 26 (4)

Wei Tang. Molding Process Simulation and Mold Design of Automotive Knuckle. Forging. 2009, 44 (2)

Yanwei Wang, Jiwei Luo, Jun Ye, etc. FEA Based fatigue analysis and Its Application. Machinery Design & Manufacture. 2008, 1

Ying Zheng. Process Design of Automobile Steering Knuckle. Economic Vision. 2013, 6

# *Computer education*

*Frontiers in Computer Education – Wang (ed.)*
 *ISBN 978-1-138-02797-8*

# Computational thinking in C++ programming language

HongBing Wang, YanRui Liang & Lin Yao
*School of Computer and Communication Engineering, University of Science & Technology, Beijing, China*

Jun Cai & LiJuan Zhu
*School of Metallurgical and Ecological Engineering, University of Science & Technology, Beijing, China*

ABSTRACT: The essences of computational thinking are abstraction and automation. Abstraction and automation in C++ programming language are discussed. Abstraction in C++ is the modelling process during which the symbols in C++ are used to describe the system accurately and strictly. Automation in C++ means that some operations are worked on the abstract modules composed of symbols, and these operations are executed in sequential or non-sequential flow structure. Finally, simple examples about abstraction, enumeration, recursion, iteration, reasoning, simulation, polymorphism, and overloading are given to show the case base about the computational thinking in C++ programming language.

*Keywords*: Computational thinking, Abstraction, Automation, C++, Simulation

## 1 INTRODUCTION

Computational thinking is taking an approach to solving problems, designing systems, and understanding human behaviour that draws on concepts fundamental to computing (Wing 2006). Computational thinking is a kind of analytical thinking (Wing 2008). Computational thinking is influencing research in nearly all disciplines, both in the sciences and the humanities (Bundy 2007).

There are three typical ways of scientific thinking: positive thinking marked by observation and induction, logical thinking marked by reasoning and deduction, and computational thinking marked by abstraction and automation (Chen 2012, Li 2012).

The principal characteristics about computational thinking are abstraction and automation. Abstraction is the process of using the symbol system to describe the problem which needs to be solved strictly and accurately. Automation means that some operation can be made on these symbols and then be automatically executed according to some flow.

There are so many methods in computer science to reflect computational thinking, such as enumeration, recursion, iteration, simulation, reasoning, polymorphism, and overloading. So it is necessary to build a case base about the computational thinking, when C++ programming language is studied.

## 2 ESSENCE OF COMPUTATIONAL THINKING

Some teachers always use simple examples for students during the teaching process of C++ programming language. They focus on how to use the programming language in the correct syntax, not on how to understand the principles hiding behind these examples. This has the advantage that students learn how to write simple programs quickly, but has the disadvantage that students consider programming to be just the process of using these tools, such as coding, debugging, building, and linking. This teaching style of instrumentalism easily leads to the habit of copying code for students who do not understand the computational thinking behind the program.

### 2.1 *Abstraction*

One essence of computational thinking is abstraction. The abstractions in C++ are extremely general because they are symbolic. For the programming using C++, the abstraction is the modelling process, during which the symbols in C++ are used to describe the system accurately. This system may be a real one like student management, or may be an abstract one like a virtual game. The program is the product of the abstraction of the system. The system has some basic characteristics and some important activities. The operands in C++ are used to describe these characteristics, and the functions in C++ are used to describe these activities. There are many activities in the system and there are some complex relationships among these activities. For example, the function to compute the definite integration maybe call the function to compute the value of the mathematical function at some point. If this mathematical function is composed of many deterministic functions such as sine

Table 1. Abstract modules in C++.

| Abstract modules in C++ | Real elements |
|---|---|
| Program | System |
| Class | Substance |
| Object | Instance |
| Function | Activity |
| Data type | Characteristics |
| Constants | Characteristic with the unchangeable value |
| Variables | Characteristic with the changeable value |
| One-dimensional array | Multiple characteristics with the same data type |
| Multiple dimensional array | Multiple characteristics with the different data type |

function, cosine function, and tangent function, these functions should be called in turn by the mathematical function. This is called the multilevel calling of the function in C++.

The operands in C++ can be divided into variables and constants. The constants are used to describe the characteristics with unchangeable values, such as the circumference ratio. The variables are used to describe the characteristics with changeable values, such as the circumference radius. Both constants and variables must have a value range and the permitted operations. For example, the score of course generally has a value range from 0 to 100 and the permitted operations include addition, subtraction, and comparison. However, the name of course can not be operated by these arithmetic operators. Data type in C++ can be used to define the value range for these characteristics and the permitted operations. The abstract modules in C++ are shown in Table 1.

C++ is an object-oriented programming language. Class in C++ is used to abstract the substance in the real world, and object in C++ corresponds to the real instance.

### 2.2 *Automation*

Another essence of computational thinking in C++ is automation. Automation means that the operations are executed automatically in sequential or non-sequential flow structure. For the execution of a program written in C++ language, automation is supported by the EIP (Extended Instruction Pointer) register, which always points to the next address of instruction to be executed, and the stack in memory, which has the last-in-first-out requirement.

When the content of the EIP register increases by itself, the instructions inside the function can be executed in sequence. With the support of the stack in memory, the jump between functions can be easily implemented. The change in the stack in memory is shown in Figure 1, when the following program in C++ is executed:

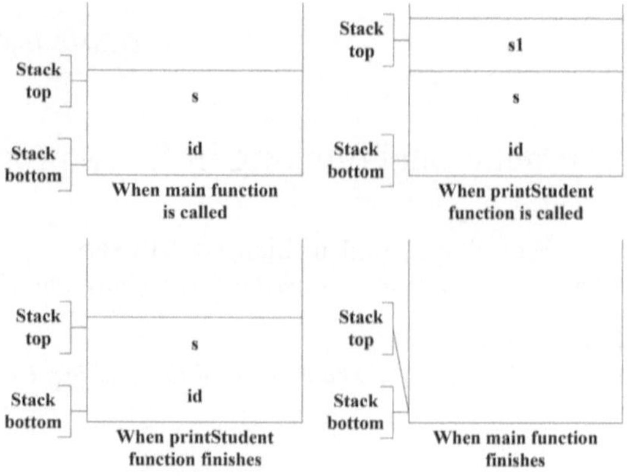


Figure 1. Change about the stack in memory.

```
#include <iostream>
using namespace std;
class Student
{
public:
    Student(int pId)
    {
      id=pId;
    }
public:
    int id;
};
void printStudent(Student s1)
{
    cout<<s1.id<<endl;
}
int main()
{
    int id=1001;
    Student s(id);
    printStudent(s);
    return 0;
}
```

When the *main* function is called, the stack in memory is allocated for this function to store its local variable *id* and *s*. When the *printStudent* function is called, the stack in memory is allocated for this function to store its local variable *s*1 at the top of stack. At the same time, the program control is transferred into the *printStudent* function. When the *printStudent* finishes, the program control returns back to the *main* function. The stack for the *printStudent* function is deleted. When the *main* function finishes, the program control returns back to the operating system and this program is over. The last-in-first-out character about the stack in memory guarantees that the top of the stack is only used for the function to get the program control.

## 3 CASE BASE ABOUT COMPUTATIONAL THINKING

The essence of computational thinking is abstraction and automation. However, computational thinking in

C++ has many different forms, such as iteration, polymorphism, and recursion. So it is necessary to build the case base in C++ to master these different forms of computational thinking.

### 3.1 *Abstraction*

Abstraction in C++ is the process where the data type is used to describe the important characteristics of the system to be developed. For example, a variable with the int data type can be used to define the number of students in a class.

```
int number;
```

An array with the float data type can be used to define the length of students in a class.

```
float length[30];
```

The class Student can define the characteristics and activities for the real students abstractly.

```
class Student
{
};
```

### 3.2 *Enumeration*

Enumeration is the process of listing all the possible results in order. Enumeration is the direct search method to find the solution. Although it is simple, it can easily be forgotten. The problem about finding the daffodil number can be solved in the enumeration method, as shown in the following:

```
for(j=100;j<1000;j++)
{
a=j%10;
b=(j/10)%10;
c=(j/100)%10;
i=a*a*a+b*b*b+c*c*c;
if(i==j)
printf(" %d\n",j);
}
```

### 3.3 *Recursion*

Recursion in C++ means that a function can call itself directly and indirectly. A classic example of recursion is the definition of the factorial function and the code is as follows:

```
int factorial(unsigned int n)
{
if (n == 0)
{
return 1;
}
else
{
return n * factorial(n – 1);
}
}
```

### 3.4 *Iteration*

Iteration is the act of repeating a process with the aim of approaching a desired goal. Unlike recursion, this repeating is often implemented in a loop control statement in C++. For example, the iteration formula to solve the square root of a positive number is as follows:

$$x_{n+1} = 1/2 * (x_n + a/x_n).$$

According to this formula, the iteration process in C++ is shown in the following:

```
x0 = a/2;
x1=(x0+a/x0)/2;
do
{
x0 = x1;
x1=(x0+a/x0)/2;
}while (fabs(x0-x1)>=1e-8);
```

Here, a is the positive number, $x0$ corresponds to $x_n$, and $x1$ corresponds to$x_{n+1}$.

### 3.5 *Simulation*

Simulation in C++ is the process which uses the probability function to enumerate the real case. The following code is to simulate the probability of head or tail appearance for a coin-tossing game:

```
srand((unsigned)time(NULL));
printf("input the test times: );
scanf("%d",&n);
i = n;
while(n–)
{
     if(x=rand()%2)
      num[1]++;
     else
      num[0]++;
}
```

Here, for every different seed value used in a call to *srand*, the pseudo-random number generator can be expected to generate a different succession of results in the subsequent calls to *rand*

### 3.6 *Reasoning*

Reasoning is the capacity for consciously making sense of things, applying logic for establishing and verifying facts, and changing or justifying practices, institutions, and beliefs based on new or existing information (Nikolas 2000). There are many kinds of reasoning. Here, just case-based reasoning is introduced. Case-based reasoning is the process of solving new problems based on the solutions of similar past problems. The calculation about the similarity between the new problem and the past case is very important for case-based reasoning. The following code is to compute the similarity, the new suspect, and the past criminals in the case library:

```
double computeSimilarity(struct criminal c1, struct criminal c2)
 {
  double similarity = 0.0;
  double distance = 0.0;
  distance+ =abs(c1.length-c2.length);
```

```
    distance+ =abs(c1.lengthOfArm-c2.lengthOfArm);
    distance+ =abs(c1.lengthOfFace-c2.lengthOfFace);
    distance+ =abs(c1.lengthOfLeg-c2.lengthOfLeg);
    distance+ =abs(c1.weight-c2.weight);
    distance+ =abs(c1.widthOfFace-c2.widthOfFace);
    if(distance==0)
      similarity = 1.0;
    else
      similarity = 1.0/distance;
    return similarity;
}
```

### 3.7 *Polymorphism*

Polymorphism means the occurrence of something in different forms. In C++, the access to different types of objects through a common base class is done by using a pointer of the type of the base object to point to the objects which derive from that base class. The following code shows that the base pointer *ppoly* can access the two different derived objects *rect* and *trgl*:

```
class Polygon {
  protected:
    int width, height;
  public:
    void set_values (int a, int b)
      { width=a; height=b; }
    virtual int area (void) =0;
};
class Rectangle: public Polygon {
  public:
    int area (void)
      { return (width * height); }
        };
class Triangle: public Polygon {
  public:
    int area (void)
      { return (width * height / 2); }
};
int main () {
  Rectangle rect;
  Triangle trgl;
  Polygon * ppoly=&rect;
  ppoly->set_values(4,5);
  cout<<ppoly->area()<<endl;
  ppoly=&trgl;
  ppoly->set_values(4,5);
  cout<<ppoly->area()<<endl;
}
```

### 3.8 *Overloading*

Overloading allows a function or a data type to be written generically so that it can handle values identically without depending on their type (Pierce 2002). Overloading is a way to make a language more expressive while still maintaining full static type safety. Overloading in C++ can be divided into function overloading, operator overloading, function template overloading, and class template overloading. The following code shows the overloading between two functions with the same name *add*:

```
int add(int a,int b)
{
  int c;
  c = a + b;
  return c;
}
float add(float a,float b)
{
  float c;
  c = a + b;
  return c;
}
```

## 4 CONCLUSION

Abstraction and automation in C++ programming language are discussed in detail. Simple examples about abstraction, enumeration, recursion, iteration, reasoning, simulation, polymorphism, and overloading are given to show the case base about the computational thinking in C++.

## ACKNOWLEDGMENT

This is financially supported by the Beijing Higher Education Young Elite Teacher Project (YETP0382), Fundamental Research Funds for the Central Universities (FRF-TP-12-086A), and the Teaching Reform Project at the University of Science and Technology, Beijing (JG2013M32).

## REFERENCES

Bundy, A. 2007. Computational thinking is pervasive. Journal of Scientific and Practical Computing 1(2): 67–69.

Chen, G.L. 2012. Computational thinking. Communications of the CCF 8(1): 31–34.

Li, L. 2012. Computational thinking – Concepts and Challenges. China University Teaching 1: 7–12.

Nikolas, K. 2000. So we need something else for reason to mean. International Journal of Philosophical Studies 8(3): 271–295.

Pierce, B.C. 2002. Types and Programming Languages. MIT Press.

Wing, J.M. 2006. Computational thinking. Communications of the ACM 49(3): 33–35.

Wing, J.M. 2008. Five deep questions in computing. Communications of the ACM 51(1): 58–60.

*Frontiers in Computer Education – Wang (ed.)*
*© 2015 Taylor & Francis Group, London, ISBN 978-1-138-02797-8*

# Discussion on the teaching methods for a computer network course

HuiTing Wu & YanMei Liu
*College of Information Engineering, Chutian College, Huazhong Agricultural University, Wuhan, China*

ABSTRACT: A computer network course is a professional, required course for computer majors. The course is wide-ranging, has abstract content, and a very strong theory content. It includes a lot of concepts, definitions, and protocol. Generally, students feel that the learning content is abstract and boring. According to the characteristics that the independent college students learn the course, the paper puts forward methods to improve classroom teaching and to cultivate quality computer application specialty talents, which aim to improve the students' computer network course learning enthusiasm and ability to solve practical problems.

## 1 RESEARCH BACKGROUND

With the rapid development of the computer network, a computer network course's requirement is the cultivation of a skill, which is a subject with strong practicability and wide application. On the other hand, a computer network course is an important manifestation of teaching and scientific research, organically combined for teachers. It is also an important conversion of passive learning and active learning for students. In the independent colleges, which mainly train applied talents, we need to advocate the setting up of new teaching ideas, reform the teaching method, and improve students' autonomous learning ability, analysis and problem solving ability, and practical ability. The teacher must carefully study teaching methods to make students better master the course. The teacher should focus on the students' thinking to inspire and stimulate students' interest in the classroom. At the same time, the teacher should pay attention to letting the students know the 'hows' but not the 'whys' when teaching knowledge points.

The practice proved that through the classroom method, students should be able to understand and master the computer network more quickly and deeply, and improve their skills.

## 2 EXISTING PROBLEMS

### 2.1 *Boring theory teaching*

Students have high expectations of a computer network course due to the rapid development of the network. However, after a few classes, the students often find that the course is not as simple as they imagined it to be. The characteristics of this course are that there are too many and miscellaneous knowledge points, with strong theory, and abstractness. Logical thinking and stereo mapping are needed. If the teacher still blindly echoes what the books say in class, and only talks about the principle of the network, it is very difficult to guide the students' learning interest.

### 2.2 *The dispersed knowledge point*

Traditional teaching uses the bottom-up teaching method. For example, the system structure of the network is explained from the physical layer to the application layer. The protocols and technology that are involved in each layer, have separate explanations. Students find this difficult to digest, and lose the learning initiative.

### 2.3 *Disconnection between experiment and reality*

For the experiment of this computer network course, students are required to study hardware knowledge of the network including the computer, the communication link, and the network equipment. Also, knowledge about the network software should be learnt, namely the network protocols and architecture. However, by the experiment's environmental restriction, the course's current experimental projects are mainly verification experiments such as simple networks or application layer protocol configuration, and comprehensive and design experiments are less. Many students still learn the low layer protocols by rote, and thus they cannot understand their practical application.

## 3 TEACHING METHODS

### 3.1 *Analogy method*

In computer network course teaching, teachers often have to explain some concept which is abstract and difficult to understand. When some very abstract concept is not easy to understand, the teacher can give living examples to study which is very similar and more familiar by people. This method can help students intuitively and vividly understand abstract knowledge.

The system structure is the key content in a computer network course. Students generally reflect that it is very abstract for understanding this part content. So the teacher can use analogy method to explain this part content. We can compare the hierarchical structure of the network to the courier company management system. Item which is sent from Beijing to Shanghai will go through the following process: After item is packaged, the courier writes the destination address, post code and telephone number in a letter. Then the item is sent to the delivery department service in Beijing. Express company staffs will baggage sorting according to the destination address. Already sorting good packages are delivered to the train station or other transport sector. The transport sector will make packages packed in bales which are ready for transport. After that item is transported the express to Shanghai. Shanghai express company inspects items and assigns to the subsidiary company. Then courier delivers the items to the people who receive them. After people who receive the item sign, the whole transport task is completed. Teacher can explain the OSI (Open Systems Interconnection) model from the bottom to the top of the function using the analogy method to make students vividly understand the concept of hierarchy. If the data encapsulation in the system structure, such as the concept of protocol data unit is not well understood, the teacher can continue to cite examples. The package is a manifestation of goods in the sender and recipient hand. The package is for the performance of bagging state in Beijing or Shanghai express company, and is shown as packed in bales of selected state both in the transportation sector. The form of package is not the same in different levels. While at the same level, the form is the same. It can help the students to understand the various layers of the PDU (Protocol Data Unit). In the architecture of the express company, the sender / recipient of this layer is the upper layer of the system. The express company is the lower layer of the system. We can see that the express company serves for its upper layer and the transport department serves for express company. This analogy method can let students better understand the relationship between different layers in the system structure. We can give an example to explain the task of each layer. A mother let the child go to wash the clothes, and the mother is waiting for airing clothes in the balcony. First of all, the mother does not care about that child washes clothes with soap or washing powder. It is OK as long as the clothes are washed. Secondly, if there is soap powder on the child's washed clothes (This layer of the task for child is not well completed), it will affect the mother to air the clothes. It can help students to understand that each layer divides the work to complete the task.

It is a difficulty for student to understand the technology of CSMA/CD (Carrier Sense Multiple Access with Collision Detection). The working process of CSMA/CD can be understood through the analogy of many students meeting in a dark room. The IEEE802.3 standard LAN in which all computers share transmission media can be likened that the students talk in a very small and totally dark room and only a person is allowed to speak in the same time. When understanding carrier sense, it can be likened that one student who will speak should check to see whether other students are speaking. If not, the student can begin to speak. When understanding multiple accesses, it is as if each of the students in the room has the same ability to speak. Conflict detection is that if there are two students speaking at the same time. Everybody cannot hear any person's speech because of the occurrence of conflict. When this happens, the two students need to stop the continued talking at once and wait for a random period of time. The students repeat the above process until speech is successful. If there are too many failures, the student, perhaps, gives up the idea of a speech.

Through the analogy method, it not only allows students to have a deeper understanding of the knowledge, but it also makes the students concretize abstract problems when understanding the knowledge points.

### 3.2 *Heuristic method*

Heuristic teaching can be reflected through the form of a question and answer, a lecture and a practice. It also can be realized through the teacher's vivid lecture to make students leave a deep impression. In computer network teaching, we must use all the available means to inspire students to broaden their thinking. For example, when the relationship between the channel utilization and time delay is talked about in the network, we can choose a question to ask which is closely related to the life of the students. Then we inspire the students to think about the problem to get the answer. For example, channel utilization means that a channel is used in the percentage of time in the computer network. If the channel utilization rate is 10%, it means this channel is occupied for 10% of the time and idle for 90% of the time. It is not better if channel utilization is higher. The relationship of delay and utilization rate is proportional. The students are often easily confused when trying to understand the problem. In our teaching, we can inspire students to think about when they use broadband access at home; the students want to browse webpages fast or often disconnect the internet. The communication channel is often shared used for the majority of users. From the user's point of view to consider the problem, the user certainly hopes that the utilization ratio of the communication channel is very low. In this case, the user can use it whenever he likes. And the user also doesn't meet that the channel is too busy to use. It is random that the user uses a public communication channel. If there are too many people using the channel at a certain time, the channel may be in the busy state. The users' complaints increase, and they are not even willing to use the services provided by the company. It will force the telecommunication companies to increase investment to expand the capacity of the communication lines for reducing the average utilization rate of the communication channel.

Through the inspiration above, it not only allows the students to deepen their comprehension and memory of knowledge, but it also explains the practical problems in life.

### 3.3 *The task-driven method*

The task-driven method requires the establishment of a task 'goal' and teaching situation. It can make the students learn in the exploration with the real task. The computer network course in this experiment, is a practical course, and it is suitable for the method. Firstly, the teacher forms groups as a unit to create situations, and puts forward the task in each experiment class. For example, a company has two departments, sales and technology, and the personal computer system of the sales department is connected to different switches. These need to communicate with each other, but for data security reasons, the sales and technology departments need to be isolated from each other. Now, appropriate configuration is needed on the switch to achieve this goal. Through the setting of this task, it enables the students to master realizing port isolation on the switch by the division of Port VLAN (Virtual Local Area Network). Then the computer systems in the same VLAN can communicate with each other across a switch, but the computer systems in a different VLAN cannot communicate with each other. Students need trial, exploration, tasks research, and correct execution in a specified time. Secondly, teachers need to explain, demonstrate, reform, and innovate according to the situation so that the students complete the task. Finally, the teacher will evaluate, check, and summarise for the completion situation of each group.

'Task-driven' is a learning method which is applicable to learning knowledge and skills of an operation class. It has formed a basic characteristic of 'task' as the main line, 'teacher' as the leading factor, and 'student' as the main body.

## 4 STIMULATION ON STUDENTS' INTEREST

The great scientist Einstein said: "Interest is the best teacher." Once students have a strong interest in something, they will take the initiative to seek knowledge, explore, and practice. A cheerful mood and experience are generated in the process of seeking knowledge, exploring, and practicing. The characteristics of a computer network course are that the network technology is complex and conceptual, and the principles are abstract to understand. However, it is unique in that the network originates from the life and the demands of people, thereby promoting the formation and development of the network. So a lot of use of network technology can be found in life. According to the characteristics of the network, teachers should combine their teaching with practical application and use a concrete example in life to connect the concept which is complex and difficult to understand in real life. For example, when the concept of a hierarchical computer network is explained, we can give examples from life: If a cook hopes his apprentice to do a delicious meal, he must be hurry-scurry. But if he assigns the task that one apprentice cooks coffee, one apprentice fries chips and one apprentice fries sandwich, this division of labour collaboration is easier to make a rich feast (with dynamic courseware of vivid colour). The idea of 'divide and rule' is introduced to solve smaller and simpler problems. So, the big problem of a computer network is solved too. Such a vivid metaphor can help students understand the abstract concept of stratification to receive more with less effort. When explaining simplex, half duplex, and full duplex concept, we can combine the remote controller, interphone, and mobile phone in life. When talking about TCP (Transmission Control Protocol) and UDP (User Datagram Protocol) protocols, the respective characteristics and difference of the two protocols need to be introduced. We can show the example of file transfer in life. Because we do not want to lose the files, TCP protocol is selected in the transport layer. UDP protocol can be selected in the transmission of video. In addition, when introducing the three handshakes in TCP protocol, we can start from a network popular joke which only a programmer can understand. Then students are gradually introduced to the principles of learning.

The abstract problem which needs to be solved in the network is introduced with a real life scenario. It can help to stimulate students' interest by making a classroom atmosphere active.

## 5 LET THE STUDENTS UNDERSTAND THE 'HOWS' BUT NOT THE 'WHYS'

When teachers explain new knowledge or review old knowledge in the introduction of relevant knowledge, students need to understand the 'hows' but not the 'whys' for learning the knowledge. When explaining the production of a twisted pair, we have to ask why twisted pair production must be conformed to the TIA/EIA586B standard, in addition to demonstrating the basic steps of conventional cable production (using pictures or video presentation). Then the 586B and 586A standard is introduced. Then we can explain that if the student is not in accordance with this standard, the cable will appear unforeseen phenomenon. Further what is the straight line and cross line can be extended to introduce. So it lets the students not only learn to make double strands, but also understand why it should be produced like that and how to use it after making it.

## 6 CONCLUSION

As the saying goes, 'Give a man a fish and you feed him for a day. Teach him how to fish and you feed him for a lifetime.' Teachers should not only teach the students knowledge and skills, but also cultivate the students to obtain a desire to learn and master the ways of learning. A computer network course contains more content, is more difficult, and has a wider application than other courses. Teachers should focus on cultivating the students' ability of analysis, software development, network management, and innovation based on network protocol.

## REFERENCES


Chatzigeorgiou, A., Tsantalis, N., & Deligiannis, I. 2008. An empirical study on students' ability to comprehend design patterns. Computers & Education, 51, 1007–1016.

Nurul I. Sarkar. 2006. Teaching Computer Networking Fundamentals Using Practical Laboratory Exercises. IEEE Transaction on education, 49, 285–291.

Paulo H.M. Gurgel, Luiz H.C. Branco, Ellen F. Barbosa, Kalinka R.L.J.C. Branco. 2013. Development of a Practical Computer Network Course through Netkit Virtualization Tool. Procedia Computer Science, 18, 2583–2586.

Vasile Dadarlat, Tom Coffe, Cosmina Ivan. 2002. A Personalized Approach for Teaching Web-Based Curriculum in Communications & Computer Networks. Proceedings of the 2002 IEEE Canadian Conference on Electrical & Computer Engineering, 32, 732–737.

*Frontiers in Computer Education – Wang (ed.)*
*© 2015 Taylor & Francis Group, London, ISBN 978-1-138-02797-8*

# The application of EXCEL in sampling

Chao Ling Qin
*School of Physical Education, Southwest China Normal University, Chongqing, China*

ABSTRACT: Sampling is an important part of scientific research, and it is related to the reliability, accuracy and scientificity of the research data. This paper lists several methods to achieve simple random sampling and systematic sampling using EXCEL software, aiming to provide technical security and reference for scientific sampling.

## 1 INTRODUCTION

The first step of scientific research is to collect information and obtain the sample data, but unrepresentative sample data is worthless for the research. The aim of sampling is to select samples in a scientific way to represent the whole in order to get reliable data and accurately determine the situation of the whole, thus studying the characteristics or regularity of the whole. It could save a lot of human and material resources and time. Theoretically speaking [1], for the whole with many units, the comprehensive characteristics of one unit are in line with the characteristics of the whole. When it is difficult to study the whole or the sample could be used to replace the whole, we use sampling to study the situation of the whole. However, how to carry out sampling and ensure its scientificity is a key problem. Probability theory studies have shown that [2] sampling should follow the random sampling principle to enable the sample to better represent the whole. Probability sampling mainly includes simple random sampling, systematic sampling, stratified sampling and cluster sampling and so on. EXCEL software can be used to achieve the several sampling and provide a technical guarantee for the scientific research.

## 2 SIMPLE RANDOM SAMPLING

If every unit in the whole has the same possibility to be selected (random sampling), and the components of the whole stay the same after the removal of one unit (independence of sampling), this sampling method is called simple random sampling.

### 2.1 *Simple random sampling implemented by EXCEL random function*

Case 1: In order to investigate the course studies of students in a particular grade (200). It is necessary to sample 20 students from 200 ones. The operating steps using Excel are shown as Figure 11.

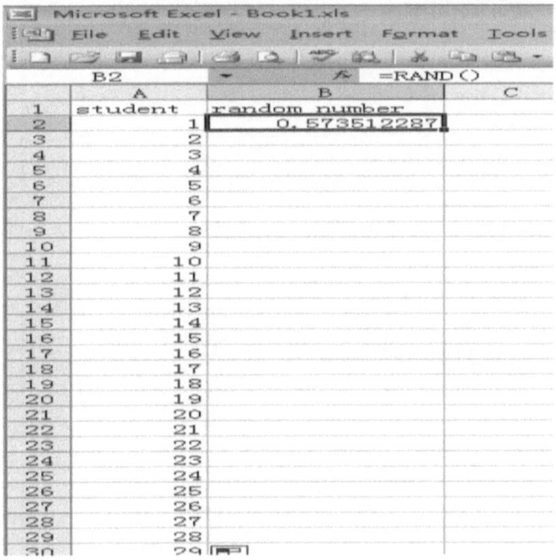


Figure 1. Generate the first random number.

Step ①: Respectively input 1 and 2 at A2 and A3. Select A2 and A3, move the cursor to the lower right corner, and the cursor changes to a solid +. Generate 1-200 integers with the filling method as the simulated number of students, as shown Figure 1; input "=rand()" function → click "Enter" at B2 to generate the first random number of the first student, as shown in Figure 1.

Step ②: Use the fill handle to generate 200 random numbers for 200 students, as shown in Figure 2. To fix the random numbers, randomly select all the numbers, and click "Edit" →"Copy"; "Edit" → "Paste selectively". The dialog box in Figure 3 appears. Paste the "Selected Value" in the dialog box; select "None" for "Operator" →click "OK", and the value is pasted in the original position.

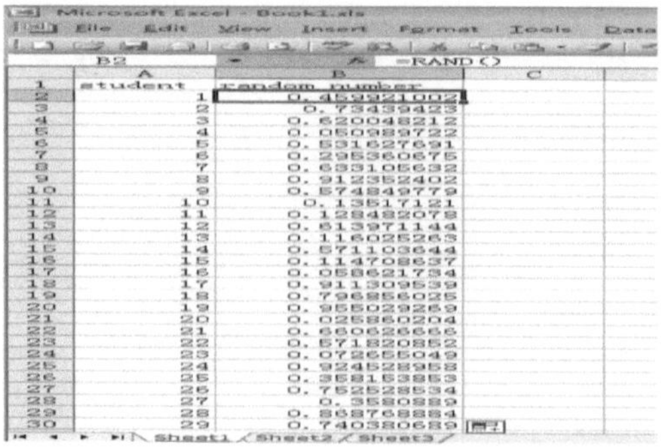

Figure 2. Use the fill handle to generate 200 random numbers.

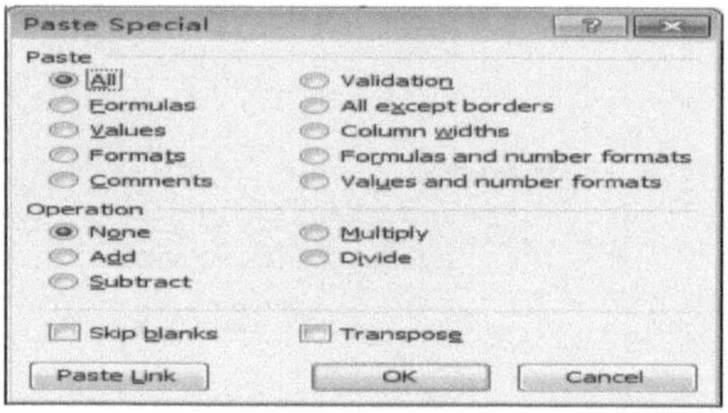


Figure 3. Paste dialog box selectively.

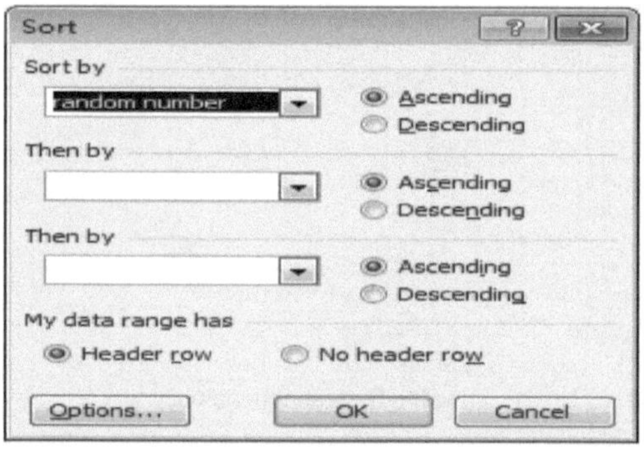


Figure 4. Sort dialog five ranking after the Sort dialog box.

Step ③: Sort the random number. Select B2: B201. In the menu "Data" →"Sort", select "Ascending" for all the key words in the sort dialog to conduct ascending sorting for the random numbers, as shown in Figure 4. Students corresponding to the top 20 numbers are selected as randomly objects, as shown in Figure 5.

## 2.2 *Use EXCEL add-ins method to achieve simple random sampling*

Case 2: In the primary school with 1100 students, where there are 100 teachers and 1000 students. Randomly sample 15 teachers and 50 students to carry out questionnaire survey. The operational steps of Excel are shown as follows:

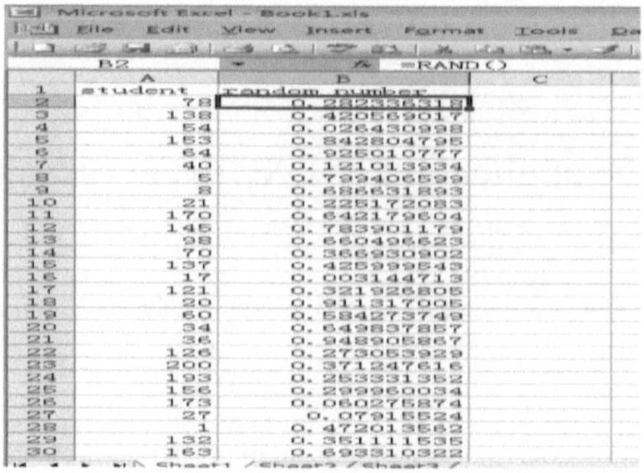

Figure 5. Random number after sorting.

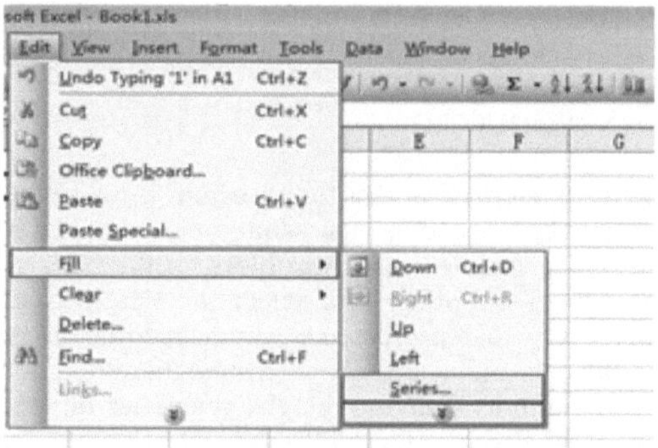


Figure 6. Click "Edit" → "Fill" →"Sequence".

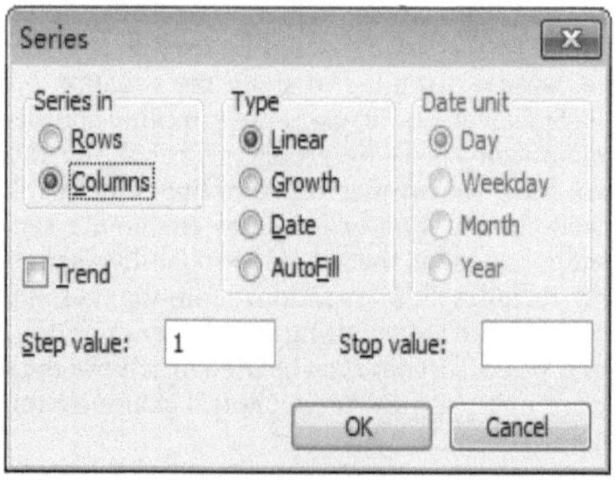


Figure 7. Dialog box of "Sequence".

Step ①: The filling is to complete the serial number of teachers and students: Enter 1 at A1, and then enter 1-1100 with the filling method. Select A1, and click "Edit" → "Fill" →"Sequence" in the menu, as shown in Figure 6; in the dialog box of "Sequence", as shown in Figure 7, select "Column" for the sequence generation, select "Arithmetic Sequence" and fill 1100 in the step termination value →click "Yes", which could generate serial number 1-1100 in column A. Enter "Teachers Sample" at B1, enter "Students Sample" at C1, which is used to enter data after sampling.

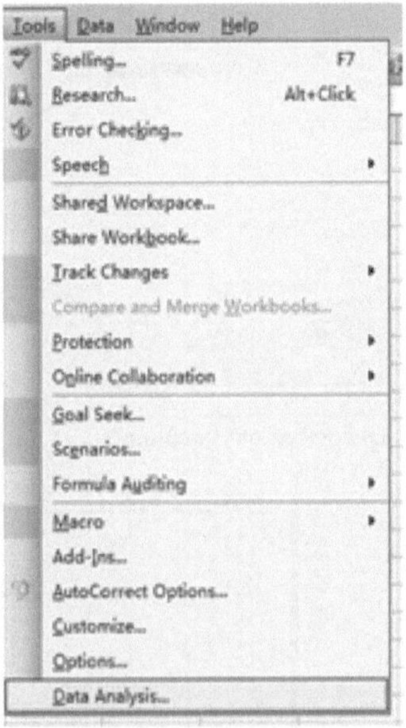


Figure 8. “Tools” → “Data Analysis”.

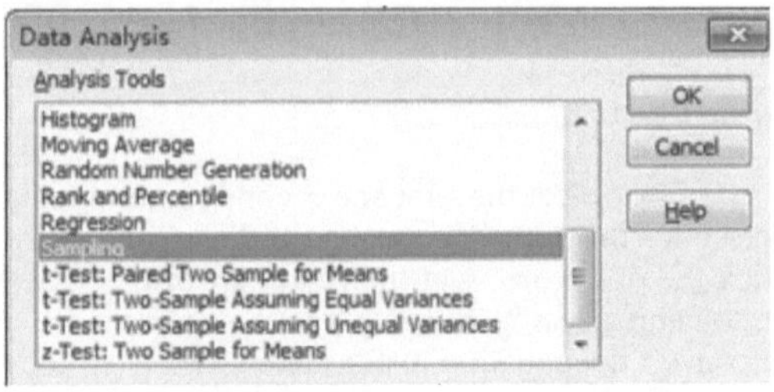


Figure 9. Dialog box of “Data Analysis”.

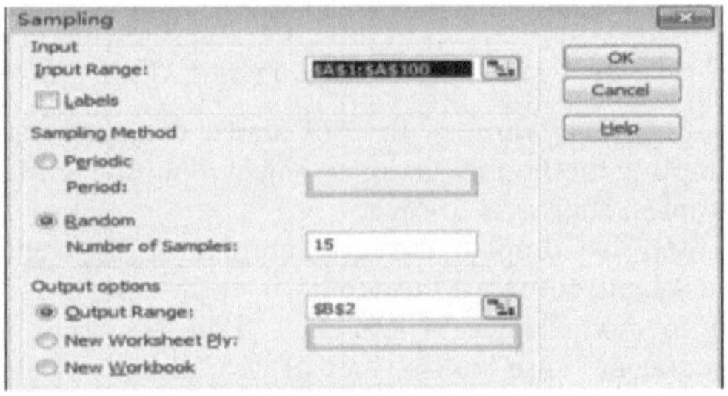


Figure 10. Dialog box of “Sampling”.

Step ②: Select “Tools” Data Analysis and then select “Sampling” in the dialog box” click “Yes” in the menu, as shown in Figure 10. The dialog box “Sampling” appears, select teacher A1 in the “Input area”: A100 area, select “Random” for the “Sampling Method”, input 15 as the number of samples, select B2 as the “Output Area”. Select “OK” to generate the sample of random numbers of these 15 teachers.

Step ③: In the same way, in the “Sampling” dialog box, select student A101 in the “Input Area”: in A1100 area, select “Random” in the menu of “Sampling Method”, input 50 as the number of samples, select C2 in the “Output Area”, →click “yes” to complete the sample of random numbers of 50 students.

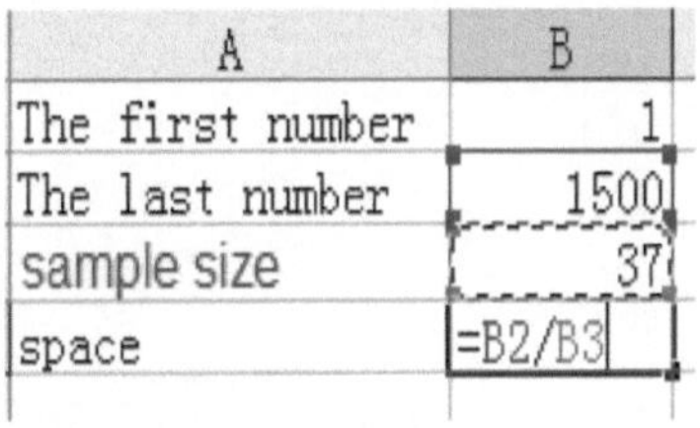


Figure 11. Steps to calculate spacing.

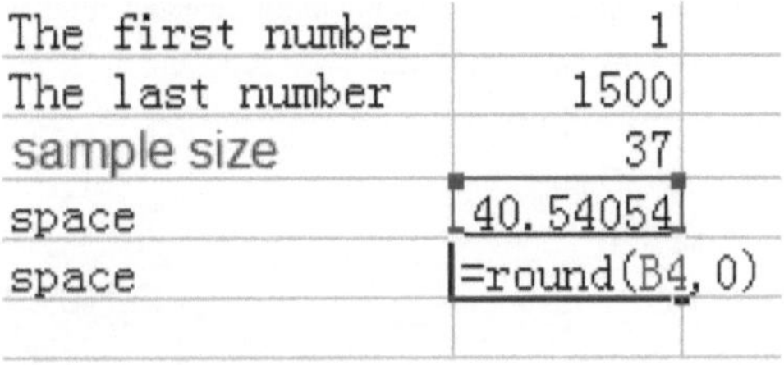


Figure 12. Rounding steps.

## 3 SYSTEMATIC SAMPLING

Systematic sampling is arranged according to a certain order. Carry out sampling based on a fixed interval which depends on the ratio of the required number of samples and the number units in the whole. The initiative number must be randomly determined. Systematic sampling has better representativeness than simple random sampling, for the sample individuals are evenly distributed among the whole. The combination of systematic sampling and simple random sampling could achieve the optimal effects.

### 3.1 *Use the EXCEL random function to achieve systematic sampling*

Case 3: In a whole with 1500 individuals, using systematic sampling method to randomly select 37 individuals as samples, shown as follows:

Step ①: input the first number at A1 in the EXCEL sheet, input 1 at B1; input the last number at A2, input 1500 at B2, input the number of samples at A3, input 37 at B3; input the interval at A4, input “= B2/B3” at B4. Click “Enter” to get the interval 40.54054, as shown in Figure 11.

Step ②: Insert a row below B4, put the integral spacing, which can be replaced by 41 after rounding or use the ROUND () function: enter “= ROUND (B4, 0)” “at B5 (Description: If it is necessary to make B4 an integer, input 0; if it is necessary to keep one decimal place for B4, enter 1, etc., as shown in Figure 12, and then click “Enter” and the spacing is changed to 41.

Step ③: Enter “= RANDBETWEEN (B1, B5)” at B6, as shown in Figure 13, click “Enter” to get the first randomly selected number.

| | |
|---|---|
| The first number | 1 |
| The last number | 1500 |
| sample size | 37 |
| space | 40.54054 |
| space | 41 |
| The first number of randomly selected | ==RANDBETWEEN(B1, B5) |

Figure 13. Generate the first randomly selected number.

| | |
|---|---|
| The first number | 1 |
| The last number | 1500 |
| sample size | 37 |
| space | 40.54054 |
| space | 41 |
| The first number of randomly selected | 36 |

Figure 14. Convert the equation to constant.

| | |
|---|---|
| The first number | 1 |
| The last number | 1500 |
| sample size | 37 |
| space | 40.54054 |
| space | 41 |
| The first number of randomly selected | 36 |
| | 77 |

Figure 15. Generate the second randomly selected constant number.

| | |
|---|---|
| The first number | 1 |
| The last number | 1500 |
| sample size | 37 |
| space | 40.54054 |
| space | 41 |
| The first number of randomly selected | 36 |
| | 77 |
| | 113 |
| | 190 |
| | 303 |
| | 493 |

Figure 16. Copy.

Step ④: For the first randomly selected number, convert the equation to constant: choose B6, click F2 and convert to edit mode, click F9, convert to the recalculating state, as shown in Figure 14.

Step ⑤: Arrange the second number "= B6 + B5" at B10, click "Enter" to get the second randomly selected number.

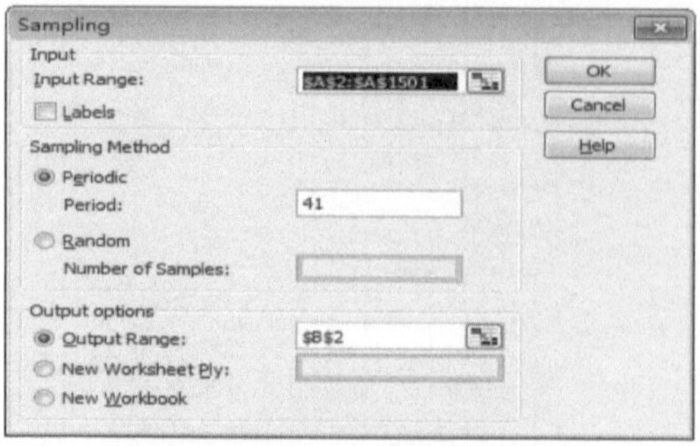


Figure 17. Dialog box of "sampling".

| number | sampling results |
|---|---|
| 1 | 41 |
| 2 | 82 |
| 3 | 123 |
| 4 | 164 |
| 5 | 205 |
| 6 | 246 |
| 7 | 287 |
| 8 | 328 |
| 9 | 369 |
| 10 | 410 |
| 11 | 451 |
| 12 | 492 |
| 13 | 533 |
| 14 | 574 |
| 15 | 615 |
| 16 | 656 |
| 17 | 697 |
| 18 | 738 |
| 19 | 779 |
| 20 | 820 |

Figure 18. Copy.

Step ⑥: Select the first and second sampling numbers, click the "Copy" button, write down the contents, click the down arrow at the right part of the "Paste" button and select "Value (V)", as shown in Figure 15, to convert the equation to constant.

Step ⑦: Put the cursor at cell B9, when it becomes solid +, move the bottom right cursor to the 37th sample, as shown in Figure 16.

## 3.2 *Achieve systematic sampling by EXCEL add-ins*

Case 4: In a whole with 1500 units, use systematic sampling method to randomly select 37 individuals as samples, shown as follows:

Step ①: Complete the serial number of 1500: enter 1 at A1 and complete the number 1-1500 using the filling method. Select A1 and click "Edit" → "Fill" → "Sequence" (use the first step of EXCEL add-ins random sample). In the dialog box of "sequence", select "column" in "sequence generation", select "arithmetic sequence" in the menu of type, fill in 1500 in the step termination value and click "OK" to generate 1-1500 numbers in column A.

Step ②: Select "Tools" → "Data Analysis", and then select "sampling" in the dialog box of "Data Analysis", and click "OK", as shown in Figure 17. There appears the dialog box of "sampling", input A1 in the "input field": in A1500 area, select "Cycle" in the "sampling method", and then input 41 for the spacing, select B2 for the "output area", as shown in Figure 17. Click "OK" to generate 37 samples with random numbers, as shown in Figure 18.

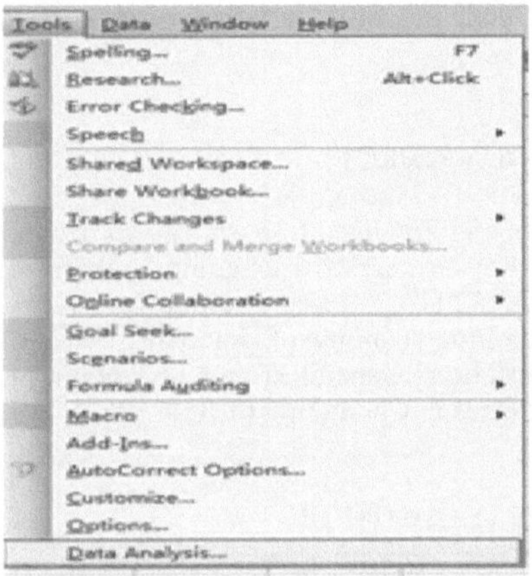


Figure 19. "Tools"→ "Data Analysis".

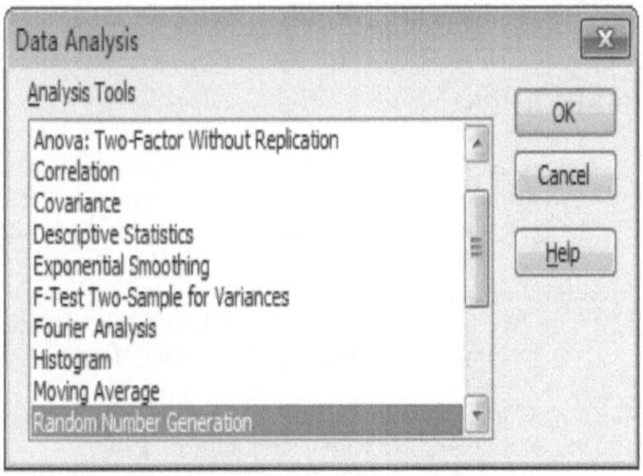


Figure 20. The dialog box of "Data Analysis".

## 4 RANDOM NUMBER TABLE

The random number table is to randomly draw 0-9 and compile these numbers as a table [1]. In any statistical book, there is a random number table in the appendix for future use. If there is no available one, EXCEL can be used to make a table, shown as follows:

### 4.1 *Make a random number table by the random number generator*

Use the EXCEL random number generator to generate a random number table composed by 5 integers, arranging 6 rows and 10 columns.

① Open the EXCEL worksheet, put the cursor in any cell, select "Tools" → "Data Analysis", as shown in Figure 19. There appears the dialog box of "Data Analysis", and then select "random number generator" →"OK", as shown in Figure 20.

② When there appears the dialog box of "Random Number Fenerator", input the number of columns of the random number table in the text box "Variable Number (V)", such as 6; input the number of rows in the text box "Number of Random Numbers(D)", such as 10; input 0-99999 in the "Parameter" (5 digits), and select A1at the output area, as shown in Figure 21. Click "OK" to generate a 10 row × 6 column random number table between 0-99999.

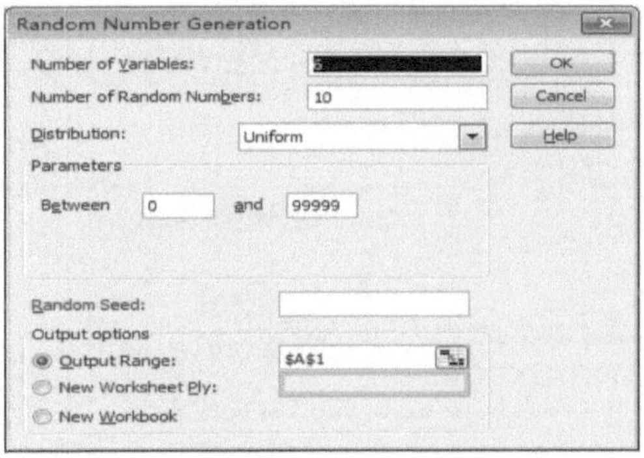


Figure 21. The dialog box of "Random Number Generator".

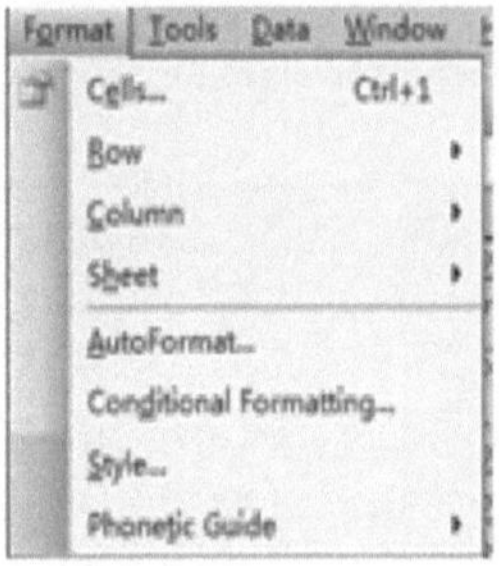


Figure 22. "Cell" steps in the menu of "Format".

③ The random number table at this time has no formatting. Select A1: A10, implement the "Cell" in the menu of "Format, as shown in Figure 22. Then go to "Digital" tab, select "Custom" in the "Classification" and input "00000" at "Type", which means to conduct automatic rounding, if the numbers are less than 5 digits, fill 0 automatically, as shown in Figure 23. Click "OK" to generate a formatted10 row × 6 column random number table between 0-99999, as shown in Figure 24.

### 4.2 *Make a random number table with the random function RAND*

Make a 10 row × 6 column 4-bit random number table with random function RAND. The steps are shown as follows:

① Input "=RAND ()*10000" at A1→ Click "Enter" to generate a random number with no format.

② Define the format of random numbers (refer to the 3nd step of random number table making by the random number generator): Place cursor at A1, select "Format" from the menu→"Cell" → the dialog box of "Number" → Select "Custom"→ Input "0000" at "Type" →Click "OK"

③ Put the cursor at lower right corner of A1 and it becomes solid +. Press the left mouse button and drag 6 columns towards the right and 10 rows downward

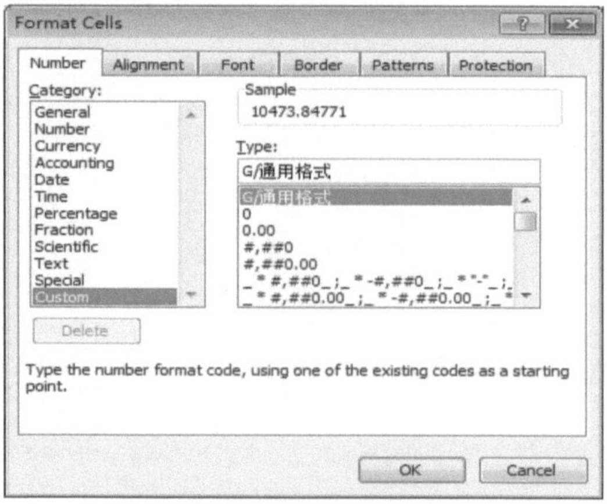


Figure 23. Numerical labels of format cells.

| | | | | | |
|---|---|---|---|---|---|
| 10474 | 63463 | 66490 | 85256 | 53956 | 97243 |
| 96138 | 32474 | 16535 | 58720 | 78307 | 43372 |
| 39927 | 35303 | 63133 | 80501 | 47679 | 72829 |
| 13755 | 02527 | 69481 | 15058 | 50285 | 44877 |
| 61805 | 54350 | 60356 | 00552 | 27933 | 84047 |
| 04416 | 16770 | 90517 | 93267 | 93871 | 73054 |
| 19815 | 94704 | 70174 | 56395 | 70735 | 26050 |
| 63072 | 99099 | 31745 | 15689 | 56028 | 94970 |
| 38868 | 53257 | 37012 | 59425 | 57823 | 55720 |
| 62794 | 49186 | 87441 | 69591 | 14783 | 07157 |

Figure 24. The formatted random number table.

to conduct filling, thus making a 10 row × 6 column 4-bit random number table.

## 5 CONCLUSION

In addition to the simple random sampling and systematic sampling, probability sampling also includes stratified sampling and cluster sampling and so on. Stratified sampling [*ii*] is to classify the individuals in a whole into several types or layers according to a certain property, and ten carry out simple random sampling in proportion to generate the research samples. Besides, the EXCEL method in the simple random sampling and systematic sampling could also be used to achieve the sampling of single type or layer, and then use these samples to compose the samples of the whole. Cluster sampling [*iii*] is to regard cluster as the sampling unit, and then sample several clusters to constitute the samples according to simple random sampling and systematic sampling. Therefore, the cluster sampling could also be achieved by EXCEL method.

As the probability sampling in scientific research involves theatrical and practical problems, research workers should constantly acquire new knowledge and technologies to improve sampling accuracy and quality.

## 6 FUND PROJECT

"Reform and Practice of Sports Talents Training Program", Key Project of Chongqing Higher Education Reform (132042); "Research on the Statistical Analysis Capabilities Training of Graduate Student of Physical Education" General Project of Teaching Reform of Graduates in Chongqing (YJG133067)

## ABOUT THE AUTHOR

Qin Chaoling (1962–), female, Dazhou, Sichuan, Professor, MA, primarily engaged in sports statistics and evaluation.

E-mial: qincl@swu.edu.cn.

## REFERENCES

[1] *Lecture Notes*, [EB / OL]. Http://www.et.pku.edu.cn/methods/textbook/class05_01_00_00_i.htm
[2] Sheng Zhou, etc., Probability *Theory and Mathematical Statistics* [M]. Beijing: Higher Education Press.1989 2nd Edition. P141–153
[3] Wang Hongru. The Application of EXCEL in statistics [M]. Beijing: China Railway Publishing House. 2004 1st edition. pp. 70–73
[4] Cong Huping. *Sports Statistics* [M]. Beijing: Publishing House. 1989 higher with a first edition. P14
[5] Cong Huping. *Sports Statistics* [M]. Beijing: Publishing House. 1989 higher with a first edition P15
[6] Pu Yingjuan, Li Xiang. The application of EXCEL in stratified sampling teaching [J]. Journal of Baoshan Teachers' College. 2008(09).
[7] Tian Bo, Hu Ruiyue, Li Zhiyou. Stratified sampling of electronic prescription by using EXCEL [J]. Chinese Prescription. 2014(07).
[8] Liu Wei, Jia Shijie. The application of EXCEL in complicated random sampling [J]. Journal of Mathematical Medicine. 2014(06).
[9] Qu Xiang. How to use EXCEL tools to do the sampling [J]. Market Research. 2010(10).
[10] Peng Weixiang. The design skills in EXCEL sampling method [J]. Statistics and Decisions. 2004(08).
[11] He Youming. How to use EXCEL to do sampling in audit [J]. Audit Monthly. 2011(11).
[12] Zhang Qingyuan, Gui Gaixia. Dynamic modeling of Gaussian sampling and probability distribution using EXCEL [J]. Journal of Mathematical Medicine. 2012(02).
[13] Yi Dong, Wang Wenchang, Zhang Wei. Using EXCEL to do the simulating test of random sampling [J]. Medical Information. 2002(05).
[14] Bai Xianchun. Methodological discussion on dividing groups based on group intervals [J]. Statistics and Decisions. 2008(18).

*Frontiers in Computer Education – Wang (ed.)*
*© 2015 Taylor & Francis Group, London, ISBN 978-1-138-02797-8*

# Design of an educational recommender system for an MOOC

Lianghui Zhao, Zhenyang Yu & Haisen Wang
*School of Economics and Management, Wuyi University, Jiangmen City, Guangdong Province, China*

ABSTRACT: From the perspective of education designing, this paper proposes ideas on constructing an educational recommender system for MOOC. The difference between an educational recommender system and a commercial one is put forward and used in designing, as well as some key issues and the design process.

*Keywords*: MOOC; Education Recommender System; context-aware

## 1 INTRODUCTION

MOOC, namely the Massive Open Online Course, can provide high quality learning resources and a complete learning experience for the majority of people on the education of learning, especially, what it is obvious of a revolutionary impact on education in higher education. As a new learning concept. An MOOC advocates autonomous learning, and provides a corresponding flexible, as well as simple, knowledge organization form, and a vivid manifestation course to learners. However, as an immature method of learning, the implementation of an MOOC also encounters many difficulties.The major obstacles include:

(1) The learners are difficult to stick to complete the course with lacking of instructors. In an unsupervised learning environment, the most difficult part of learning is "information overload" that is facing too much knowledge, which making leaeners do not know how to choose it. At the same time, the teacher can easily give students guidance in the traditional learning environment, however, the number of learners can easily reach over tens of thousands in an MOOC environment, and teachers' guidance becomes impossible. Therefore, the number of learners can easily shrink causing a high dropout rate, such as the online courses in artificial intelligence from Coursera which attracted 160,000 students at the initial launch, but in the end only 14% of these learners finished the course[1].
(2) An MOOC is difficult for learners to gain the skills providing with the interaction, such as critical analysis and learning to put forward its own ideas and discussions and high level thinking etc[2]. Some MOOC organizers (such as Coursera) introduced mutual learning to deal with it, such as peer assessment, and discussion groups etc, but students also faced difficulties to choose an appropriate peer and build trust network, as well as other practical difficulties.

The introduction of an educational recommender system in the process of an MOOC can help students learning, and it is hoped that the above problem will be solved satisfactorily, as shown below:

(1) It can provide learning guidance with the help of an education recommended system. The recommender system based on peer to peer technology, that is, study the behaviour of other peer learners, which is based on the learning characteristics and learning progress of the current learner to recommend the learning resources. Thus, the recommender result is personalized and consistent with the actual needs of learners. In addition, the recommender existence creates a direction of the learning environment to reduce the difficulty of courses completed and increase the possibility of courses completed.
(2) The recommender system can help students to build Interactive Web Personalization. Based on the characteristics of the learners, with their own as well as the information of other learners, the recommender system can quickly compare the similarities and differences between learners to recommend the most suitable learners to form a partnership, which is much better than the user with blinding study, and also can overcome the users shyness and timid to interactive others.

This paper, based on the above analysis, explores the design of a personalized educational recommender system for the MOOC, which can greatly enhance the quality of the implementation and improve the user experience of the MOOC. Compared with the traditional network video courses, the MOOC has a very important change what provides resources provided services in a single such as certification courses and test and discuss etc, which those given the static course with a new vitality. An MOOC-based educational recommender system can make these services richer and more delicious. Othersides it paves the way for this innovative learning form the brilliant learning style.

## 2 THE RESEARCH STATUS AT HOME AND ABROAD

2012 was known as the "MOOC year", and it was immediately popular begin from its introduction. Nowadays, many international first-class universities and domestic well-known colleges and universities have joined the MOOC camp. The current research on the MOOC remains in the warm-up phase[1−5], and needs further exploration.

The research of an educational recommender system, although the number of less than the burgeoning business recommender systems, have begun to form scale; Among them, Buder believes recommendation systems in e-commerce have been very widely applied, but the majority of applications for education development go no further than the prototype system stage; and commercial recommender system should be reformed to deep transformation for education recommender system[6]. In recent years, the educational recommender system have the Linton's recommender systems of description for "software function learning", which a recommender method design to long-term observe of the conditions of using a desktop software, and it can provide help for the use of software[7]. Drachsler describes a learning environment that is based on the hybrid network technology, where the environment is based on a Web2.0 learning network to provide useful information for the user; and the user can specify the Web2.0 framework service to point the information provided with training the recommender system[8]. Hsu proposed a use for ESL(English as a Second Language) teachers' teaching recommender systems that can help teachers to position difficulties in the learning process of students[9]. The analysis of educational recommender systems have also published, such as Aehnelt etc are presented for a social technology model of use of collaborative learning in workshop[10]; Femandez analysis for the health educational recommender system facing opportunity and challenge[11]. Kardan presents the recommender reference tool using concept maps and labels[12]; Liu Xudong describes a basic framework of a learning information recommender system[12−13], and Wang Zhimei uses a learning state evaluation vector to calculate the similarity between the learner, and introduce a recommender system of Agent Technology to achieve the learning content[14]. Wang Yonggu, Sun Xin, and so on, used collaborative filtering technique to realize recommender resources and described the corresponding system architecture[15−16]. SHI Lin compared the similarities and differences of online shopping and online learning recommender systems and applicable recommender technology[17].

The most literature on the research of educational recommender systems takes a purely technical point of view, ignoring the special of what educational recommender system. In fact, the educational recommender system, unlike the commercial recommender system in many aspects, so it is not suitable for education recommender system with only considering the application environment and the user experience and will simply commercial recommended technology and algorithm of system copy. This article is written from the educational designer's perspective, rather than the algorithm designer's perspective, and fully considers the educational environment of recommender systems, and combines the special requirements of the MOOC to design an educational recommender system.

## 3 THE DESIGN OF AN EDUCATIONAL RECOMMENDER SYSTEM FOR THE MOOC

### 3.1 *System design*

Starting with the learning environment's characteristics of massive open online courses and the actual needs of learners, research into the design methods of personalized education has recommended a system, which closely combined with MOOC, can, according to the current situation of learners, recommend the learning resources which form a learning guide to reduce the difficulty of learning.

Research ideas take three ways to do many things to shoulder that educational theory as the guide and the MOOC system as the structural basis as well as the commercial recommended algorithm, driven as an engine prototype, designed to meet the educational concept and meet the MOOC's actual needs and the leading level of the operation's efficiency educational recommended system.

(1) By the theory of pedagogy as system design guidelines

The fundamental purpose of the recommended system is to promote learning. Therefore it must be based on an educational theory as the design guide. WANG Ping[5] believes that most of the current MOOC is focused on the traditional teaching mode of xMOOC, which is different from the early cMOOC that is 'based on the view of Knowledge Construction Association'. But whether based on the connectivism learning theory, or the currently popular learning theory that emphasizes the construction of learning significance, so the design of the recommender systems must meet and promote such meaningful construction, and some principles such as learner centred and emphasizing the creation of a learning environment as well as an emphasis on collaborative learning and attention to feedback etc will be carried out. For example, constructivism holds that the construction of new knowledge is based on a previously completed knowledge structure network, so the recommended system of learning resources should be fully take into account this principle. It and similar or homologous learners' knowledge

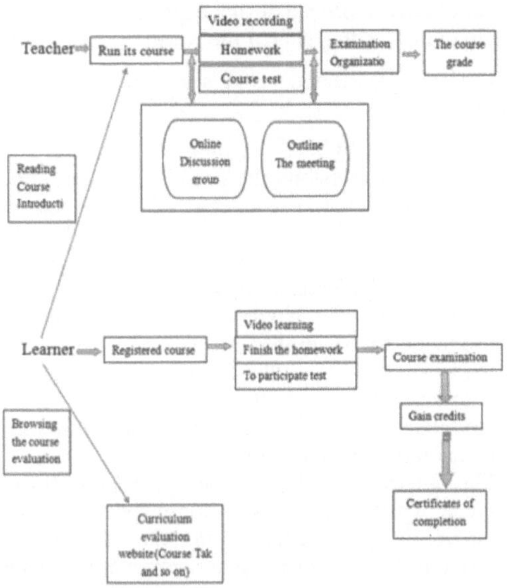


Figure 1. The current mainstream MOOC curriculum model.

background will as one of the standards recommended learning resources to help learners from the original knowledge experience in the "growth" of new knowledge and experience.

(2) Design the recommended system structure based on the structure of MOOC

Wang Ping described the current mainstream MOOC curriculum model, see Figure 1:

In Figure 1, the learning resources use the video instruction, but it can contain other forms as an auxiliary. We believe that any course should have its own knowledge structure, and the video shows the structural fragments. In addition, each video is corresponding to a point of knowledge, and all the knowledge points from a concept map. For the introduction of an auxiliary learning process, this knowledge can continue to diffuse outwards and extend from other knowledge points which can be the basic original knowledge of the principle and the original historical as well as an example and the recent development situation etc. Similarly, it also can be in parallel with the original knowledge. All knowledge points constitute an extension of the concept map to combine into the core of curriculum knowledge system. A collection of knowledge points constitute the recommended space of the recommender system (or a part of the recommender space), accordingly. As an important basis for making the recommended decision, the structure of the concept map also needs to be absorbed in the recommended system. In addition, the logical structure of MOOC course and the link connection with the course Web page and the curriculum content of the logic distribution in the database will also become the basis for system's design.

(3) With context-aware recommender technology to drive the engine's system.

Traditional recommender system in User – Item in two space dimensions try to on those who have not yet been scoring "user-item on" calculation of score function R:

$$R : User \times Item \rightarrow Rating$$

As to the education recommended system, the "goods" is the knowledge points that recommended by the system to the user. This score is an ordered digital symbol of user preference degree. R can be regarded as a partial function, namely the combination of some user * item that has not yet been given a determined score (that is to say not know the evaluation of the corresponding items of the user), which need to use collaborative filtering (or other technologies) to estimate. Once the R of the entire user * article space are established, then some items of the highest score can be a recommender to each user.

Context aware recommender technology will be the environmental information as the third party to join at the recommender space, Verbert[19] argues that the context aware recommender system, used for modelling and predicting user preferences through contextual information, and combined effectively, can be more accurately inferred using behaviour to improve the recommender satisfaction. The corresponding score function can be expressed as

$$R : User \times Item \times Context \rightarrow Rating$$

In the MOOC environment, objects can be mapped into knowledge points, and the scores can be directly or indirectly the score of users for learning resources (such as whether to watch, watch, watch long times etc converted into digital score), in addition, the situation can be mapped into a hierarchy or multidimensional variables, of which the contents should include the use of learning characteristics and including from the start the description of the learning model and the concept's map content.

At present, the context-aware recommender technology is one of the key research subjects in the personalized recommender system, although, it is subject to form a unified algorithm for the diversity of its context. There are some mainstream algorithm structures, such as context pre filtering, context after filtering, and context modelling, etc [20]. Based on those algorithm building the recommender system can improve the user's satisfaction.

### 3.2 *System design and planning*

Based on the above ideas, the design planning of the educational recommended system is as follows:

(1) Research in the MOOC is of particular significance and for the user needs from the perspective of education;

Compared with other learning environments, the characteristics of MOOC is very obvious, such as learner differences being very large, with learning content fragmented, and a high degree of freedom to learn, and teacher exchange opportunities small, etc. This leads to the needs of personalized learning support and the learning construct as well as interaction skills in difficulties with learners. With the learner at the centre improvement methods of education for the above difficulties can be put forward, thus forming an analysis of educational recommender system requirements.

(2) The construction of an educational recommender system framework, based on a mature e-commerce recommender system.

In the field of electronic commerce, the personalized recommender system technology is mature and widely applied, which can be used as a source of technology for the educational recommender system. However the special educational environment makes it impossible to copy the e-commerce recommender system as an application and a profound transformation must be carried out. The current research trend of the personalized recommender system, context-aware recommender system based on user context to recommend, gives full consideration to the outside position of a user and recommender items of environmental information. Thus, it is suitable for the technology as a basis for educational recommender system; since the definition of the context itself was too broad, so context awareness technology also presents the characteristics of diversification in view of the teaching situation, especially the MOOC context to explore the corresponding recommender technology, which will become the core content of the subject.

(3) Combined with the form of design, the educational recommender system and MOOC completes the system's scheme;

The educational recommender system must be embedded into the MOOC Web page, with a seamless integration into the network courses. Between the MOOC curriculum and recommender system are to promote mutual relations, such as the recommender system requires MOOC to record the learning process perfectly, providing for basic data for calculating similarity of users; otherwise the curriculum team learning tasks and mutual evaluation and mutual aid activities also needs that the recommender system provides the best members combination and score with partner. The implementation of the educational recommender system needs to apply the network technology. In addition, we must consider the user experience, speed, recommender time, and recommender user interaction, which are closely linked issues. This article was readying to referent context teaching design skills. In the meantime, it referent commercial recommender system human-computer interaction experience structural prototype system. Finally, it gradually improved design experience by the user to complete this part of the work.

The specific design process can be divided into three steps:

a. Research on MOOC teaching characteristics, are needed to understand the needs of learners. From the point of view of pedagogy education research the recommender system should achieve the effect and take the form of recommendations, in order to format the demands of the analysis report.
b. As the goal of the current mainstream courses products, for example Coursera, edX, and Udacity, to analysis of MOOC system architecture and combining form with the recommender system. To complete the overall design and implementation of educational recommender system.
c. Through trial operation for evaluation and improvement of the educational recommender system, a mature design scheme and implementation of the program for the educational recommender system can be formatted.

## 4 THE KEY PROBLEMS OF THE DESIGN OF THE EDUCATIONAL RECOMMENDER SYSTEM FOR MOOC

### 4.1 *Grasp the particularity of an MOOC learning situation, seek an innovative design that make a breakthrough in traditional recommender thinking*

In detail, it can be divided into three aspects as follows:

a. Using a novel recommended way complete a recommendation of learning partners. In the environment of MOOC, learning partners have an 'accompany' function, in addition, it must also be borne in part a 'guide' function; So it results in the significant difference that in the user is found between the educational recommender system and the traditional commercial recommender system; First of all, the commercial recommender system hope to find the most similar to the current user, and predict the current user purchase behaviour; However, the educational recommender system wants to find the learning level close to but slightly higher than the user to provide "scaffolding" guidance and help with the user, which conforms to the educational concept of 'the zone of proximate development'. Courses in a Shanghai sharing centre's data show that students in online learning will lead to polarization, some people feel just like a fish in water, others are not just as one wishes [18]; For the latter, which is obviously inappropriate to recommend a learning partner who also backward.
b. In contrast to the traditional viewpoint, the learners will be integrated into the recommender process, which integrate into the design pattern and the degree needing to elaborate. The commercial recommender system generally does not require the

user to participate in the recommender process (that is to make the evaluation results recommender to help improve the quality of recommendation), and the user also did not participate in the power. However, the educational recommender system require learners to participate actively in the recommender system, and it has two advantages: The first is to increase the system interaction, and "interactive" is what MOOC has been advocating the behaviour, so it is in line with the learning as a building and cooperation activities of spirit; The second, the performance of the recommender system can be evaluated explicitly. Different from the commercial recommender system can evaluate by the user of purchase rate, what is difficult to obtain similar evaluation, however, learners' feedback to fill this gap.

c. Using the characteristic of pedagogy "situation" as the recommender foundation. The subject will use the "context aware recommender technology" as the core algorithm, and the "situation" can generally be understood as any information depicting the entity characteristic. The recommender systems context contain the user's educational background, age and other demographic information. in addition, the educational characteristics of learners' learning style and learning object concept map will constitute the core content of insinuation.

### 4.2 *Gets the user fully trust in the recommender system*

Education recommendation system as a freshman affairs for the majority of people are familiar with it, and the active service characteristics are similar to push advertising, so it may be causeed the user rejection, especially when the recommended time error. At present there is no good way to improve user trust degree of ascension, but we can proceed from the following several aspects:

a. To determine the recommender principles and recommend depth by studying the recommender experience of the mature recommender system; and recommender criteria of existing commercial recommender system can be used for reference, such as a useful recommendation (mufti recommender only in the alternative items) and one reason to recommend (given a suitable reason of recommendation to improve user trust) as well as based on the social network of friends recommendation (user-based recommendation), those recommender criteria all can be transplanted to the research.
b. To determine the recommender time on the basis of MOOC curriculum content difficulty distribution. Obviously we should be based on the characteristics of the course to determine using the recommender plan. With the help of course designers to provide a "degree of difficult or easy map" to trigger preset recommender; in addition, adopt the way of looking for the recommender time with dynamic monitoring, such as learners stay repeatedly in one knowledge point, or long time no mouse operation, which may be the best time to recommend.
c. To adjust the recommender mechanism basis of learner feedback to dynamically. That is the learner containment strategy. Form of elaborate design recommender system to adjust learners of positive emotion with lively, humanized page and language of expression.

## 5 CONCLUSIONS

This article has discussed the MOOC educational recommender system's design, and explained the idea of the design and the important points that need attention and discussion. Although that are not provided a consummately recommended system case., based on those basis, it can achieve close corresponding recommendation system practical combined with curriculum combined with the specific MOOC courses.

A new form of teaching needs a new tutoring style to match, which is the starting point for designing this recommended education system. Because of the need to learn the powerful auxiliary functions, the design of the control system in the technical level is more complex, if appropriate to reduce the requirements, then the technology can greatly reduce the difficulty; and the education recommended system need Closely integrated with the lecture content and teaching style of teacher, so it can be process of modular design of the system, frame good system structure, by the MOOC to complete the construction of the main teaching content filling and knowledge network to "design guide" form, to complete the implementation process of the system.

## ACKNOWLEDGMENT

This work is supported by the project of the "Twelfth – Five" plan of educational science for universities in Guangdong, China (name of the project: The design of e-commerce professional program aligned by "Outcome-based Education" theory), and teaching quality project of Wuyi University (Grant No. JG2013008).

## REFERENCES

[1] Wang Wenli. The development of MOOC and its influence on Higher Education [J]. The Higher Education in Jiangsu, 2013, (02): 53–57.
[2] Bartholet J. Hope and Difficulties of MOOC [N]. Guangming Daily, 2013-08-31.
[3] Fan Wenqiang. The massive open online course connectivism based (MOOC) and learning support [J]. Journal of Distance Education, 2012, (03): 31–36.
[4] Wang Ying, Zhang Jinlei, Zhang Baohui. MOOC typical characteristics of the project and Its Enlightenment [J]. Journal of Distance Education, 2013, (04): 67–75.
[5] Wang Ping. The new development and application of large-scale online open Courses: from cMOOC to

xMOOC [J]. Research on Modern Distance Education, 2013, (03): 13–19.
[6] Buder J, Schwind C. Learning with personalized recommender systems: A psychological view [J]. ACM Comput Surv ACM Sigplan Notices ACM Trans comput syst, 2012, 28(1): 207–216.
[7] Linton F, Schaefer Hp. Recommender Systems for Learning: Building User and Expert Models through Long-Term Observation of Application Use [J]. User Modeling and User-Adapted Interaction, 2000, 10: 181–208.
[8] Drachsler H, Rutledge L, Van Rosmalen P, et al. ReMashed – A Usability Study of a Recommender System for Mash-Ups for Learning [J]. International Journal of Emerging Technologies in Learning, 2010, 5: 7–11.
[9] Hsu M-H. Proposing an ESL recommender teaching and learning system [J]. Expert systems with applications Expert Syst Appl, 2008, 34(3): 2102–2110.
[10] Aehnelt M, Ebert M, Beham G, et al. A Socio-technical Approach towards Supporting Intra-organizational Collaboration [M] // Dillenbourg P, Specht M. Times of Convergence Technologies Across Learning Contexts; Springer Berlin Heidelberg. 2008: 33–38.
[11] Fernandez-Luque L, Karlsen R, Vognild LK. Challenges and opportunities of using recommender systems for personalized health education [J]. Studys in Health Technology Informatics, 2009, 150: 903–907.
[12] Kardan AA, Abbaspour S, Hendijanifard F. A Hybrid Recommender System for E-learning Environments Based on Concept Maps and Collaborative Tagging [M]. 2009: 300–307.
[13] Liu Xudong, Zhang Mingliang. Personalized E-learning design and implementation of learning system information recommendation [J]. Chinese Adult Education, 2013, (10): 106–107.
[14] Wang Zhimei, Yang Fan. Based on similar learners find Resource Recommendation System [J]. Journal of Zhejiang University (Engineering and Technology Edition) 2006, (10): 1688–1691+1791.
[15] Wang Yonggu, Qiu Feiyue, Zhao Jianlong, et al. Research on personalized recommendation of learning resources based on Collaborative Filtering Technology [J]. Journal of Distance Education, 2011, (03): 66–71.
[16] Sun Xin, Wang Yonggu, Qiu Feiyue. Online Personalized Learning Resources recommendation system based on collaborative filtering [J]. Distance Education Chinese, 2012, (08): 78–82.
[17] Shi Lin, Wang Kanliang. Comparison of online shopping and online learning in personalized recommendation system [J]. Journal of Tsinghua University (Natural science edition), 2006, (S1): 1030–1035.
[18] Liu Shiyu, Lu Fang: How to develop the "Mu course" in Fudan University [J/OL] 2014, http://www.shedunews.com/zixun/shanghai/gaodeng/2014/01/17/607593.html.
[19] Verbert K, Manouselis N, Ochoa X, et al. Context-Aware Recommender Systems for Learning: A Survey and Future Challenges [J]. IEEE Transactions on Learning Technologies, 2012, 5(4): 318–335.
[20] Adomavicius G, Tuzhilin A. Context-aware recommender systems [C] // proceedings of the Proceedings of the 2008 ACM conference on Recommender systems, Lausanne, Switzerland: 1454068: ACM, 2008: 335–336.

*Frontiers in Computer Education – Wang (ed.)*


# Creating a new computer education system in engineering higher education

Yuan Liu & Lihua Guo
*School of Information Science and Engineering, Fujian University of Technology, Fuzhou, China*

ABSTRACT: This article reviews the origin of engineering education and the progress of engineering education in China. Then, by taking the civil engineering specialty in the Fujian University of Technology as an example, the importance of utilizing computer technology to create distinctive instructional resources in engineering education is illustrated, and a new computer education system for non-computer majors is presented. The new all-round college computer education reform requires efforts from all sides.

## 1 GENERAL INTRODUCTION

### 1.1 *Origin of engineering education*

The origin of engineering education can be traced back to the "Apprenticeship" implemented by western countries and their associated industries in the 11th century. Since the Bourbon dynasty of France created a civil engineering school in 1747, engineering education has been developed for more than 260 years up until now. The purpose of engineering education established by a country is to cultivate talents for industry development in one's own country, so that the comprehensive national power can be strengthened by economic development.

International higher engineering education went through three phases, the original phase, oriented to engineering practice, the transition phase deviating from practice, and the surpassing phase returning to engineering practice. Nowadays, international engineering education is situated in the "returning to engineering practice" phase. However, it is not simply returning back, instead it spiral developed to modern engineering practice.

### 1.2 *The training goal of higher engineering education*

Engineering is the extension and recreation of science; contemporary engineers must be capable of transforming science into technology and then transforming technology into productivity. By taking the subject of technology as the principal subject, and catering to the practice application of a project, higher engineering education aims for training engineers that are adept in transforming science into productivity to adapt to the developing industries of society and science technology. The US National Academy of Engineering suggested the following qualities for being an engineer by the year 2020:

- Analytical skills
- Practical ingenuity
- Creativity
- Business and management principles
- Leadership
- High ethical standards
- Sense of professionalism
- Dynamism, agility, resilience, and flexibility

### 1.3 *Development of engineering education in China*

At the beginning of the foundation of the PRC, China followed the Soviet Union model, and then engineering education developed quickly by connecting theory with practice. The "combining education with productive labour" policy was brought in in 1958, which promoted the development of engineering education.

In the 1960s, Chinese college students who majored in engineering took part in not only experimental tuition, but also took part in enterprise productive recognition, metalworking practice, enterprise productive practice, etc.

A long time after China carried out its reform and open policy, engineering education had been leap-forward developed. In 1980, the Ministry of Education modified the training objectives for undergraduate education from cultivating engineers to cultivating students that obtain the basic training of an engineer. The transformation emphasized on solid subject foundation other than specialty. It was the signal of deviating from practice.

### 1.4 *The deviating and the returning of engineering education in China*

The "deviating from practice" phase in China was later than in Occident. In 1990s, Chinese engineering education found itself still in the "deviating from practice" phase, when the international engineering education had been developed to the "returning back to practice" phase.

In 2010, "China's Medium- and Long-Term Talent Development Plan (2010-2020)" and "Plan for Educating and Training Outstanding Engineers" (PETOE) were published by the Communist Party of China (CPC) Central Committee and the State Council. PETOE was a reform programme implemented in selected units. This plan aimed to explore the ways in which colleges cooperate with enterprises to train talents, and it expanded the spread of engineering education.

After the start of PETOE, many industry associations, local governments, enterprises, and colleges worked together and made huge progress in the quality of China's engineering education. The new school-enterprise cooperation training mode was established. It improved students' engineering practice ability, engineering design ability, and engineering innovation ability. As a result, the graduates were well received at various enterprises. At the International Engineering Alliance Meeting held on June 19th 2013, China was approved to become the Washington Accord's 21st contracting party.

### 1.5 *Various international recognition agreements*

In order to satisfy the needs of economic globalization and the mobility of international talents, some countries such as America began to build an international recognition system of engineering education and engineering technicians in the 1980s. It involved the standardizing of engineering education and continuing education, the certification of engineer qualifications, and authentication of educational background, etc. The six agreements included in the system can be divided into two levels. The "Washington Accord", the "Sydney Accord", and the "Dublin Accord" aim for the mutual recognition of engineering education and technique education. The "Engineer Mobility Forum", the "Asia-Pacific Engineer Agreement", and the "Engineering Technologist Mobility Forum" aim for a certified qualification of the related engineering technique.

The Washington Accord is the most influential mutual recognition agreement for engineering undergraduate degrees. It aims to promote engineers work in different countries by the mutual recognition of an engineering education or engineering qualification certification. The standard of engineering education and engineering qualification certification proposed by the accord is the authoritative requirement for an engineering graduate and engineer. The participation of China in the Washington Accord illustrated the approval of China's engineering education quality in an international society.

Participation in the Washington Accord will promote standards for the cultivation of talents in China which are equal to the standards of the accord. After that, colleges and enterprises will be connected tightly. So it is helpful for improving China's engineering education level and the overall strength of China's engineering manufacture.

## 2 UTILIZE COMPUTER TECHNOLOGY IN ENGINEERING EDUCATION

### 2.1 *Civil engineering specialty at the Fujian University of Technology*

We will take the civil engineering specialty at the Fujian University of Technology as the example in the following sections. The Fujian University of Technology (FJUT) is a new provincial college created in 2002, which is now devoted to becoming an outstanding and distinctive engineering-based Applied University. The civil engineering major of the FJUT is a national-level characteristic specialty, and also is the initial national-level pilot reform specialty. By utilizing computer technology and following progressive engineering education principles, an in-depth school-enterprise cooperation model was created during the talents-training reform. The model was carried out successfully during the 2010-2013 graduate education major in civil engineering, achieving a full employment rate and specialty counterpart rate.

The "Innovation and practice of in-depth school-enterprise cooperation model for civil engineering specialty" programme achieved the second class reward of the national teaching achievement reward. The civil engineering specialty passed the undergraduate education appraise assessed by the Higher Education Appraisal Committee of the Housing and Urban-Rural Development Ministry. The validity period is three years, from May 2014 to May 2017.

### 2.2 *Utilize computer technology to create distinctive instructional resources*

The four-year training programme for civil engineering specialty offers students comprehensive development of moral qualities, intellectual ability, and physical fitness to serve for the modern civil construction of the "Economic Zone on the West Side of the Straits", and also offers students general training of being an registered engineer and innovative practical advanced engineering specialty talents. Graduates can be engaged in construction and management of civil engineering projects with primary project planning and designing ability. Graduates are also qualified for project construction and management jobs in building construction, municipal engineering, road bridge engineering, railway track engineering, etc.

The multimedia courseware developed by the school-enterprise cooperation integrates pictures and video with 3D animation. It is more advanced, visual, and diverse, because it combines the course content with engineering practice to solve the obscurity of some specialty courses such as construction technology, and it also solves the problems of outdated textbooks. The "Building Construction" courseware was demonstrated at many national conferences with a favourable reception. The "Construction Organization of Civil Engineering" teaching courseware won the first-class award in the National Multimedia Materials Contest for Civil Engineering Education.

The software developed by schools and enterprise together can be generally used in tuition and construction production, including templates optimization design, scaffold design, temporary electricity design, construction plan design, quality acceptance, schedule control, investment control, safety management, civilized construction management, etc. The software can be helpful for students to solve the problem of project construction scheme designing, and work out solutions during the curriculum design or the graduation project. In this way, students can master the skills for project construction scheme designing and implementation in order to get a better job later. The software has been applied in the tuition for seven years and also in hundreds of construction enterprises. The software developed for tuition and production both narrows the gap between theory and practice. This achievement won the third prize for technology award in Fujian.

A quality and safety supervision simulation system was developed to simulate the multilevel examination and approval process in different roles. It demonstrates advanced construction techniques by 3D animation, which encourages students to learn obscure techniques through vivid animation. By using this system, students can simulate an on-the-spot inspection of an observation record in the construction site, and estimate deviation according to the regulations. Also students can virtually simulate construction workers to check the observation records, simulate a project manager to examine and approve construction quality or safety examination materials, simulate party A to review, and simulate designers to review the engineering project or conduct a sign procedure.

## 3 CREATE A NEW COMPUTER EDUCATION SYSTEM FOR NON-COMPUTER MAJORS

### 3.1 *Motivation of the new computer education reform*

Since the 1980s, computer education for non-computer majors has been developed over the three decades. After the study and practice of generations of experts and instructors, computer education theory for non-computer majors has been developed step by step, and a curriculum system of computer education for undergraduates has been created as a consequence.

The initial computer foundation education mainly treated the computer as a tool for mastering computer basic knowledge and techniques. The concept of a computer culture was brought up in the 1990s and increased the universality of computer education. Nowadays, with the popularization of information technology in elementary and secondary schools, many first-year university students know computer operation techniques well, so it is necessary to promote a new computer education reform.

According to the "Washington Accord", a certification validity period can be provided. For example the validity period for the civil engineering specialty in FJUT is from May 2014 to May 2017. Various techniques will be updated during the three years. Therefore, we have to train college student majors in engineering to be more flexible, and prepare them for the next engineering education assessment.

### 3.2 *The features of new all-round college computer education reform*

Computer education for undergraduates has been developed with continuous reform. The new round college computer education reform proposed here shows significant breakthrough and evolution compared to the current education model. Following the principle of cultivating students' abilities, the essence of abilities can be changed and developed greatly. It breaks through the computer science field, and achieves universal computing thinking and practical abilities to solve problems.

The functions of a college computer education have significantly changed, along with the essence of changing abilities. They are not to be limited to solving the problems within computer fields, but can also solve problems of various specialties and solve issues in real life by computing thinking and practice. It is a novel feature of the new all-round college computer education reform. And the developing of science thoughts including computational thinking was also a novel contribution to computer education.

"Computational Thinking" was a concept put forward by American scholars in the 1990s. Professor Jeannette M. Wing, in America, gave a comprehensive definition of computational thinking in 2006; it is the basic quality with which to solve a problem. So, computational thinking can be integrated into a new all-round college computer education to improve students' thinking ability.

### 3.3 *Create a curriculum system oriented to computer application ability*

Computer education for engineering education must be categorized, not only by different subjects, but by different college directions in China, as well.

Computer education for undergraduates must be application oriented, and meet the specialty needs. The

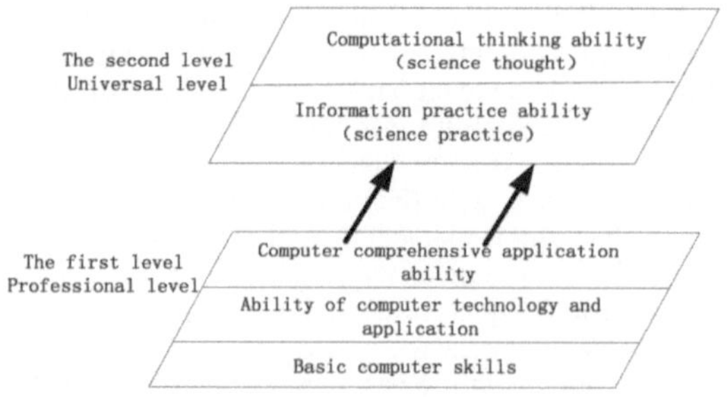


Figure 1. Structural system of college computer application ability.

computer related courses must be provided continuously during the four-year college study, and must be deepened step by step.

The new curriculum system may include a professional level and a universal level.

The Professional level (first level) can be divided into the foundation, technique, and application levels. The Universal level (second level) is a level to help the integration of thought and action. Courses are non-essential in the second level; we can only train students by integrating some courses of the first level. From the non-computer major's perspective, computer techniques can be divided into calculation techniques, data techniques, network techniques, and design technique domains. Utilizing the latest computer techniques to update the course materials is very important.

The civil engineering specialty at the FJUT provide a "3.3 plus 0.7" training programme which requires students to perform computer-aided design and computer-aided management tasks. Therefore, we can create a curriculum system as follows. In the first year, computer foundation and object-oriented programming language courses can be provided for students to learn about the development of computer techniques, master information retrieval skills, solve practical problems with programming language, and pass the second-level of the Computer Rank Examination for universities in Fujian. In the second year, combined with civil engineering cartography and building architecture courses, basic drawing and editing functions of the popular computer graphics software, AutoCAD, can be taught. Students can learn how to draw engineering graphs, and master several mobile applications. In the third year, courses such as Advanced Computer Aided Design (CAD) of Building, Principles of Concrete, Structure Design courses etc, can be offered. Students can learn the popular functions of several CAD soft wares and master several simulate software. The skilled students or those who are interested may also join the school-enterprise team to develop software together.

Finally, in the fourth year, students will be encouraged to take part in engineering practice, so that the application of computer-aided design in graduate programmes will be enhanced, and students' computer application levels will be improved.

To satisfy the needs of the curriculum system, the supporting textbooks have to be provided. Textbooks are important for the achievement of education thought and the tuition requirements, and also are an important fundamental construction of education reform. Besides bringing in some outstanding foreign computer textbooks, the textbook system has to be thoroughly designed according to the needs of Chinese students who are non-computer majors. The textbooks have to be scientific, gradually improved, progressive, and accessible for students' active learning.

The Massive Open Online Course (MOOC) was an education revolution launched by Silicon Valley in 2011. The course videos of the MOOC are short, lasting from a few minutes to more than ten minutes, according to the fragmentation feature of reading in the network era. The courses in the MOOC have many quizzes; students have to answer these to continue with the course. If students have questions about the course, they can post on the course forum to ask for help from anyone all over the world.

Learning from the model of the MOOC, the University Open Online Courses (UOOC) Alliance was created by local colleges in May 2014. Fujian University of Technology was invited to join in the alliance in September 2014. The UOOC Alliance integrates the high-quality tuition resources of local colleges so that superior courses can be shared, tuition and learning methods are enriched, higher education is developed and balanced, and the training level and serving ability of college talents are improved. There are sixty colleges joined in the alliance so far, involving thirty-eight cities all around China. The "College Computer" course provided by Shenzhen University is one of the initial online courses.

In brief, the computer industry is developing rapidly nowadays; the urgent need for outstanding engineering talents is increasing in countries gradually, so the creation of a new college computer education system for non-computer majors is not only an opportunity, but also a challenge.

## 4 CONCLUSIONS

Engineering education has been developing for years, leading to more rational hierarchical structures, various training objectives for talents, and a more optimized training system for engineering technique talents. The application of computer techniques to engineering education practice helps interdisciplinary engineering talent to possess an international perspective. The new all-round college computer education reform requires effort from all sides. Not only do all the related teachers need to upgrade their perspectives, improve their abilities, and become actively involved in the reform, but also the leaders and administrative departments of colleges, the related governments, and industries have to pay more attention to the reform. In this era of rapid information development, by strengthening the researches and formulating the related policies, the new college computer education reform will play a more important role in engineering education progress.

## REFERENCES


Li, Guoqian & Xiong, Haibei. 2013. Engineering educational assessment and international professional accreditation. *Journal of architectural education in institutions of higher learning* 22(1).

National College Computer Education Seminar & China Railway Publishing House. 2013. Opinions of New Round College Computer Education Reform. *Computer Education* 20.

Wang, Ling & Lei, Huan. 2008. The Characters of Engineering Education Accreditation of Signatories of Washington Accord and their Impact on China. *TSinghua journal of education* 29(5).

Xie, Zuoxu. 2001. *Researches in Popularization of Chinese Higher Education*. Fuzhou: Fujian Educational Press.

Zha, Jianzhong & He, Yongshan. 2009. *The Three Strategies to reform engineering education in China*. Beijing: Beijing Institute of Technology press.

Zhu, Y. 2009. Computational Thinking: Its Scientific Position, Basic Principles and Innovation Methods. *Computer Science* 36(4): 53–55.

Yan, Dainian. 2009. *The Wonderful Work of Engineering Education–The Industrial Center of Hong Kong Polytechnic University*. Nanjing: Southeast University Press.

*Frontiers in Computer Education – Wang (ed.)*
*© 2015 Taylor & Francis Group, London, ISBN 978-1-138-02797-8*

# Correlations between learning networks and academic performance

Q. Zhao, K. Zhao, D.S. Zhao & J. Yang
*College of Modern Educational Technology, Northwest University for Nationalities, Lanzhou, Gansu, China*

ABSTRACT: This paper investigates the correlations between learning networks (network topology, network structure, and network properties) and academic performance. We analysed a learning network comprised of 204 students at the Northwest University for Nationalities. We found that indegree is positively related to English performance, and betweenness is positively related to computer performance. We came to the conclusion that in learning networks, the students who have a high-level of indegree or betweenness, play a very important role in communication between students, they can affects academic performace of students each other.

## 1 INTRODUCTION

Social networks have attracted considerable interest in recent years, largely because of their contribution to human understanding of various social structures and processes, such as information processing (Dodds, Watts & Sabel, 2003), distributed search (Kleinberg, 2000), and diffusion of social influence (Valente, 1995). There are mainly two branches of the studies of social networks: the static features of network topological structures, and the dynamical evolving processes measured from some network generation models. Social learning theory (Vygotsky 1978; Wenger 1998) suggests that learning takes place during social interaction and in social contexts. Learning networks, are one kind of social network composed by students in the process of learning and there is a relatively small amount of research on learning networks so far.

To realize the relationship between the topological features of a learning network and academic performance, can improve students' academic performance and optimize the structure of learning networks. Social Network Analysis (SNA) (Serrat, 2010) is a method with increasing application in the social sciences and has been applied in areas as diverse as psychology, health, business organization, and electronic communications. More recently, interest has grown in the analysis of leadership learning networks to sustain and strengthen their relationships within and across groups, organizations, and related systems. In addition, in this paper we also use SNA to study the correlation between topological features (betweenness closeness, clustering coefficient indegree and outdegree) of learning networks and academic performance (common courses in English and computer studies).

## 2 RELATED WORKS

The social network approach holds that the behaviour of an individual is affected by the kinds of relations, or technical ties, and networks, more than by the norms and attributes that an individual possesses. The social, informational, or material resources, that two individuals exchange characterize their ties. In social network analysis, these resource exchanges are termed "relations." Some positive and negative relations are assumed to be related to an individual's performance. Researchers empirically demonstrated that friendship and advice relations were positively related to a student's academic performance and an employee's job performance. On the other hand, the effects of an adversarial network were negatively related to performance (Baldwin, Bedell & Johnson, 1997; Sparrowe, Liden & Kraimer, 2001).

SNA has also been applied in community studies to investigate the social relationships and patterns that evolve through member interactions (Cho, Lee, Stefanone & Gay, 2005; Haythornthwaite, 2002; Reffay & Chanier, 2002). The application of SNA to community studies typically involves an investigation into the types of resources exchanged, their frequency, quantity, and flow, in order to describe the elements comprising the social system (Haythornthwaite & Wellman, 1998). For example, Reffay and Chanier (2003) incorporated SNA to describe relationships and interactions occurring among students and staff participating in computer-supported collaborative learning. The authors maintained that the use of SNA provides an opportunity to gauge the communicative interactions that take place, and to assess the degree of cohesion within the group. Girvan and Newman (2002) have drawn similar conclusions regarding the applicability of SNA to ascertain the degree and strength of social ties developed within a social system in developing models of community structures.

A great deal of work in SNA focuses on the performance of the individual in the social networks, the evolution of the social network, network structure, and so on. In addition, aside from a comparatively small number of studies are about, network topology

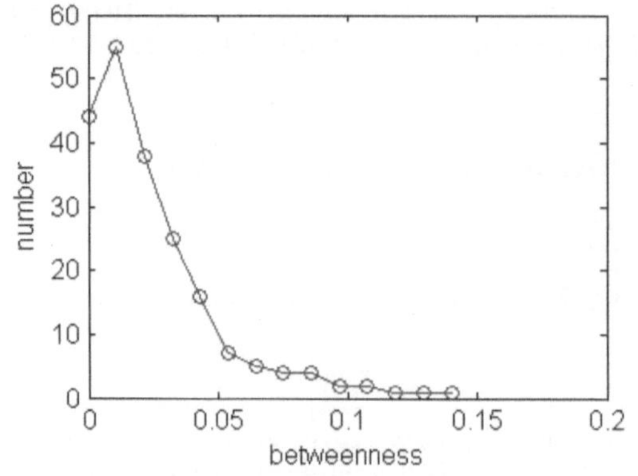


Figure 1. Betweenness distribution of learning network l1. The x-axis is the number of nodes with degree k.

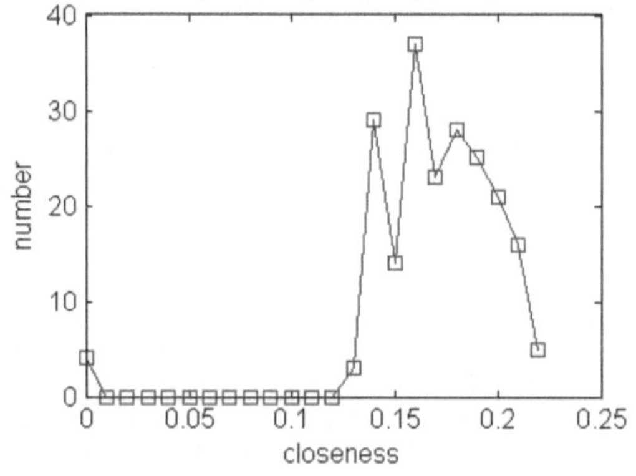


Figure 2. Closeness distribution of learning network l1. The y-axis is the number of nodes with degree k.

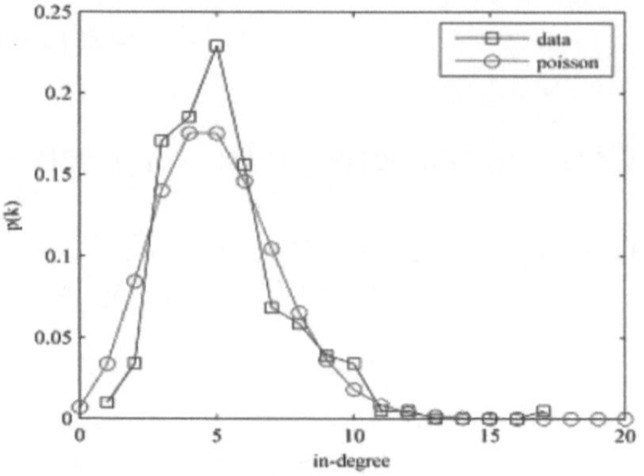


Figure 3. Indegree distribution of p(k) of learning network l1. p(k) is the probability of nodes with indegree.

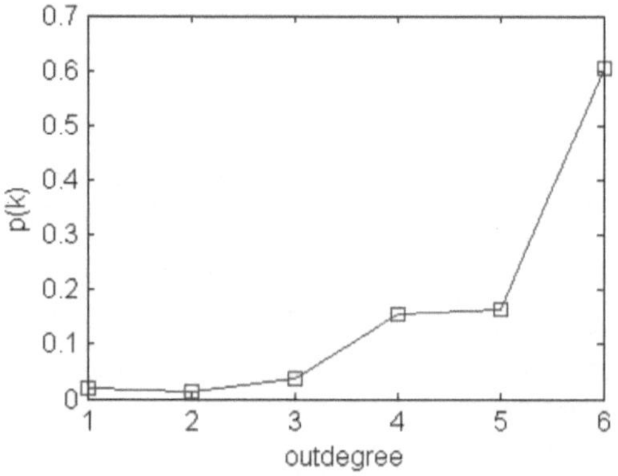


Figure 4. Outdegree distribution of p(k) of learning network l1. p(k) is the probability of nodes with outdegree.

structure characteristics, such as betweenness, closeness, clustering, and degree distribution, which relate to academic performance, that prompts our passion for the related study.

## 3 DATA DESCRIPTION

### 3.1 *Data*

In this study, we gathered data on the learning networks of all first-year undergraduate students in the College of Modern Educational Technology at the Northwest University for Nationalities during the first semester of the academic year 2011–2012. The group of selected students consists of 204 students with 3 majors, and each major contains 2 classes.

All students were asked to complete the learning network questionnaire, which provided a student roster containing a list of their names. The students choose them with two items, "schoolmate", and "non-schoolmate." The students were asked to fill questionnaires out in 45 minutes. Questionnaires were presented to the students after a month of them entering the university. The corresponding learning networks were indicated by L1. We also linked the learning network data to teaching administrative data in order to acquire the academic performance of the first semester academics. This allowed us to analyse the correlations between learning network and academic performance (English performance and computer performance).

### 3.2 *Topological features*

We measured L1's betweenness, closeness, degree distribution, and clustering spectrum. The results gave us the clues as to how the learning network is formed.

1) Betweenness: betweenness of the learning network is shown in Figure 1. The blue curve is drawn according to the observed learning network data.
2) Closeness: closeness of the learning network is shown in Figure 2. The blue curve is drawn according to the observed learning network data.
3) Degree distribution: in our study, the learning network link is directed, therefore the degree distribution of the learning network has InDegree distribution and OutDegree distribution, respectively, shown in Figure 3 and Figure 4. In Figure 3, the blue curve is drawn according to the observed learning

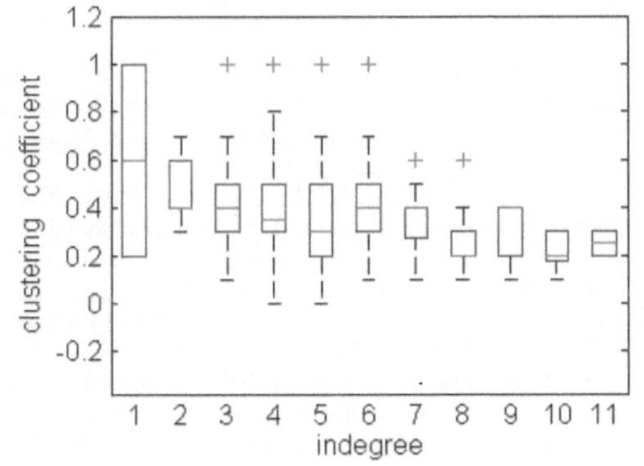


Figure 5. Clustering spectrum of 11. The y-axis is the clustering coefficient of nodes with indegree.

Table 1. Correlation matrix.

| | english | computer | betweenness | closeness | clustering. | indegree | outdegree |
|---|---|---|---|---|---|---|---|
| english | 1.00 | 0.28 | 0.04 | −0.03 | −0.03 | 0.21 | 0.05 |
| computer | 0.28 | 1.00 | 0.17 | 0.01 | −0.07 | 0.09 | 0.00 |
| betweenness | 0.04 | 0.17 | 1.00 | 0.34 | −0.42 | 0.36 | 0.24 |
| closeness | −0.03 | 0.01 | 0.34 | 1.00 | −0.26 | −0.11 | 0.32 |
| clustering | −0.01 | −0.07 | −0.42 | −0.26 | 1.00 | −0.16 | −0.05 |
| indegree | 0.21 | 0.09 | 0.36 | −0.11 | −0.16 | 1.00 | 0.05 |
| outdegiee | 0.05 | 0.00 | 0.24 | 0.32 | −0.05 | 0.05 | 1.00 |

network data. A Poisson degree distribution is also drawn in Figure 3.

4) Clustering spectrum: clustering spectrum uses boxplot to depict the relation between nodes in degree k and the clustering coefficient. In Figure 5, it produces a flat clustering spectrum, whose value is higher than 0.2 in most cases.

# 4 DATA ANALYSIS

In this section, we use R to compute the correlation between topology characteristics and academic performance.

## 4.1 *Correlation matrix*

To analyse the level of closeness between objective things with appropriate statistical indicators, that is correlatively analysed. In order to more accurately describe the level of linear correlation between the variables, the correlation coefficient for correlation analysis can be counted. The correlation coefficient is an important indicator which can indicate the level of correlation. The correlation matrix in our statistics is shown in Table 1.

## 4.2 *Probability values*

P-value is probability. It reflect the probability of an event occurring. In statistics, p-value is the important basis of hypothesis testing. p-values in our statistics are shown in Table 2.

Table 2. Probability values.

| | english | computer | betweenness | closeness | clustering | indegree | outdegree |
|---|---|---|---|---|---|---|---|
| english | 0.00 | 0.00 | 0.60 | 0.72 | 0.64 | 0.00 | 0.50 |
| computer | 0.00 | 0.00 | 0.02 | 0.85 | 0.34 | 0.22 | 0.96 |
| betweenness | 0.60 | 0.02 | 0.00 | 0.00 | 0.00 | 0.00 | 0.00 |
| closeness | 0.72 | 0.85 | 0.00 | 0.00 | 0.00 | 0.14 | 0.00 |
| clustering | 0.64 | 0.34 | 0.00 | 0.00 | 0.00 | 0.03 | 0.49 |
| indegree | 0.00 | 0.22 | 0.00 | 0.14 | 0.03 | 0.00 | 0.44 |
| outdegiee | 0.50 | 0.96 | 0.00 | 0.00 | 0.49 | 0.44 | 0.00 |

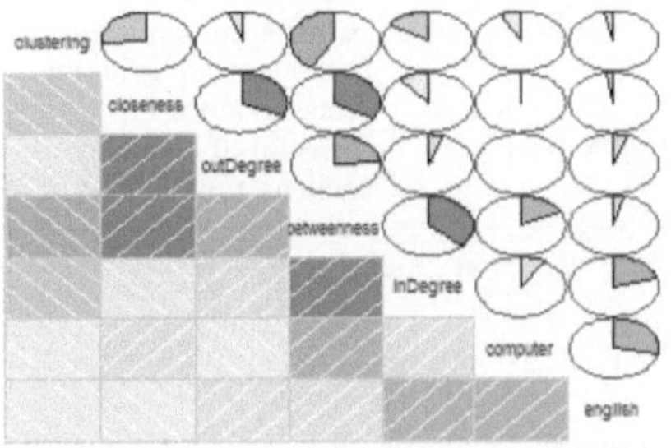


Figure 6. Correlation between topology features and academic performance.

## 4.3 *Correlation*

The correlation between topological features and academic performance is shown in Figure 6. We begin to explain these graphics from the lower triangular cell (below the main diagonal cells). In the blue cell, the slash where the lower-left cell points to the upper-right cell, indicates that the two variables were positively correlated. In turn, the red cell, upper-left point to the lower-right representing a negative correlation. The deeper the colour, the greater the correlation between the variables. Pie chart of triangular cell (upper the main diagonal cells) shows the same information. Colour function is the same above, the bigger the size of the pie, the greater the correlation between the variables. Positive correlation fill pie clockwise from 12 o'clock, while a negative correlation fill pie the counter-clockwise. In these graphics, we can seeindegree and English have a positive correlation. Betweenness and computer also have positively correlation with each other.

## 4.4 *Students in the real world*

In order to study the correlations between topological features and academic performance better, we investigated the real social relationship with a high-level of indegree in the collegeSurvey result is shown in Table 3. The students with a high-level of indegree are mostly cadres of the Student's Union or cadres of a class, who have a significant effect on communication between students. They maybe have an influence on various aspects of the learning network, including academic performance.

Table 3. Student relationship.

| Student ID | indegree | Social relationship |
|---|---|---|
| P11*****30 | 10 | class monitor |
| P11*****52 | 16 | Psychological counselling work office |
| P11*****08 | 9 | commissary in charge of studies |
| P11*****46 | 9 | The student union minister of learning department |
| P11*****21 | 9 | class monitor |
| P11*****54 | 9 | institute president of student union |
| P11*****93 | 7 | Class commissary in charge of organization |
| P11*****72 | 8 | class monitor |
| P11*****25 | 8 | director of office of college students |

## 5 CONCLUSION

In the university campus, a learning network may be the most basic relationship of their social networks. In this paper, we research the correlations between topological features of learning networks, which are composed of students from the Northwest University for Nationalities and academic performance (English performance and computer performance). Accurately, we respectively calculate the correlations between betweenness, closeness, clustering coefficient, indegree, and outdegree, and English performance and computer performance. According to the correlations analysis, we found that indegree and English show a positive correlation, and betweenness and computer also positively correlated with each other. It suggests students, who have a high-level of indegree or betweenness, play a very important role in the academic performance of learning networks.

## ACKNOWLEDGMENT

This work is supported by "the Fundamental Research Funds for the Central Universities" under NO. zyz2011085 and "Middle-aged and Youth Funds for Northwest University for Nationalities" under NO. x2010-24, Z.Q., Z.K., Z.D., Y.J. gratefully acknowledge the financial support from Northwest University for Nationalities.

## REFERENCES

Baldwin, T. T., Bedell, M. D. and Johnson, J. L. (1997). The social fabric of a team-based M.B.A. program: network effects on student satisfaction and performance. Academy of Management Journal, 40(6): 1369–1397.

Cho, H., Lee, J.-S., Stefanone, M., & Gay, G. (2005). Development of computer-supported collaborative social networks in a distributed learning community. Behaviour & Information Technology, 24 (6), 435–447.

Dodds, P. S., Watts, D. J. & Sabel, C. F. Proc. Natl. Acad. Sci. U.S.A. 100, 12516 (2003).

Girvan, M., & Newman, M. E. (2002). Community structure in social and biological networks. Proceedings of the National Academy of Sciences of the USA, 99 (12), 7821–7826.

Haythornthwaite, C., & Wellman, B. (1998). Work, friendship, and media use for information exchange in a networked organization. Journal of the American Society for Information Science, 49 (12), 1101–1114.

Haythornthwaite, C. (2002). Building social networks via computer networks: Creating and sustaining distributed learning communities. In K. A. Renninger & W. Shumar (Eds.), Building virtual communities: Learning and change in cyberspace (pp. 159–190). New York: Cambridge University Press.

Kleinberg, J. M. Nature 406, 845 (2000).

Reffay, C., & Chanier, T. (2002). Social Network Analysis used for modelling collaboration in distance learning groups. Lecture Notes in Computer Science, 2363, 31–40.

Reffay, C., & Chanier, T. (2003). How social network analysis can help to measure cohesion in collaborative distance-learning. Paper presented at the Computer Support for Collaborative Learning Conference, June 14–18, 2003, Bergen, Norway.

Serrat, O. (2010). Social network analysis. Washington, DC: Asian Development Bank.

Sparrowe, R. T., Liden, R. C. & Kraimer, M. L. Social networks and the performance of individuals and groups. Academy of Management Journal, 44(2): 316–325, 2001.

Valente, T. W. Network Models of the Diffusion of Innovations (Hampton Press, Cresskill, NJ, 1995).

Wenger, E. (1998). Communities of Practice, Cambridge, USA. In Wenger, E., McDermott, R., and Snyder, W. (2002). Cultivating communities of practice: A guide to managing knowledge, Boston, Harvard Business School Press.

Vygotsky, L.S. (1978). Mind in Society: The development of higher psychological processes. Cambridge, MA: Harvard University Press.

*Frontiers in Computer Education – Wang (ed.)*
*© 2015 Taylor & Francis Group, London, ISBN 978-1-138-02797-8*

# Research and application on electronic magazine in an antenna professional teaching course

Xiaoying Ran & Zhen Yu
*North China Institute of Science and Technology (NCIST), Langfang, China*

ABSTRACT: The subject combines electronic magazine, which often appears in entertainment media, with boring science courses, innovatively, so that changes are made to the traditional teaching methods with images, text, sound, and video, disseminating rich information in a lively, fun, and artistic way. Zinemaker software is used to design and develop the magazine in the field of antennas. Electronic magazine contains images, text, sound, video, and other multimedia elements, and displays the evolution, classification, and application of antennas in a lively and informative way. By way of example, electronic magazine's cover, table of contents, and various components and features are introduced.

*Keywords*: antenna course; E-zine; teaching method; research and application

## 1 INTRODUCTION

Electronic magazine, abbreviated as E-zine, also called network magazine or interactive magazine, has the characteristics of plane media and internet. E-zine combines text, images, sound, and video information and contains hyperlinks and timely interaction network elements as well, in a very enjoyable reading mode[1].

An antenna and radio wave propagation course is a very important practical professional course with abstract and complex theory in communication engineering and electronic information engineering specialties. The course covers the theoretical research and design application of antennas and radio waves, and dozens of different categories of antennas that are widely used in the field of wireless communication are discussed. In order to enrich teaching contents and expand the horizons of students, the electronic magazine media form is combined with the antenna curriculum firstly, which changes the traditional teaching methods in science and engineering courses, with a large number of images, texts, sounds and videos, and that disseminates rich information in a lively and artistic way to improve the interest of students[2–3].

## 2 E-ZINE FEATURES

E-zine is a kind of machine reading magazine, which can give full play to the advantages of computing speed and mass storage of the computer to increase the amount of information greatly, and find contents in the ocean of information with the computer specific query functionality, quickly. Moreover, with more forms of multimedia people cannot only see text and pictures, but can also hear all kinds of sounds and see moving images. E-zine has good extension that can be transplanted to a variety of personal terminals such as PDA, MOBILE, MP4, PSP, and TV (digital TV, set-top box), etc. The page turning effect of E-zine is more lifelike than an e-book, and the appearance of E-Magazine's user interface is more stereo than a PDF and webpage interface.

## 3 APPLICATION OF E-ZINE IN TEACHING

E-zine has the characteristics of exciting content, page turning effect, navigation buttons, link addresses, and more rich audio-visual elements to attract readers and promote active attention into passive attention. The advantages of E-zine can be applied to design teaching practice and enrich the traditional teaching method with new technological elements, in order to attract learners using the wealth of learning resources to participate in thoughtful discussions efficiently[4].

E-zine broadens the horizons of learners, and attracts learners to participate with the vivid and rich learning environment. With the popular PPT (PowerPoint) software in teaching, E-zine can produce a very good complementary effect aiming at the writing on the blackboard. When classroom-oriented, E-zine creates vivid scenes and promotes interaction between teachers and students. Table 1 gives the comparison of similarities and differences between E-zine and PPT.

## 4 DESIGN OF AN ANTENNA E-ZINE

### 4.1 *E-zine development software*

The development software commonly used for E-zine are IEBook, Pocomaker, Zmaker, and Zinemaker, where Zinemaker is standard Windows-based

Table 1. Comparison of similarities and differences between E-zine and PPT.

| | Items | E-zine | PPT |
|---|---|---|---|
| Similarities | 1. Order of Presentation | Flip effect | Pointer clicks |
| | 2. Transition effects | Transition effects (increase special effects such as fade, shear, switching the window) | Custom animations (increase special effects such as blinds, checkerboard, dissolves, erasure) |
| | 3. Link to audio and video | Import audio and video | Inserting audio and video |
| | 4. Navigation effect | Set the directory navigation, jump pages difficult to trek | Set the link navigation, easy to trek |
| Differences | 1. Progress indicators | Can prompt progress page | Unable to prompt progress page |
| | 2. Aggregate effect | Aggregating multimedia elements, visual harmony, sound and picture combinations, and leaves no trace of multimedia factors combined | Can achieve sound and picture combinations, but leaves traces of multimedia factors combined |
| | 3. Writing style | Require the selection of illustrations, text template, or modify the XML, time-consuming | Directly change font style, saving time |
| | 4. Directory Setup | Use directory templates, set the directory that corresponds to the page, and page setup is clear | Link-directory, without any idea of jump |
| | 5. Add annotations and modify | Works cannot be modified when there is no project file | Show annotations can be added, the work can be modified |

Figure 1. Magazine cover page.

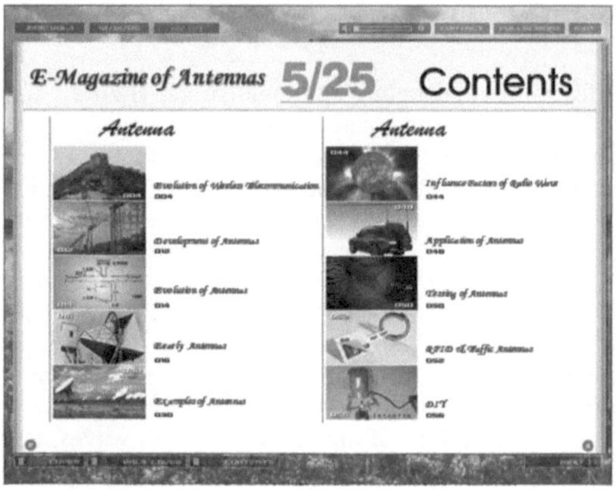

Figure 2. Magazine directory page.

with 128-bit strong encryption technology that can protect the user's Flash files from being maliciously cracked. The produced E-zine file is an independent EXE file with built-in FLASH 8 player, which can be opened directly to watch without needing other platform or plugin support, or changing the system or registry information of the user's computer. It supports the new Flash file format fully with beautiful text, video, animation, and a special web page effects template to show the best effects. At the same time, with multiple sets of fine Flash cartoon templates and a large number of Flash page effects, it can help more ordinary users to produce their own electronic magazine and provides online release function as well. E-zine generally contains a cover, a back cover, text, feedback, a scaling effect, a turning effect, and a directory link to the text elements.

### 4.2 *Overall structure and application examples of an antenna e-zine*

The magazine takes antennas as the theme and includes several aspects such as the application development of wireless communication and evolution of antennas, early famous and classic antennas, application and testing of antennas, wave propagation, etc. The magazine presents almost one hundred types of civil and military type of antenna, and is almost sixty pages in total. In order to show different types of antennas, it adapts different performance styles to enhance its art and appreciation. The magazine takes a famous antenna as the cover and designs the graphic layout, as shown in Figure 1.

In the magazine directory page, click the small figure to enter the related child directory for more reading, or click the lower right "PREVIOUS" and

a) b) c) d)

Figure 3. Page screenshot.

"NEXT" bottom to flip, also can drag magazine's margins to flip just like reading the paper magazine, as shown in Figure 2. Click the lower left bottom of "COVER", "BACK COVER", and "CONTENTS" to jump to the page number. Drag the top right volume control slide  to control the sound effect. Click "FULLSCREEN" and "EXIT" to achieve full screen and exit the magazine.

The E-zine begins with the wireless communications evolution. Click the dynamic signage of "ANCIENT", "DISCOVERY", and "DEVELOPMENT" in the upper-left of the page to pop up the background knowledge introduction corresponding with the text description at the bottom of the page. There are some small scrolling pictures below the text; click them to zoom out for more reading at the right side of the page, as shown in Figure 3-a).

The military antennas section is categorised by land, sea, and air. In the upper area of the page, loops playback small antenna pictures. Click the picture to study the antenna background knowledge at the left side of the page, and enlarge the image information at the right side meanwhile, as shown in Figure 3-b).

The civilian antenna part uses different forms of pictures. Click them to pop up an enlarged picture of the antenna. In addition, video media is contained in the page, as shown in Figure 3-c.d). Through the transformation of the art form, the boring knowledge becomes lively and entertaining.

## 5 CONCLUSION

E-zine breaks through the pure combination of words and pictures in a paper journal, and uses new technology to combine multimedia elements to form a unique style. It can give students a multi-faceted, audio-visual experience in classroom teaching.

## REFERENCES

Liu, R. X. The service of electronic journal resources and electronic information online. *Knowledge of Library and Information Science*, 1999(2).

Yu, Z. & Ran, X.Y. Study and teaching practice of the course of antenna and radio wave propagation. *Journal of Langfang teachers college*, 2009. 9(6): 118–119.

Ran, X.Y. Research and application on network and media technology in antenna professional courses. *Journal of*

*North China Institute of science and technology*, 2014. 11(6): 96–100.

Zhang, L.M. & Sui, Y. Reform and exploration of electronic information comprehensive system experiment teaching. *Experimental technique and management*, 2006. 24(10): 118–120.

Jiang, Z.C. & Zhai, L.j. S.C. Zhang, Experiment teaching practice and exploration of microwave technique and antenna course based on virtual reality technology. *Experimental technique and management*. 2010. 27(6): 85–88.

Fang, S.J. & Li, H.B. Research on Antenna experiment teaching simulation system. *Laboratory research and exploration*, 2010. 29(1): 50–52.

Xie, Z.M. Application on network teaching in microwave technique and antenna course, *Journal of electrical and electronic teaching*, 2008. 30(3): 113–115.

Ren, L.L. & Yi, Y.C. Study on interactive virtual antenna experiment courseware, *Journal of Beijing broadcasting Institute*, 2000. 30(2): 20–26.

*Frontiers in Computer Education – Wang (ed.)*
*© 2015 Taylor & Francis Group, London, ISBN 978-1-138-02797-8*

# Practical innovation about the Minimum Spanning Tree (MST) in data structure teaching

Jing Yi, Sheng'en Li, Xiaobing Tang & Xiaofei Niu
*School of Computer Science and Technology, Shandong Jianzhu University, Jinan, China*

ABSTRACT: In the teaching of data structure, there are lots of problems including those of vague concepts, abstract ideas, and non-concrete algorithms. How to solve these problems from a theoretical and practical point of view is one of the troublesome tasks of teaching. Taking the Minimum Spanning Tree (MST) as an example, this paper introduces a method of case teaching with professional characteristics. During the learning process, as well as learning how the algorithms work, the students have to also improve their original knowledge and ability structure. This innovative practice teaching mode played a very good teaching effect.

## 1 INTRODUCTION

Trees and graphs are the keystones and difficulties of data structure teaching. Specially, it is very hard for students to acquire knowledge about the concept and application of graphs. To solve these problems to attain teaching objectives, the basic theories and engineering during the teaching process were combined By taking case teaching method and graphic teaching method; It makes the students to think positive under the guidance of teacher. This article takes the Minimum Spanning Tree (MST) of a graph as an example and combines the graphic teaching and practical teaching. Meanwhile, it provides a teaching case with their professional characteristics. It can not only make students learn the content of MST, but also it can improve their quality of overall problem-solving on a professional level as well. It will improve their cognitive level of professional knowledge and make the structure of knowledge perfect so as to achieve a better teaching effect.

## 2 ANALYSIS OF CURRENT TEACHING CASE

### 2.1 *Definition of the minimum spanning tree*

In all the spanning trees with weighted graphs, the spanning tree with the sum of the edges for the minimum cost is called the minimum spanning tree. Its characteristics are as follows: (1) The edge number of the minimum spanning tree edges is $|V| - 1$ (V is the vertices number of the weighted graph), (2) the minimum spanning tree is a tree structure with no ring, and (3) the minimum spanning tree contains all the vertices.

### 2.2 *Analysis of current teaching case about the minimum spanning tree*

The theory of the minimum spanning tree is very important in the teaching of data structure, which is a typical application of combining the theory and practice. There are many teaching cases about the minimum spanning tree in documents, eg, literature [1] enumerates some typical teaching cases in the course of data structure, documents [2–4] uses the road network planning problem as a teaching case for the minimum spanning tree, that is, how to design a minimum cost network which connects n cities.

Establishing a transportation network between n city needs $n - 1$ minimum edges. Possible lines between n city traffic are up to $1 + 2 + 3 + \cdots + (n - 1) = n^*(n - 1)/2$. How to find the $n - 1$ lines saving money most in the $n^*(n - 1)/2$ lines is the core issue of solving this case. The basic idea to solve the problem is as follows: Find $n - 1$ lines costing the least from possible lines, which is also equivalent to looking for a minimum spanning tree network; choose $n - 1$ edge (does not constitute a loop) in all the possible $e(n - 1 \leq e \leq n^*(n - 1)/2)$ weighted edges and make the weights of all chosen edges minimum.

The teaching case where highway network planning saves money the most has the two following features:

(1) Easy understanding. The problem is clear at a glance, easy to understand, and can guide students to think actively, analyse, and then find a possible solution to the problem;
(2) High pertinence. Abstract the possible link between n cities for a network, use vertex to represent the city, the edge in the connected net presents the boundary line between two cities, and the weight on the edge of the net represents

the communication cost; such a transportation network planning problem is equivalent to a even lton-us minimum spanning tree problem.

This case can meet the basic conditions required for teaching case, combined with the graphic method, and can be easily understood and grasped by students. However, through the teaching results, students will still have questions after solving problems, for example, a real traffic network, in order to bring convenience to transportation, is often designed and established in accordance with the network mode, not in accordance with the minimum spanning tree finally obtained. This phenomenon illustrates the teaching case does not have good applicability. So, seeking better teaching case becomes an important task in order to strengthen the teaching effect.

## 3 TEACHING CASE WITH PROFESSIONAL CHARACTERISTICS

### 3.1 *Selection of teaching case*

Minimum spanning tree is important and difficult in the graph teaching of the data structure. The choice of case is the core for the case teaching method. A excellent teaching case can not only combine the practical environment, but also can be understood easily. At the same time, in the process of collecting case materials, teaching case that is related with the professional characteristics of students can make students combine the learned data structure knowledge, and relate professional knowledge effectively. As a result, students can improve their original recognition structure in the process of analysing and solving problems.

In accordance with the purpose of case selection, we can find a teaching example based on campus network planning: Taking the Shandong Jianzhu University area network planning as the example, it takes the XINXI academic building (Network Center) as the centre, connects several big teaching offices in the campus, and discusses how to plan the campus network.

### 3.2 *Analysis of the teaching case*

Taking the situation into account where students have mastered the basic network knowledge before learning the data structure, two kinds of solutions about design of campus network can be obtained, that follows: tree structure (as shown in Figure 1) and graph structure (as shown in Figure 2).

The two campus network planning methods are compared, and their characteristics are described in Table 1.

So, the method of solving the planning campus network is as follows: we can deploy the campus network in redundancy wires, and can make use of the spanning tree of the network to run the network. the problem of planning campus network is, this is to say, a problem to work out a minimum spanning tree in a connected graph (shown in Figure 3).

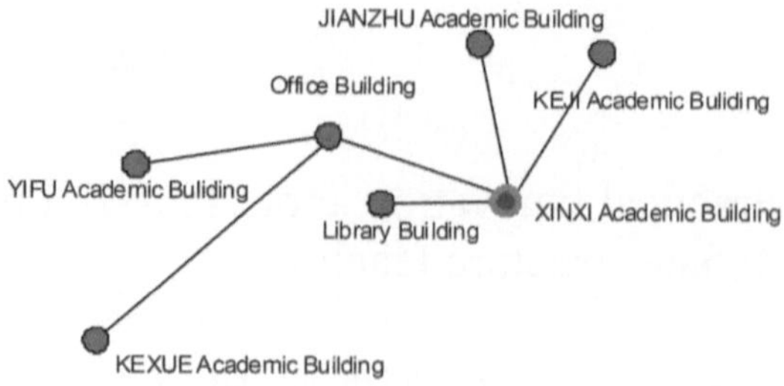


Figure 1. Tree structure about the campus network.

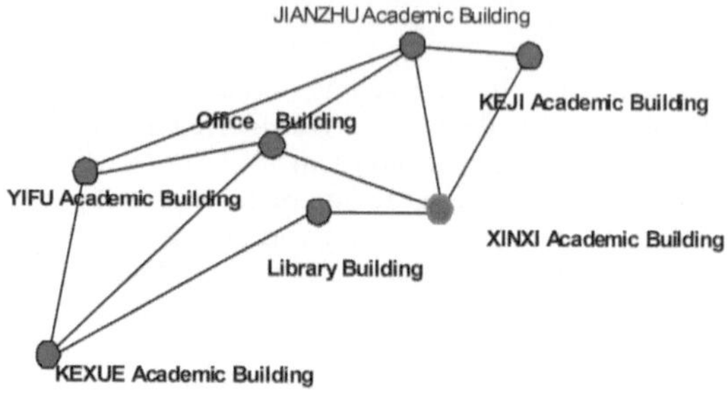


Figure 2. Network structure about the campus network.

Table 1. Contrast of two kinds of blueprint about the campus

| | Model | |
|---|---|---|
| Characteristic | Tree mode | Graph mode |
| advantage | simple structure connectivity | complex structure ring |
| disadvantage | single fault point | the broadcast storm |
| the method of solution | redundant wiring | using tree structure to start network |
| conclusion | redundant wiring and enable network with the minimum spanning tree | |

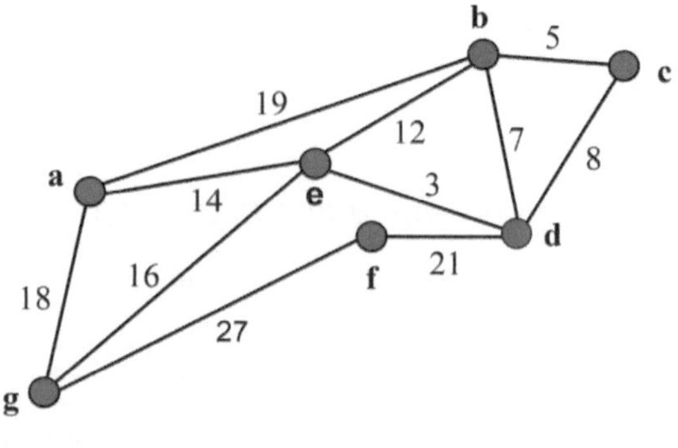


Figure 3. Abstract graph about campus.

### 3.3 *Solution for the teaching case*

There are two ways of obtaining the minimum spanning tree in a connected graph: the Kruskal algorithm

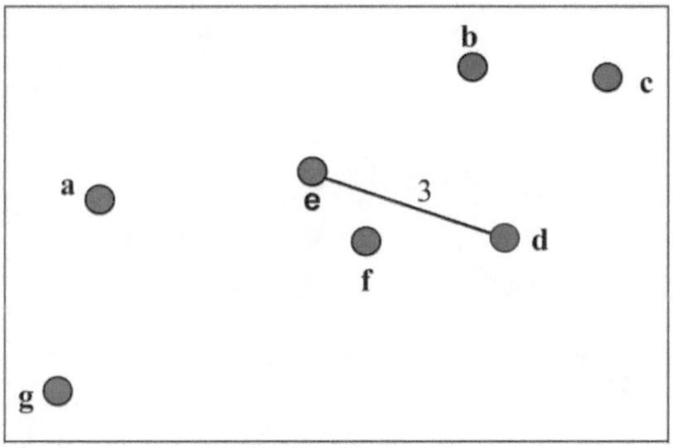


(4.a) add edge <e,d>

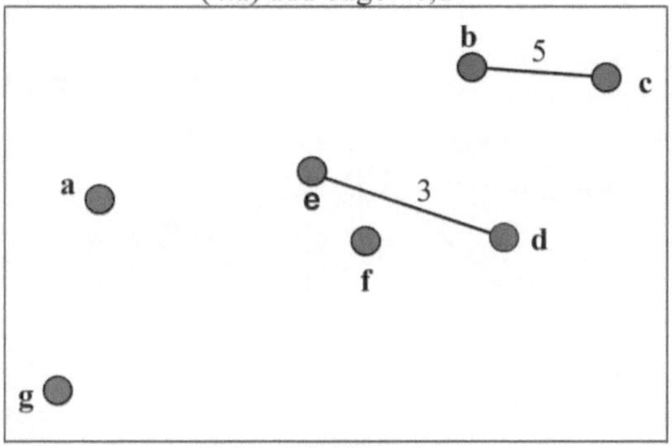


(4.b) add edge <b,c>

and Prim algorithm[2],[3],[4],[5],[6]; in the course of teaching the algorithms, documents [5], [6] adopt the illustration method. The author with the experience of many years of teaching, explains the solving process of the minimum spanning tree well using illustration and multimedia animation and, at last, provides two thoughts for the two algorithms.

The minimum spanning tree obtained by the Kruskal algorithm is shown in Figure 4, and its core thought is to add edges but not constitute a loop. Figure 4 shows the process for the minimum spanning tree using the Kruskal algorithm (4.a–4.f).

The pseudocode description for Kruskal's algorithm is as follows:

```
CreateGraph(ST, n, 0)
 k=0; i=1;
 while( k<n-1 ) {
    select a minimum-weight edge (u,v);
    if (!isCircle(ST,u,v) {
       addEdge(ST,u,v);
       k++;
    }
    i++;
 }
```

The minimum spanning tree obtained by Prim's algorithm is shown in Figure 5 and its core thought is adding vertices and keeping the net connected. Figure 5 shows the process for the minimum spanning tree using Prim's algorithm (5.a–5.f)

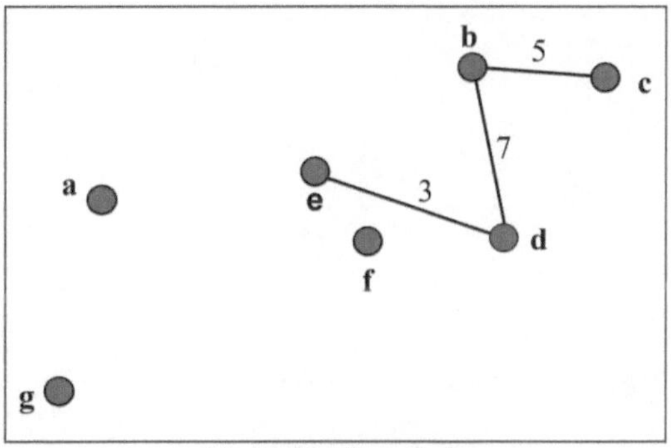


(4.c) add edge <b,d>

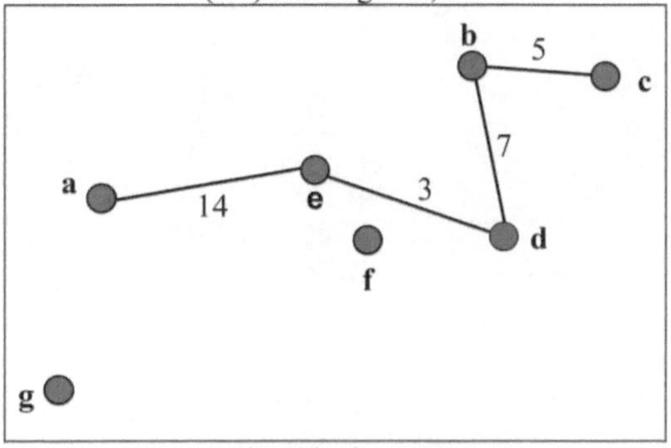


(4.d) add edge <a,e>

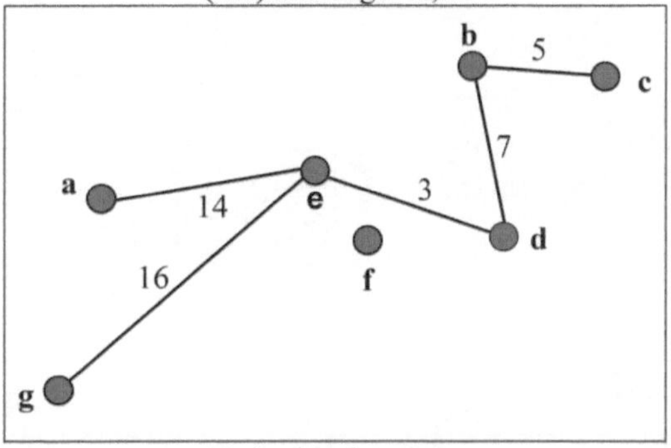


(4.e) add edge <e,g>

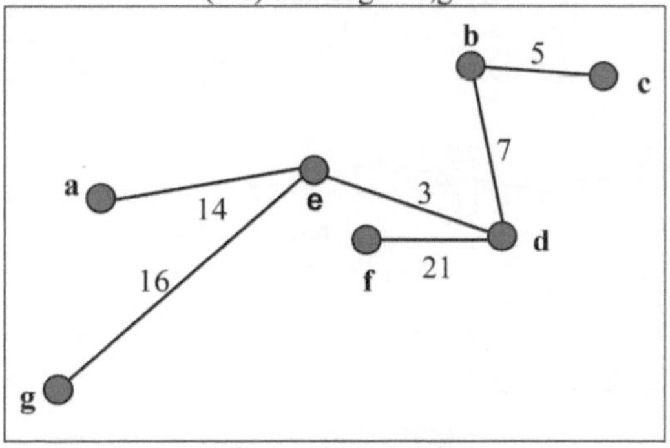


(4.f) add edge <e,g>

The pseudocode description for Prim's algorithm is as follows:

```
addVerToCollection(v0,U);
while V-U≠ Φ {
```

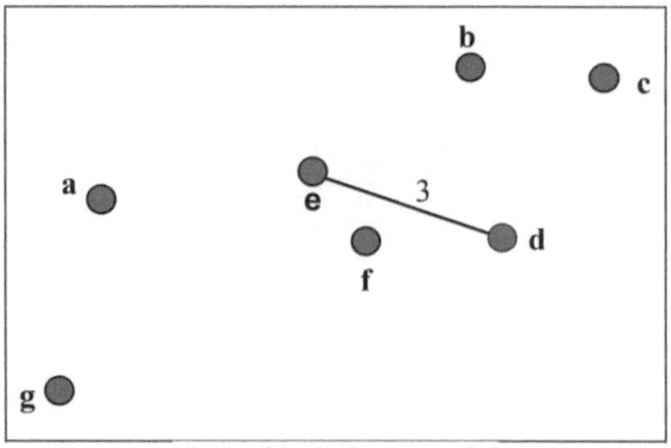


(5.a) add vertices e and d

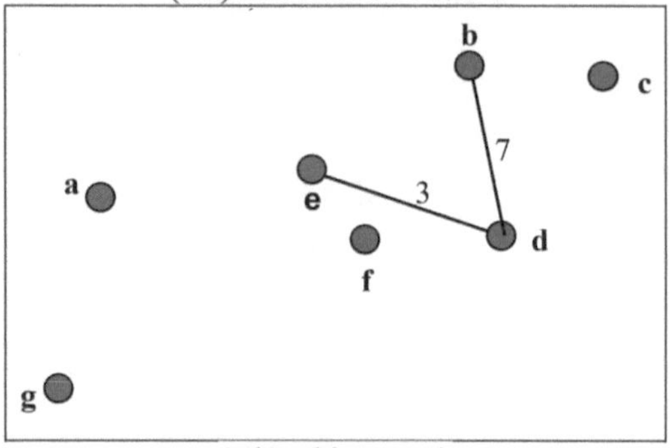


(5.b) add vertex b

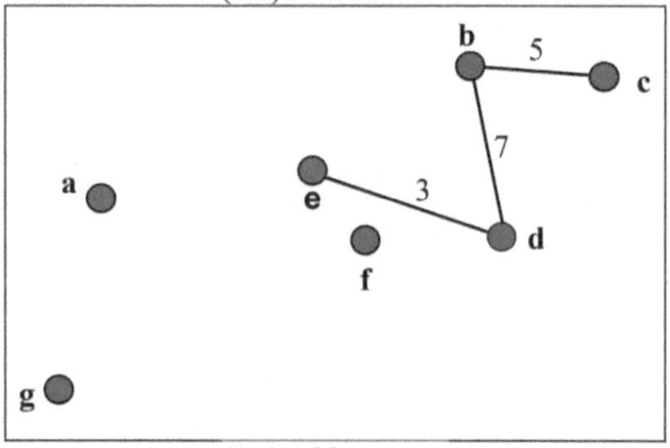


(5.c) add vertex c

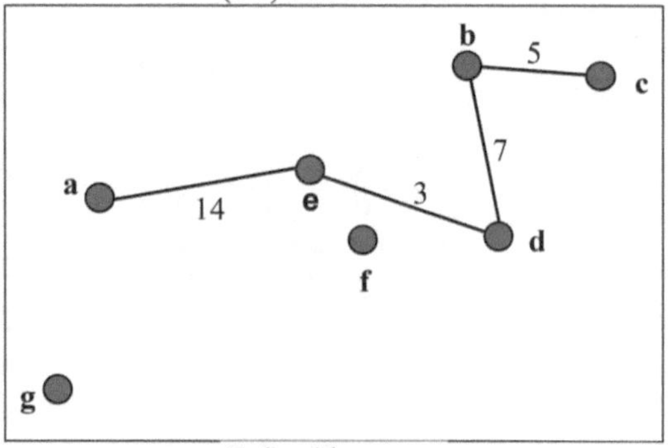


(5.d) add vertex a

```
select a minimum-weight edge (u,v);
//u comes from U, and v comes from V-U
   add VerToCollection(v,U);
};
```

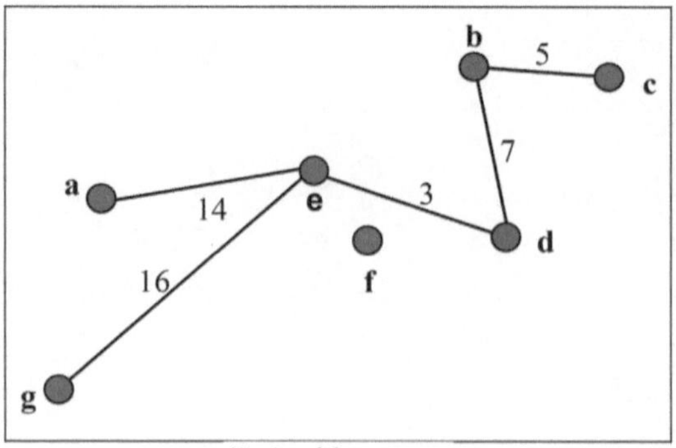


(5.e) add vertex g

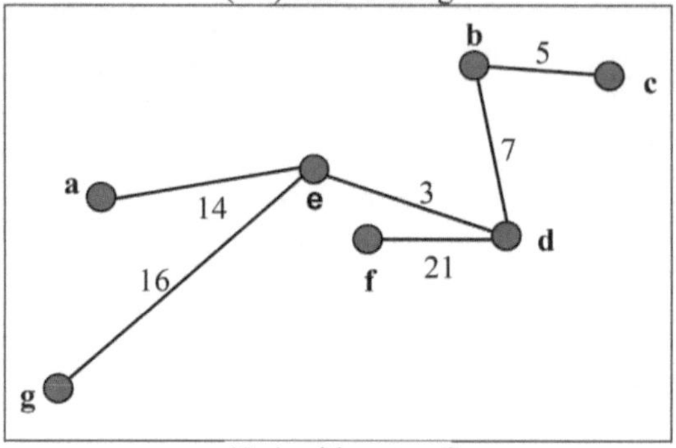


(5.g) add vertex f

And finally, we conclude the shape of the minimum spanning tree is not unique.

### 3.4 *Summary and expansion about the teaching case*

We can obtain the detailed process for two algorithms (Prim's algorithm and the Kruskal algorithm) of the minimum spanning tree in the teaching process by introducing the teaching case of "the campus network; how to plan and wire", proposing the concept of a minimum spanning tree, and combining the teaching by way of illustration and multimedia animation. The tested outcome of classroom has proved that 99% of students easily understand and accept this problem, and master it well.

We can conduct the case extension on the basis of understanding and grasping the issues of the minimum spanning tree, and then ask the question that "if a circuit (such as the edge <d, e>) breaks down, how will the network run?" in order to guide students to draw the conclusion that they can work out the new minimum spanning tree through the algorithm of the minimum spanning tree from the new figure after deleting the edge <d, e>, so that the students can fundamentally understand the rules of network operation and enrich their original knowledge systems. Such a knowledge point is the Spanning Tree ProtocolSTP) of the "Computer Network", which will be taught in the following courses, and the extension of the teaching case plays a connecting role in the teaching of these following courses.

## 4 SUMMARY

Data structure is a very important specialized elementary course, which emphasizes both theory and practice. Moreover, it plays an important role in teaching the computer system. During teaching of the students, we combined the student characteristics with their major, and we introduced the method of case teaching into network engineering, which showed the importance of case selection during the teaching. In the process of the application of the graphic teaching method, the program is presented in language. The design of the case teaching mode not only makes students enhance their understanding of the algorithm process, but also makes them realize the practical significance of this algorithm in the course. Besides, it can expand the student's understanding domain through the study of the subsequent courses, make their knowledge structure perfect, and improve the quality of teaching.

## REFERENCES


[1] Cerui Fu. The Research in Teaching for Data Structure Courses Based on Java [J], Microcomputer Information, 2011, 27 (4) 56–59.

[2] Xiaofei Niu, Sheng'en Li, etc. Discussion on the recursion in data structure teaching [J], Journal of Shandong Jianzhu Uninversity, 2010, (06) 656–661.

[3] Qifang Luo. The Practical Application of the Case Teaching in The Course of Data Structure [J]. China Science and Information. 2010, (10) 101–106.

[4] Danlin Cai, JunjieLiu. Teaching Reform and Practice Innovation in Data Structure, Journal of Chifeng University (Natural Science Edition), Vol. 27, No. 10, Oct. 2011.

[5] Zongyao Sha, Fuling Bian. Application of Graphic Method in Teaching of Data Structure and Algorithm in Education [J], Computer Education, 2009. (18) 18–23.

[6] Chunfang Xing, Application of Graphic Teaching Method in Teaching Data Structure [J], Journal of Xitai College. 2010.25(2) 114–116.

[7] Xiaofei Niu, Sheng'en Li, Dongmei Zhang. Study on Single Linked List Language. Data Structure Teaching [J], Computer Education, 2008, (20) 40–44.

[8] Mark Allen Weiss. Data Structure and Algorithm Analysis: Described by Java Language (the second version) [M], Beijing: China Machine Press, 2008.

[9] Sartaj S. Data Structures, Algorithms and Applications in C++[M].Beijing: China Machine Press, 2008.

[10] Weimin Yan. Data Structure (Described by C Language) [M], Beijing: Tsinghua university press, 1997.

*Frontiers in Computer Education – Wang (ed.)*
*© 2015 Taylor & Francis Group, London, ISBN 978-1-138-02797-8*

# An Empirical Mode Decomposition (EMD)-based steganographic method for radar images

Y.H. Li
*Huangpi NCO School, Air Force Early Warning Academy, Wuhan, China*

Y.L. Gao
*No. 5 Department, Air Force Early Warning Academy, Wuhan, China*

J.H. Yang
*Huangpi NCO School, Air Force Early Warning Academy, Wuhan, China*

ABSTRACT: In order to use radar images as cover images, whose noise level is higher than that of natural images, we propose a steganographic method for radar images based on BEMD (Bi-dimensional Empirical Mode Decomposition). First of all, we divide a cover radar image into many blocks, and we utilize the BEMD to decompose every block into a set of IMFs (Intrinsic Mode Function), plus the residue. Then, we calculate the texture strength of a block via all the IMFs' coefficients. Namely, the hiding capacity is evaluated by the texture strength. Finally, we embed the secret information into the cover radar image with modulus function-based steganography. The simulation results demonstrate that the proposed steganographic method provides a larger hiding capacity and maintains a better visual quality of the stego-image.

## 1 INTRODUCTION

Since the 1990s, steganography has been a technique for data protection, which is mainly used in the hiding of communication[1]. Because there are many images on the Internet, some people utilize certain steganographic methods to hide secret information within images, such as LSB (Least Significant Bit). LSB is the most common method, but it embeds only one bit into one pixel. That is to say, for a 256*256 image, only 65536 bits are embedded into this image. Therefore, it has a small embedding capacity. To enlarge the hiding capacity, modulus function-based steganography[2] was proposed in 2003, which can embed more bits into a pixel. However, not all pixels in an image can tolerate equal amounts of changes without causing a noticeable distortion. The changes, which occur in the texture and edge areas, can hardly be noticed by human eyes, so scholars designed steganographic methods to make use of HVSs (Human Visual System), such as BPCS (Bit-plane Complexity Segmentation)[3], PVD (Pixel-value Differencing)[4], Side Match[5], and WCLSB[6] (Wavelet Contrast Least Significant Bit). Using these steganographic methods, more secret information is embedded in the texture and the edge areas in images; less information is embedded in smooth areas. For example, these methods embed 2–4 bits into one pixel. That is to say, for a 256*256 image, about 131072-262144 bits are embedded into this image. Fortunately, the PSNR (Peak Signal to Noise Ratio) of these methods are about 38–40 db. Thus, we see that the hiding capacity of information is greatly improved while the quality of visual imperceptibility is maintained. However, these methods utilize the differences of neighbouring pixels to estimate the human eye sensitivity, which does not make full use of HVSs. Besides, these methods use the natural image as the cover image, whose noise lever is lower and zero at some times, and in these methods, the noise is not considered. However, in radar images, the noise lever is higher and can destroy the cover image at the same time. Thus, these methods do not adapt to the radar images.

At the end of the 1990s, Huang et al. developed a novel data decomposition method as a novel signal analysis method for non-stationary and non-linear signals, named the EMD (Empirical Mode Decomposition). Different from the wavelet, the EMD can decompose signals into a set of IMFs and obtain the residue. Each IMF is the detail with a certain scale, and the residue represents the approximation. Thus, the EMD can be considered as a multi-resolution analysis method of a signal. Being an expansion of the EMD method in the two-dimensional field such as the image, the BEMD[7] (Bi-dimensional Empirical Mode Decomposition) is a novel analytical method for image multi-resolution. As we known, the multi-resolution analysis is closely related to the human visual system. Thus, we can use BEMD to build the HVS evaluated function in this paper.

To embed more secret information into cover radar images and make the radar images effectively robust to noise, a novel steganographic method based on BEMD and HVS is proposed in this paper, where the radar images are adopted as the cover images. Since in the radar images, the magnitude of the IMF's coefficient can reflect the texture strength of images, we can estimate the texture strength of each block according to the sum of absolute value of all the IMF coefficients. Considering that human eyes are less sensitive to the noise in a strong texture area than in a weak one, we embed more secret information into stronger texture areas. Besides, as we know, the performance of the modulus-based steganography is best in all steganographic methods where the HVS is not being considered; we utilize it as the embedding method in this paper.

## 2 STEGANOGRAPHY METHOD BASED ON BEMD

Considering the fact that the radar image's noise level is higher than that of the natural images, we develop a novel radar image steganography based on BEMD and modulus function in this paper. First of all, a cover radar image is divided into many blocks, and every block is decomposed into a set of IMFs and the residue with the BEMD. Then, we estimate the texture strength. The texture strength of a block is evaluated by the sum of the absolute value of all the IMF coefficients. As we know, the stronger the texture is, the larger the hiding capacity is. Finally, the secret information is embedded by modulus function-based steganography, whose performance is very good. The process includes the embedding method and extracting method. The details are as follows:

### 2.1 *Embedding method*

The detailed secret data hiding steps are shown in Figure 1. The process is as follows:

*Step 1*: The cover image is divided into blocks of equal size, $P * Q$ and then each block is decomposed into a set of IMFs and the residue by BEMD.

*Step 2*: Calculate the texture strength of a block. In these letters, we utilize the sum of the absolute value of all the IMF coefficients in the block as the texture strength of the block:

$$S = \sum_{j=1}^{J} \sum_{x=1}^{P} \sum_{y=1}^{Q} | C_j(x, y) | \tag{1}$$

where $C_j(x, y)$ is a coefficient of the $j$-th IMFs in the block, and $J$ is the number of IMFs.

*Step 3*: Obtain the modulus $n$ according to the texture strength. Determine the parameters $n_{\min}$, $n_{\max}$, and $\Delta$, then calculate modulus $n$ as Equation (2):

$$n = \min(floor(S / \Delta + n_{\min}), n_{\max}) \tag{2}$$

where $n_{\min}$ and $n_{\max}$ are the minimum and maximum of $n$, respectively, and $\Delta$ is the quantization coefficient. From Equation (2), we can see that the larger $S$ is, the larger the modulus $n$ is and $n$ adopted by modulus function-based steganography is proportional to the hiding capacity. This suggests that more secret information is embedded into stronger texture areas.

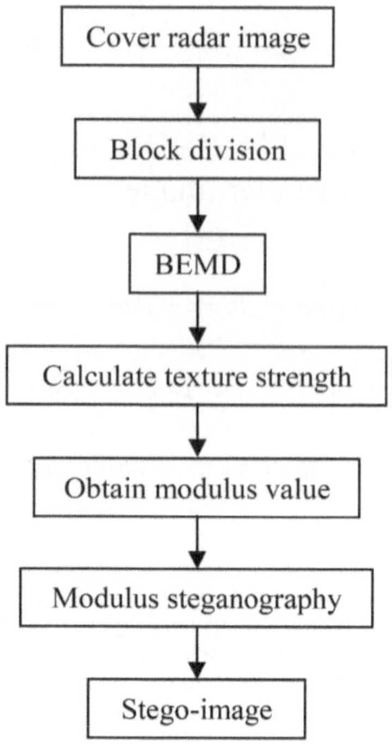


Figure 1. Flow chart of the embedding algorithm.

*Step 4*: Embed additional and secret information into a block using modulus function-based steganography. According to a secret key, we choose a pixel from the block, and embed the value of modulus $n$ (the additional information) into it by modulus function-based steganography with modulus $n_{\max} + 1$. Then, by modulus function-based steganography with modulus $n$, we embed $floor(\log_2 (n^{P*Q-1}))$ bits secret information into other pixels in this block. However, before we embed secret information into the block, we should transform 2-base notational secret information into an m-base notational one.

*Step 5*: Go to Step 2, until all blocks are processed.

### 2.2 *Extracting method*

We can retrieve secret information from the stego-image as shown in the following steps:

*Step 1*: The stego-image was divided into blocks of $P * Q$.

*Step 2*: According to a secret key, find the pixel embedded with modulus $n$ and extract the additional information for a block.

*Step 3*: Secret information is extracted as Equation (4) with modulus $n$. Then, secret information is converted into a 2-base notational system. The pixel embedded with modulus $n$ need not to be extracted.

*Step 4*: Go to Step 2, until all blocks are processed.

## 3 REVIEW OF MODULUS STEGANOGRAPHY

The modulus function-based steganographic method has the following virtues: (a) its SNR (Signal to Noise Ratio) is high in all steganographic methods where the HVS is not considered; (b) its embedding load is large; it can embed 2–4 bits of secret information into one pixel of the cover image. However, it has a weak point. Namely, it does not make use of the HVS effectively, and is sensitive to noise. We will use this method as a step in the proposed method. It is described as follows.

Assume that $k$ kinds of symbols are used in the secret information. The digital system representing the embedded secret information is: $\{0, 1, 2, \ldots, k-1\}$. Moreover, assume that the i-th data $x_i$ $(0 \le x_i < k)$ is embedded into the pixel $y_i$. Thus, the pixel $y_i$ is changed into $y_i^*$ as follows:

$$y_i^* = y_i - k \bullet sign(d_i)\left|round\left(\frac{d_i}{k+1}\right)\right| + d_i \tag{3}$$

where the value $d_i$ can be presented as:

$$d_i = x_i - \mathrm{mod}(y_i, k) \tag{4}$$

In a few cases, the $y_i^*$ in Equation (3) may fall outside the range of [0, 255]. If that occurs, $y_i^*$ is reset using the following equation:

$$y_i^* = y_i^* - k \bullet round\left(\frac{y_i^* - 1}{255} - \frac{1}{2}\right) \tag{5}$$

Extracting the data digit $x_i$ from $y_i^*$ is very easy. Just calculate as shown in Equation (6):

$$x_i = \mathrm{mod}(y_i^*, k) \tag{6}$$

## 4 SECURITY ANALYSIS

Steganography security includes the perceptibility security and the statistical security. The perceptibility security is obtained when the human eye cannot notice the distortion of the stego-image, while the statistical security means that the steganalysis cannot detect the secret information embedded within the image.

Next, we analyse the statistical security. On one hand, the reader can determine five parameters. As a result, steganalysis against our method must estimate five parameters. As with BPCS, PVD, and SM, each method has only one parameter to determine, and designing steganalysis methods against these methods is easy. However, our method has five parameters, and designing steganalysis against our method becomes difficult. On the other hand, the pixel into which the modulus value of a block is embedded, is chosen by the secret key. Thus, anyone who does not know the parameters and the secret key can hardly extract or detect the secret information, and the steganalysis against our method is defeated.

In this paper, the perceptibility security is analysed through the simulation results.

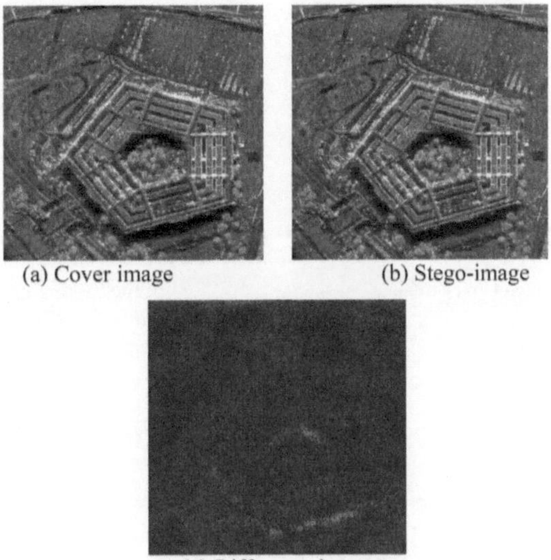

(a) Cover image (b) Stego-image

(c) Difference image

Figure 2. Simulation results for the Pentagon, Virginia, USA.

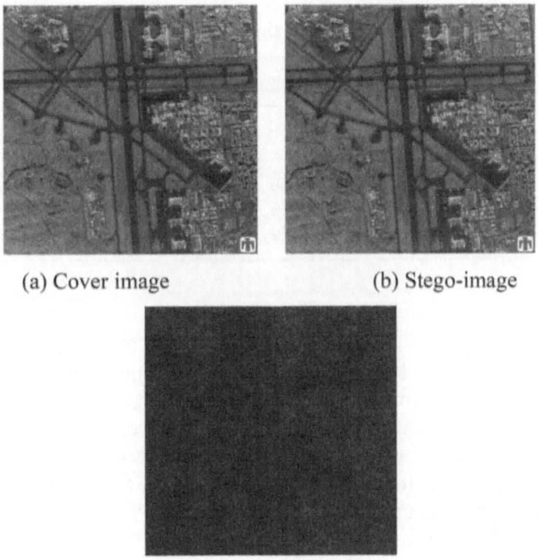

(a) Cover image (b) Stego-image

(c) Difference image

Figure 3. Simulation results for abq_apt.

## 5 EXPERIMENTAL RESULTS

In this paper, we utilize fake stochastic sequence as the secret information, and adopt 8, 8, 4, 11, and 3000 as the parameters $P$, $Q$, $n_{\min}$, $n_{\max}$, and $\Delta$, respectively. To enhance steganographic security, the reader can choose the parameters himself. Figure 2, Figure 3, and Figure 4 show the simulation results for a radar image of the Pentagon, abq apt, and isleta, respectively. We cannot find the difference between the cover radar image and the radar stego-image with our human eyes.

(a) Cover image

(b) Stego-image

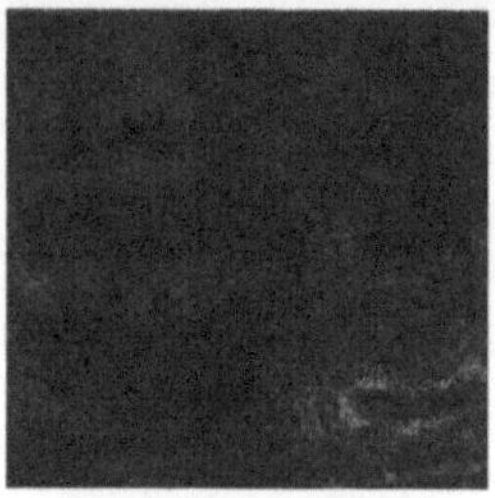

(c) Difference image

Figure 4. Simulation results for isleta.

Table 1. Simulation results for the proposed method.

| Image | Performance Capacity (bit) | PSNR (db) |
|---|---|---|
| Pentagon | 641 650 | 42.5734 |
| abq_apt | 608 369 | 42.1698 |
| isleta | 585 489 | 43.1584 |

Figure 2(c), Figure 3(c), and Figure 4(c) are the difference images between the radar stego-images and the cover radar images (with the differences of gray values being scaled 30 times). It can be seen that major changes lie in the strong texture areas and the minor changes lie in the weak texture areas. These results suggest that good visual imperceptibility is obtained.

Then, we choose three test radar images ($496 \times 496$) for simulation in this paper, and the hiding capacity and PSNR are shown in Table 1. From Table 1, we can see that the hiding capacity is greater than 30% of the size of the image (2.4 bits of secret information in one pixel of the cover radar image), and all value-sof PSNR (Peak of Signal-to-Noise Ratio) are higher than 38 db. As we know, when the PSNR is higher than 38 db, the human eye can hardly notice the distortion of the stego-image. Thus, the quality of visual imperceptibility is maintained. These results suggest that the hiding capacity of our method is great, while good visual imperceptibility is maintained.

## 6 CONCLUSION

In this paper, we have proposed a novel radar image steganographic method based on modulus function-based steganography and BEMD. The hiding capacity of our method is improved, by about 2.4 bits of secret information in one pixel of the cover radar image. Fortunately, good visual imperceptibility is maintained; the PSNR is higher than 42 db. Therefore people cannot find the existing of the secret information.

In this method, the reader can determine five parameters. As a result, statistical analyses against our method must estimate five parameters. As we know, only one parameter can be hard to estimate. The estimation of five parameters becomes very difficult. In addition, the pixel into which the modulus value of a block is embedded is chosen by the secret key. Thus, steganographic analysts can not obtain parameters or secret key. That is to say, anyone who does not know the parameters and the secret key can hardly extract or detect the secret information.

## REFERENCES


A. Cheddad, J. Condell, "Digital image steganography: Survey and analysis of current methods," Signal Processing, vol. 90, pp. 727–752, 2010.

C. C. Thien, and J. C. Lin, "A simple and high-hiding capacity method for hiding digit-by-digit data in images based on modulus function," Pattern Recognition, vol. 36, pp. 2875–2881, 2003.

E. Kawaguchi, and R. O. Eason, "Principle and applications of BPCS steganography," Proc. of SPIE: Multimedia Systems and Applications, Boston, MA, USA, SPIE-Int. Soc. Opt. Eng, 1998, pp. 464–473.

D. C. Wu, and W. H. Tsai, "A steganographic method for images by pixel-value differencing," Pattern Recognition Letters, vol. 24, pp. 1613–1626, 2003.

C. C. Chang, and H. W. Tseng, "A steganographic method for digital images using side match," Pattern Recognition Letters, vol. 25, pp. 1431–1437, 2004.

J. Liu, Z. W. Kang, and Y. G. He, "A steganographic method based on wavelet contrast and LSB," Acta electronica Sinica, vol. 35 pp. 1391–1393, 2007.

J. C. Nunes, S. Guyot, and E. Delechelle, "Texture analysis based on local analysis of the bidimensional empirical mode decomposition," Machine Vision and Applications, vol. 16, pp. 177–188, 2005.

# *Education and teaching reform*

*Frontiers in Computer Education – Wang (ed.)*
*© 2015 Taylor & Francis Group, London, ISBN 978-1-138-02797-8*

# A strategy generation method based on a Trace Substitute (TS) – Strengths Weakness Opportunity Threats (SWOT) model

XianMing Shi & Mei Zhao
*Shijiazhuang Mechanical Engineering College, Shijiazhuang, P.R. China*

ChunLiang Chen & YanHua Cao
*Academy of Armored Force Engineering, Beijing, P.R. China*

JianJun Zhang
*Troops, Shijiazhuang, P.R. China*

ABSTRACT: SWOT is a technique for making a strategic plan based on the analysis of internal and external factors. The classic SWOT analysis technique needs very high personnel diathesis and its results are comparatively random. Therefore, the internal and external factors with weak correlation can be taken as the break point. Through rationally dividing the associated factors, substituting low level factors for high level factors, and increasing the correlation of internal and external factors, a TS-SWOT analysis model is established based on the trace substitution principle. Through research on the strategy generation principle, grammar, and steps, the classic SWOT analysis is improved, and a strategic planning method is proposed based on TS-SWOT. This method provides a basis for generating more scientific strategic planning.

## 1 INTRODUCTION

The rapid development of society and science depends on advanced and feasible strategic planning. The internal and external environmental analysis is requested in strategic planning (Research Department of National Defense University, 2009). SWOT analysis is an analysis of Strengths (S), Weaknesses (W), Opportunities (O) and Threats (T) of the system, thus generating a system development strategy and providing guidance for system development. When making a SWOT analysis, it should firstly be known exactly what the strengths, weaknesses, opportunities, and threats of the research object are. Then, through comprehensively evaluating and analysing the strengths and weaknesses of internal factors and opportunities and threats of external factors, strategic planning can be made according to the studied conclusions (Gao, 2010). The system strategy generating from SWOT analysis is concluded in the following steps:

(1) Comprehensively analysing the internal conditions, history, current situation, and future trends of the external environment of the research object by means of investigation, literature analysis, etc. The internal conditions include the organizational management, personnel technology, etc., and the external environment mainly includes the national policy, the social environment, etc.
(2) According to the influence degree of the research object, ordering all factors through the survey analysis, and constructing a SWOT matrix.
(3) By using the method of system analysis, matching each two of the four kinds of influence factors of the strengths, weaknesses, opportunities, and threats together. Then, based on principles of exerting strengths, overcoming weaknesses, seizing opportunities, and avoiding threats, a series of alternative development strategies are generated, such as Strength-Opportunity strategy (SO), Weakness-Opportunity strategy (WO), Strength-Threat strategy (ST), and Weakness-Threat strategy (WT).

A primary advantage of the SWOT analysis is its comprehensiveness and systematic way of studying problems. The comprehensiveness means that it considers all influencing factors of the research object, including internal conditions, external environments, and the correlation of these two kinds of factors. Based on this, we can make the alternatives of strategic planning to choose. The systematicness means that it undertakes research according to a logical procedure of analysing the present situation, finding out the problems, and dealing with the problems. However, the classic SWOT analysis still has the following problems:

(1) The classic SWOT analysis requires higher personnel quality.
(2) Conclusions gained from the classic SWOT analysis are often different from person to person.
(3) The classic SWOT analysis fails to consider the order of the strategic priority.

One important reason for these problems is that different researchers analyse and match the internal and external factors differently; especially, when the correlation of internal and external factors is not strong, this is always ignored by some researchers. Even though some researchers have considered the correlation, it still leads to inaccurate research conclusions due to their different understanding of the two factors.

In order to solve the problems above, the internal and external factors with weak correlation are taken as the break point. Through rationally dividing the associated factors, substituting low level elements for high level elements, and increasing the correlation of internal and external factors, a TS-SWOT analysis model is established, based on the trace substitution principle. Based on the strategy generation principle, grammar, and steps, the classic SWOT analysis is improved, and a strategic planning method is proposed, based on TS-SWOT, which provides a basis for generating more scientific strategic planning. By using more rigorous ideas and quantitative analysis methods, the arbitrariness of the proposed strategy can be avoided to some extent, and the strategy can be described quantitatively, based on the correlation degree and influence degree, which improves the scientific feature of the SWOT analysis.

## 2 TS-SWOT MODEL

### 2.1 *Description of TS-SWOT model*

Definition: a TS-SWOT model can be obtained by analysing internal factors, external factors, and their associations, which contains 6 elements:

$$\mathbf{T}_r = (\mathbf{E}, \mathbf{N}, \mathbf{R}, \mathbf{r}, \mathbf{d}, \mathbf{a}) \tag{1}$$

where,

$\mathbf{T}_r$ is a TS-SWOT object;

$\mathbf{E} = \mathbf{O} \cup \mathbf{T}$ is the set of external factors;

$\mathbf{O} = \{O_i, i = 1, 2, \ldots, I_O\}$ is the opportunity set, that is, the factors which belong to external opportunities of the object system, and $I_O$ is the number of opportunity factors.

$\mathbf{T} = \{T_i, i = 1, 2, \ldots, I_T\}$ is the threat set, that is, the factors which belong to external threats of the object system, and $I_T$ is the number of threat factors;

$\mathbf{N} = \mathbf{S} \cup \mathbf{W}$ is the set of internal factors;

$\mathbf{S} = \{S_j, j = 1, 2, \ldots, J_S\}$ is the strength set, that is, the factors which belong to strength conditions of the object system, and $J_S$ is the number of strength factors.

$\mathbf{W} = \{W_j, j = 1, 2, \ldots, J_W\}$ is the weakness set, that is, the factors which belong to weakness conditions of the object system, and $J_W$ is the number of weakness factors.

$\mathbf{R} = \{\mathbf{E} \times \mathbf{N}\}$ is a relation pair composed of external environmental factors and internal condition factors, where $E_i \times N_j$ is just the correlation between the external factor $E_i$ and the internal condition $N_j$. According to the definition of $\mathbf{E}$ and $\mathbf{N}$, we know that $\mathbf{R} = \{\mathbf{R}_{OS}, \mathbf{R}_{OW}, \mathbf{R}_{TS}, \mathbf{R}_{TW}\}$, where $\mathbf{R}_{OS} = \{\mathbf{O} \times \mathbf{S}\}_{I_O \times J_S}$, $\mathbf{R}_{OW} = \{\mathbf{O} \times \mathbf{W}\}_{I_O \times J_W}$, $\mathbf{R}_{TS} = \{\mathbf{T} \times \mathbf{S}\}_{I_T \times J_S}$, and $\mathbf{R}_{TW} = \{\mathbf{T} \times \mathbf{W}\}_{I_T \times J_W}$;

$\mathbf{r} = \{r_{ij}\}$ is the correlation set.

$r_{ij} = r(E_i \times N_j)$ is the correlation of the external factor $E_i$ and the internal condition $N_j$, where $0 \le r_{ij} < 1$. If there is no relation between $E_i$ and $N_j$, then $r = 0$; otherwise, the supposed $E_i$ can be equivalently substituted by $k_i$, sub elements of the same level, $N_j$, can be substituted by $k_j$ sub elements of the same level, and if the number of these two kinds of elements with strong correlation is $b$, then:

$$r_{ij} = \frac{b}{k_i k_j} \tag{2}$$

$\mathbf{d} = \{d_{ij}\}$ is the layer distance set of factors;

$d_{ij} = d(E_i \times N_j)$ is the layer distance extent of the external factor $E_i$ and the internal condition $N_j$. If $E_i$ and $N_j$ are all the first level influence factors, then $d_{ij} = 2$; if the external factor is the next level of the $m$ level influence factors, and the internal factor is the next level of the $n$ level influence factors, then:

$$d_{ij} = 2 + m + n \tag{3}$$

$\mathbf{a} = \{a_{ij}\}$ is the set of the strategy importance;

$a_{ij} = a(E_i \times N_j)$ is the importance of strategy comprised of the external factor $E_i$ and the internal condition $N_j$, which is a measure of strategy availability or its influence on the system development, where $0 \le a_{ij} \le 1$. If the strategy is not feasible or necessary, that is to say, this strategy is not attractive, then $a_{ij} = 0$.

### 2.2 *The principle of TS-SWOT analysis*

Theorem: If $r(E_i \times N_j) \neq 0$ and $r(E_i \times N_j)$ is smaller, then $\exists(E_{ik} \in E_i, N_{jl} \in N_j)$; it will make $r(E_{ik} \times N_{jl}) > r(E_i \times N_j)$, and $r(E_{ik} \times N_{jl})$ is bigger.

Proof: Suppose $E_i = \{E_{i1}, E_{i2}, \ldots, E_{iK}\}$, $N_j = \{N_{j1}, N_{j2}, \ldots, N_{jL}\}$, and if $r(E_i \times N_j) \neq 0$, it means that $E_i$ is associated with $N_j$. Assuming the number of the factors with strong correlation is b, according to Formula (2), we know that:

$$r(E_i \times N_j) = \frac{b}{KL} \tag{4}$$

If $r(E_i \times N_j)$ is smaller, it means that $E_i$ and $N_j$ are not closely related, then $\exists(E_{im} \in E_i)$, it makes $r(E_{im} \times N_j) = 0$ or $\exists(N_{jl} \in N_j)$, and it makes $r(E_i \times N_{jl}) = 0$. To the first situation, suppose $E_i'$ is a set without this factor, then:

$$r(E_i' \times N_j) = \frac{b}{(K-1)L} > \frac{b}{KL} \tag{5}$$

To the second situation, suppose $N_j'$ is also a set without this factor, then:

$$r(E_i \times N_j') = \frac{b}{K(L-1)} > \frac{b}{KL} \tag{6}$$

Through continuous subtraction, we can find a relatively small number of substitution sets or substitution set combinations, which makes the association among the sets remains unchanged after subtraction, and the value of $r(E_{ik} \times N_{jl})$ can become bigger to a larger extent.

Conclusion: In the TS-SWOT analysis, through rationally dividing the internal and external factors, and substituting low level elements for high level elements, the correlation degree can be effectively increased. This conclusion is called the trace substitution principle.

With regard to the division of factors, there are mainly three situations, as follows:

(1) division of the internal condition factors;
(2) division of the external environmental factors;
(3) division of the internal conditions and external environmental factors at the same time.

In the process of generating a strategy, the classification method needs to be chosen according to actual situations.

Thus, in the case of weak correlation, the description and application of internal and external factors is the primary cause of the randomness of strategic analysis. According to the above theorem and conclusion, as long as the factors with weak correlation can be divided rationally and the superior set can be substituted by the next subset, the correlation of factors can be increased and the situational analysis results will be relatively definite.

The above process of dividing influence factors with weak correlation into the next level is called the trace process, and the process of substituting the next level sub factors for the original is called the substitution process. The SWOT situation analysis which introduces these two processes is called TS-SWOT analysis.

## 3 STRATEGY GENERATION PROCESS BASED ON TS-SWOT ANALYSIS

### 3.1 *Analysis of dividing TS-SWOT level*

Based on trace substitution, TS-SWOT analysis can resolve the problem of strategy generation under the circumstances of little association between internal conditions and the external environment, which can be achieved through further division of internal conditions and the external environment.

As is seen above, internal conditions or external environmental factors can be decomposed through level division. It is a good way to remove irrelevant factors and increase the correlation degree by substituting the next level factor for the superior. However, while factors are traced and substituted, it also leads to a narrow scope and low importance of the strategy generated. For example, as for the problem of the army equipment support transformation strategy, "high-tech" is an external environmental factor and "information construction" is an internal factor. When analysing the correlation of these two kinds of factors, we need to trace and substitute the "high-tech" for "information technology" to increase the correlation, thus "information technology" and "information construction" will have a strong relationship. However, as "high-tech" factor is more important, its weight will be reduced if replaced with "information technology." If it is 0.2 times more than the original, the importance degree of the strategy generated by "information technology" and "information construction" should be 0.2 times the product of "high-tech" weight and "information construction" weight. If it is further decomposed, the strategy importance degree will be 0.04 times the original. The importance of the strategy declining indicates that the strategy effectiveness reduces. Therefore, to make sure the usefulness of the strategy, factor division levels should be fewer.

### 3.2 *Strategy generation grammar of TS-SWOT*

By using the method of system analysis, each two of the four kinds of influence factors of the strengths, weaknesses, opportunities, and threats, together can be matched, thus forming a series of optional development strategies of Strength – Opportunity strategy, Weakness-Opportunity strategy, Strength-Threat strategy, and Weakness- Threat strategy, etc. This paper will propose a method of strategy generation based on the combined syntax, which is shown in Formula (7):

$$R_{ij} = \langle V_E, E_i, V_N, N_j, V_V \rangle \tag{7}$$

where, $R_{ij}$: strategy generated by $E_i$ and $N_j$;
$V_E$: external environmental factor verbs;
$E_i$: external environmental factors;
$V_N$: internal condition factor verbs;
$N_j$: internal condition factors;
$V_V$: direction of strategic application.
$V_E$, $V_N$ and $V_V$ are defined as follows:

$$V_E = \begin{cases} \text{grasp, } E_i\text{, opportunity} & \text{if } E_i \in \mathbf{O} \\ \text{avoid, } E_i\text{, threat} & \text{if } E_i \in \mathbf{T} \end{cases} \tag{8}$$

$$V_N = \begin{cases} \text{exert, } N_j\text{, strength} & \text{if } N_j \in \mathbf{S} \\ \text{overcome, } N_j\text{, weakness} & \text{if } N_j \in \mathbf{W} \end{cases} \tag{9}$$

$$V_V = \begin{cases} \text{utilize} & \text{if } E_i \in \mathbf{O} \text{ and } N_j \in \mathbf{S} \\ \text{improve} & \text{if } E_i \in \mathbf{O} \text{ and } N_j \in \mathbf{W} \\ \text{monitor} & \text{if } E_i \in \mathbf{T} \text{ and } N_j \in \mathbf{S} \\ \text{eliminate} & \text{if } E_i \in \mathbf{T} \text{ and } N_j \in \mathbf{W} \end{cases} \tag{10}$$

The strategy can be generated according to the grammar above.

### 3.3 *Steps of TS-SWOT analysis*

In actual correlation analysis, we can divide the correlation according to {0, less, 1}, thus distinguishing whether the external factors are related with the internal conditions, and whether trace substitution is required. On this basis, the process of TS-SWOT analysis is established. The concrete operation steps are as follows:

Step (1): analysing internal condition factors;

Ways: it is carried out by studying the internal conditions of the system. Internal condition factors are positive or negative factors existing in the process of the system development and evolution, which need to analyse the resources, organization, system, and mechanism, etc. When analysing, the status of the object system, as well as its future trends, should be considered; defining the positive factor of the internal conditions as strength, and denoting $\mathbf{S}=\{S_j,\ j=1,2,\ldots,J_S\}$; defining the negative factor as weakness and denoting $\mathbf{W}=\{W_j,\ j=1,2,\ldots,J_W\}$; thereby the condition factor set is $\mathbf{N}=\mathbf{S}\cup\mathbf{W}$.

Step (2): analysing external environmental factors;

Ways: by means of investigation and analysis of the external environment, we can obtain all external environmental factors of the object system; considering factors of economy, politics, society, science, technology, and competition, etc., and dividing them into two kinds of factors of advantages and disadvantages; defining the advantages as opportunities and denoting $\mathbf{O}=\{O_i, i=1,2,\ldots,I_O\}$; defining the disadvantages as threats and denoting $\mathbf{T}=\{T_i,\ i=1,2,\ldots,I_T\}$; thereby the external environmental factor set is $\mathbf{E}=\mathbf{O}\cup\mathbf{T}$.

Step (3): ordering internal condition factors;

Ways: ordering the internal condition factors obtained from Step (1) according to its influence on the development and evolution of the system or the priority improvement degree (all are called the 'influence degree'). We can gain the internal condition factors after ordering, which are denoted as: $\mathbf{N}=\{N_j|N_j\in\mathbf{S}\cap\mathbf{W}, A(N_k)>A(N_l)$ while $k<l$, where $j,k,l=1,2,\ldots,J_S=J_W\}$, where $A(N_j)$ is the influence degree of internal condition factor $N_j$ on the object system, which can be obtained by analysing the dependence degree of the system development goals on internal factors with the Quality Function Deploy (QFD) method.

Step (4): ordering external environmental factors;

Ways: ordering the external environmental factors obtained from Step (2) according to its influence on the development and evolution of the system. We can gain the external environmental factors after ordering, which are denoted as: $\mathbf{E}=\{E_i|E_i\in\mathbf{O}\cap\mathbf{T},\quad B(E_k)>B(E_l)$, while $k<l$, where $i,k,l=1,2,\ldots,I_O+I_T\}\}$, where $B(E_i)$ is the influence degree of external environmental factor $E_i$ on the object system. Similarly, $E_i$ can also be obtained in the same way as Step (3).

Step (5): give an initial value $j=0, i=0$; define the strategy set $\mathbf{R}=\Omega$, $k=0$;

Step (6): let $j=j+1$; if $j\leq J_S+J_W$, then go to Step (7), otherwise go to Step (19);

Step (7): let $i=i+1$, if $i<I_O+I_T$, then go to Step (8), otherwise, go to Step (6);

Table 1. External environmental factors.

| Types | External environmental factors | Sign | Weight |
|---|---|---|---|
| opportunities | The domestic political situation is stable | O1 | 0.044 |
| | The new military revolution thoroughly develops in the world | O2 | 0.088 |
| | Equipment support has successfully transformed in developed countries | O3 | 0.077 |
| | The national economy is soundly developing | O4 | 0.154 |
| | Defense spending is continuously growing | O5 | 0.066 |
| | The 'high-tech' is rapidly developing | O6 | 0.110 |
| | Technology investment is bigger | O7 | 0.088 |
| | Resources for transformation work is abundant | O8 | 0.054 |
| | …… | … | …… |
| threats | Security situation is more serious | T1 | 0.055 |
| | Combat style has changed | T2 | 0.044 |
| | Combat mission requirement is diverse | T3 | 0.110 |
| | Weapons and equipment develops rapidly | T4 | 0.044 |
| | The technological level lags behind overall | T5 | 0.066 |
| | …… | … | …… |

Step (8): determine the correlation of $E_i$ and $N_j$, if $r(E_i\times N_j)=0$, then go to Step (7); if $r(E_i\times N_j)$ is bigger, then go to Step (9); if $r(E_i\times N_j)$ is smaller, then go to Step (15).

Step (9): if $N_j\in\mathbf{S}$ then go to Step (10), otherwise, go to Step (12);

Step (10): if $E_i\in\mathbf{O}$ then generate the strategy in the mode of SO, and denote the strategy as $S_jO_i$. Let $R_k=S_jO_i$ and take it into the strategy set $\mathbf{R}$, then go to Step (14);

Step (11): if $E_i\in\mathbf{T}$ then generate the strategy in the mode of ST, and denote the strategy as $S_jT_i$. Let $R(k)=S_jT_i$ and take it into the strategy set $\mathbf{R}$, then go to Step (14);

Step (12): if $E_i\in\mathbf{O}$ then generate the strategy in the mode of WO, and denote the strategy as $W_jO_i$. Let $R(k)=W_jO_i$ and take it into the strategy set $\mathbf{R}$, then go to Step (14);

Step (13): if $E_i\in\mathbf{T}$ then generate the strategy in the mode of WT, and denote the strategy as $W_jT_i$. Let $R(k)=W_jT_i$ and take it into the strategy set $\mathbf{R}$, then go to Step (14);

Step (14): calculate the importance of the strategy $R_k$, that is $a_k=A(N_j)B(E_i)$, the layer distance of the strategy is $d_k=2$, let $k=k+1$; go to Step (7).

Table 2. Internal condition factors.

| Types | Internal condition factors | Sign | Weight |
|---|---|---|---|
| strengths | The channel of equipment supply is unobstructed | S1 | 0.060 |
| | Personnel quality and ability is higher | S2 | 0.143 |
| | Executive ability of the system is stronger | S3 | 0.122 |
| | Operating mechanism is flexible | S4 | 0.105 |
| | …… | … | …… |
| weaknesses | Theoretical research lags behind | W1 | 0.114 |
| | Accessory of equipment is not enough | W2 | 0.116 |
| | The foundation of technical transformation is weak | W3 | 0.085 |
| | Information construction is not systematic | W4 | 0.130 |
| | Organization system is not flexible | W5 | 0.124 |
| | …… | … | …… |

Step (15): let $l=0$, $m=0$.

Step (16): process of the trace substitution. Decompose the external environmental factor $E_i$ according to its maximum influence discrepancy on the internal condition factor $N_j$. Find out the primary influence factor, and define it as $E_{il}$, $l=1,2,\ldots,L_i$. Meanwhile, we can determine its corresponding $B(E_{il})$, where: $\sum_{l=1}^{L_i} B(E_{il}) \leq B(E_i)$. Analyse the correlation $r(E_{il} \times N_j)$, if more than one of $r(E_{il} \times N_j)$ is bigger, let $d(E_{il} \times N_j)=3$; go to Step (17); if $r(E_{il} \times N_j)$ are all smaller, then decompose the internal condition factor $N_j$ according to its maximum influence discrepancy of $E_{il}$, define it as $N_{jm}$, $m=1,2,\ldots,M_j$, and determine $A(N_{jm})$, where $\sum_{m=1}^{M_j} A(N_{jm}) \leq A(N_j)$. Then analyse each correlation $r(E_{il} \times N_{jm})$, and find out the bigger factor pair. Then let the factor layer distance $d(E_{il} \times N_{jm})=4$; go to Step (18).

Step (17): generating the strategy set according to Step (10)–Step (13), and after every strategy generation, let $a_k=a(E_{il} \times N_j)=A(N_j)B(E_{il})$ and $d_k=d(E_{il} \times N_j)=3$; $k=k+1$; go to Step (7);

Step (18): generating the strategy set according to Step (10) ~ Step (13), and after every strategy generation, let$a_k=a(E_{il} \times N_{jm})=A(N_{jm})B(E_{il})$ and $d_k=d(E_{il} \times N_{jm})=4$; $k=k+1$; go to Step (7);

Step (19): let $k=k-1$, we can obtain the strategy set **R** and the responding importance set **a**;

Step (20): classify the strategies according to SO, ST, WO, and WT, and summarize the importance of every kind of strategy. Then, use ordering and analysing to get the current situations.

For the strategy set generated from the above steps, we can abstract and conclude further to obtain feasible strategy planning.

Table 3. TS-SWOT analysis.

| | S2 | W4 | W5 | S3 | W2 | W1 | S4 | W3 | S1 |
|---|---|---|---|---|---|---|---|---|---|
| O4 | 0.2 | 1 | 0 | 0 | 1 | 0 | 0 | 1 | 0.2 |
| O6 | 0 | 0.2 | 0 | 0 | 0.2 | 1 | 0 | 1 | 0.1 |
| T3 | 0 | 0 | 1 | 1 | 1 | 0.1 | 1 | 0 | 0 |
| O2 | 1 | 0 | 0 | 0 | 0 | 1 | 0 | 0 | 0 |
| O7 | 0.2 | 0.1 | 0 | 0 | 0.2 | 1 | 0 | 1 | 0.1 |
| O3 | 0 | 0 | 0.1 | 0 | 0 | 1 | 0.1 | 0 | 0 |
| O5 | 1 | 1 | 0 | 0 | 1 | 1 | 1 | 1 | 1 |
| T5 | 1 | 1 | 0 | 0 | 1 | 0 | 0 | 1 | 0 |
| T1 | 0 | 0 | 0 | 1 | 1 | 0 | 0 | 0 | 0 |
| O8 | 0 | 0 | 0 | 0 | 0 | 0 | 0 | 0 | 0.01 |
| O1 | 0 | 0 | 0 | 1 | 0 | 0 | 0 | 0 | 0 |
| T2 | 1 | 0 | 1 | 1 | 1 | 1 | 1 | 1 | 1 |
| T4 | 1 | 0 | 0 | 0 | 1 | 1 | 0 | 1 | 0 |

## 4 STRATEGY GENERATION INSTANCES OF THE ARMY EQUIPMENT SUPPORT TRANSFORMATION

### 4.1 *Backgrounds*

The army equipment support transformation is a complicated system of engineering. It has characteristics of wide ranges, many uncertain factors, enormous cost, long duration time, and strong correlation (Liu, 2011, Cao, 2012). A strategy for the army equipment support transformation can provide guidelines for equipment support transformation. Due to its many influencing factors, it is very difficult to generating strategies. Below, we will use the method proposed in this paper to generate a transformation strategy.

### 4.2 *Analysis of internal conditions and the external environment*

We need to analyse the internal conditions and the external environment of the equipment support system effectively (Zhang, 2011), including fully grasping the opportunities brought by the external environment for the transformation work, clearly defining the threats which the transformation work may be faced with, understanding the current situations of the equipment support system, and determining the strengths and weaknesses of an army equipment support system; all of these will lay the foundation for making a scientific transformation strategy. Through expert scoring, the factor weights can be determined by the method of Analytic Network Process (ANP) and rough set theory. The external environmental factors and their weight are shown in Formula 1, the internal condition factors and their weight are shown in Table 2.

### 4.3 *Analysis of correlation of the army equipment support transformation factors*

Through correlation analysis of the internal condition factors and the external environmental factors above, the results are gained and shown in Table 3.

Table 4. Denotation of trace substitution factors.

| Factors pair | Trace substitution | |
|---|---|---|
| | Factors | Description |
| S2O4 | O4 | Education investment |
| S2O7 | S2 | The innovation ability |
| W4O6 | O6 | Information Technology develops rapidly |
| W4O7 | O7 | Innovation investment of Information Technology |
| W5O3 | O3 | Successful transformation of foreign organization |
| W2O6 | W2 | Promoting the equipment development |
| ... | ... | ... |

Values in table 3 indicate the correlation $r(E_{ik} \times N_{jl})$ of the internal conditions and the external environmental factors.

$r(E_{ik} \times N_{jl}) = 1$ indicates that there is strong correlation between its internal conditions and external environmental factors.

$r(E_{ik} \times N_{jl}) = 0$ indicates that there is no correlation between its internal conditions and external environmental factors.

$0 < r(E_{ik} \times N_{jl}) < 1$ indicates that more than one of the internal conditions and external environmental factors needs trace substitution to determine their correlations.

### 4.4 *Analysis of trace substitution factors*

After analysing the factors with little correlation in Table 4, the trace substitution goes along with substituting sub factors for original factors to increase the correlation degree of factors. The results of trace substitution are shown in Table 4, the factors which are traced and substituted are shown in brackets.

By substituting the traced and substituted factors for the original factors, we can further get the strategy planning.

### 4.5 *Analysis of effect*

From the process above, we can see that as a result of trace substitution, the correlation of the factors is obviously increased; therefore, the strategy finally generated will be more concrete, scientific, and feasible.

## 5 CONCLUSIONS

The scientific feature of the strategic planning is a basic requirement for strategy generation. In original SWOT methods of generating strategies, different researchers often have had different understanding of the correlation, as led to the difference in strategies generation. However, through the trace substitution of factors, researchers can consider a problem from the nature of the correlation, and thus be able to understand the system in an analytic perspective, as can still improve the factors correlation and minimize the randomness of the strategic planning. And it is absolutely essential for scientifically understanding the system development trend and effectively planning the strategy.

## REFERENCES


Cao Yanhua 2012. Research on Autonomic Logistics Key Technology of Armored Equipment. Beijing: A Dissertation Submitted for the Doctoral Degree Academy of Armored Force Engineering.

Gao Zhenguo 2010. Empirical Study of Development Strategy of Coal Enterprise Group: Taking Zhengmei Group as an Example. Journal of China University of Mining (4): 56–60.

Liu Zhongpeng 2011. Requirement of Transformation of the Way of Equipment Support. Value Engineering (2): 209–210.

Research Department of National Defense University of PLA. 2009. Road mapping-A new means of strategy management. Beijing: National Defense University Press.

Zhang Jian 2011. Analysis of SWOT in SEAN for Weapons and Equipment Development. Journal of the Academy of Equipment Command & Technology 22(01):21–25.

*Frontiers in Computer Education – Wang (ed.)*
*© 2015 Taylor & Francis Group, London, ISBN 978-1-138-02797-8*

# A case-driven 4S teaching model

Ming Huang & Shu Yu Hu
*Hunan Radio and TV University, Changsha, China*

ABSTRACT: As educational reform goes further and deeper, a growing number of schools have recognized certain teaching ideas, such as "combination of working and learning", "task-driven", "experimental training", and "project teaching based on working process", which are put into active practice. In response to the curricular reform and the idea of Personal Finance, it is especially important to rebuild the knowledge system, adjust and reinforce the teaching contents, and explore the proper teaching model accordingly. In the principle of "satisfying the demand, exhibiting the teaching characteristics, focusing on quality development, and simplifying the teaching content", it is required to incorporate new teaching concept and practice into the course, and precisely reflect the newest policy, law and regulations, industrial standards, technological methods, and development findings in the financial industry. By doing so, the "case-driven 4S teaching model" with "updated concept, system, content, and method" may be discovered, which could be well-suited to the development of application-oriented talent.

*Keywords*: Case-driven; 4S Teaching Model; Personal Finance

Teaching Fund Projects: one of the key programmes of teaching reform under the support of Hunan provincial-level vocational college (ZJA2013030); one of the key programmes of Hunan provincial-level soft science research (2014ZK2041); one of the research programmes of Hunan provincial situation and policy consulting Committees (2013ZZ62).

## 1 INTRODUCTION

China's economy is on the track to continuous development, which is the impetus for the development of family finance. Under such a context, Personal Finance has been set up. With a large group of students, it is worthwhile to probe into how to cater to their financial demand, and strengthen their financial concept and ability by improving the course teaching. To that regard, we should think from the angle of students, and conduct the teaching of personal finance through discussing the case-driven 4S teaching model, according to the practical characteristics of personal finance. It is intended to promote the improvement of teaching and the transformation of learning, and enhance class efficiency and student quality, thereby attaining the goal of the personal finance course.

## 2 THE GUIDELINE OF TEACHING MODEL REFORM IN PERSONAL FINANCE

According to constructionism, knowledge can only be acquired with the help of others (including teachers and learning companions), and by tapping into the necessary learning materials through the sense-making model, rather than the transfer of knowledge from teachers. Constructionism calls for a learner-centred study under the instruction of teachers, or in other words, it not only emphasizes the learner's role as a cognitive subject, but also stresses the predominant role of teachers. The learning environment of constructionism consists of four elements, which are situation construction, collaboration construction, dialogue construction, and meaning construction.

As personal finance is one subject with good applicability, its reform of the teaching model should highlight the practical demand, and emphasize the student's practical and operational capability. Under the conditions of modern information and technology, the case-driven personal finance 4S teaching model is guided by humanism and constructionism, which adheres to the human-based and learning-centred principle, and emphasizes the stimulation of the student's internal demand for learning, thereby encouraging proactive learning; it also concentrates on the social and situational nature of knowledge, which enables students to learn the way of study, collaboration, dialogue, and application. In this sense, this teaching model conforms to the demand of time and the requirement of social development.

## 3 THE CONTENT AND FRAMEWORK OF THE CASE-DRIVEN 4S TEACHING MODEL

The case-driven 4S teaching model combines teaching with practice, makes the teaching task typical, and

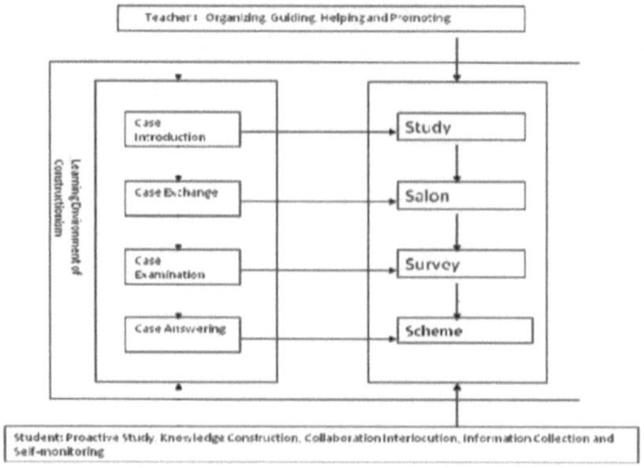


Figure 1. The basic framework of the case-driven 4S teaching model.

simulates the case for the student to consider, analyse, and judge; it develops the student's capability for self-study, cognition, and finance through the 4S method, the elements of which are "Study, Salon, Survey, and Scheme", thereby building a case that drives teaching organization throughout the teaching processes, according to the teaching objective and requirement of personal finance. The basic framework can be shown in Figure 1 below:

From the above, this case-driven 4S teaching model is made up of four parts or steps, Study—Salon—Survey—Scheme, and these four case-driven parts collectively create an open and constructionist learning environment for students. In such an environment, the teacher can organize and guide the teaching processes, and help or encourage students to complete the meaningful construction. In contrast, students are the active participators of study, the active constructors of knowledge, collaborative interlocutors of study, sharers of information collections, and self-test monitors of process throughout the teaching processes.

## 4 THE MAIN CHARACTERISTICS OF THE CASE-DRIVEN 4S TEACHING MODEL

The case-driven 4S teaching model for Personal Finance is oriented at financial demand and driven by feature case, while its creativity, smoothness, and distinctive features can be exhibited through the teaching organization - "Study—Salon—Survey—Scheme."

1. Core: "Study" is the core part. In the four parts under this model, the role of the teacher is only limited to "guiding", but students are the practitioners of activities, thus greatly highlighting the student's subjectivity and self-study. For instance, "Financial Study" requires the students to actively study and construct the knowledge; "Financial Salon" demands the students to collaborate in dialogue, and actively involve themselves in the group study; "Financial Survey" requires the students to collect the financial information by themselves; while "Financial Scheme" demands the students to carry out the meaning construction by performing the acquired theories and skills. These parts can not only drive the students to reinforce the learning process, but also stimulate the students to fully use and absorb the effective study.
2. Objectives: It is intended to develop the student's learning capability and financial capability. Each part emphasizes the need to develop the student's learning capability. "Introduce the case and learn the finance", for instance, guides the students to find out the problems by illustrating the cases that are close to reality; "Verify through the case and survey the finance" can lead the students to realize the problems; "Financial Scheme" and "Financial Salon" can encourage the students to examine and solve the problems. In the meanwhile, the four parts are always targeted at developing the financial capability gradually, which requires the students to grasp the key points of finance, exchange their financial experience, keep up with their financial status, and work out the financial scheme, etc.
3. Symbols: The feasibility of this teaching model is marked by three symbols: first, the guiding role of the teacher is fully displayed; second, teaching organization and designing could effectively improve the student's learning; third, the student's teaching objective and requirement can be greatly met. The models such as "case-driven", "expert financial salon" and "financial scheme" not only exercises the guiding role of the teacher but also greatly promotes the student's self-study. The other four parts are linked one another, guiding the students to master the knowledge and expertise of personal finance in an logical, purposeful, and strategic manner.
4. Strategies: Guiding students to aid their study, encouraging students to facilitate their study, leading students to practice their study, and evaluating the students according to their practices. "Guiding students to aid the study" is the basic strategy that we are always adopting; "encouraging students to facilitate the study" is the fundamental strategy throughout the case-driven teaching model; "leading students to practice the study" requires the students to be aware of the learning objective of Personal Finance, which can be described as learning the financial scheme, encouraging the action by study, and practicing the study by action; "evaluating the student according to their practices" teaches the students to have a down-to-earth learning attitude towards personal finance, while students must put the acquired knowledge into use, and promote the study according to their financial demands.
5. Focuses: 1. The student's position. This model follows closely the characteristics of HNRTU(Hunan Radio and TV University) students as "general professionals", which focuses especially on the family and personal financial demand, combines their professional capability with the instruction, and encourages the students to accumulate more

experience of practices in daily work and life; 2. The action orientation. This model has the action-oriented "how to self-manage finance", which emphasizes the student's role as the learning subject, thereby creating an atmosphere where teaching interacts with learning and the teacher interacts with the student. Through their specific financial situations, students are guided to acquire knowledge and expertise, and complete the task by themselves, thus enhancing their execution capability; 3. The case-based drive. Whenever the teacher introduces a financial instrument, he or she can drive the student's craving for knowledge by involving a great many of financial stories and scheme cases, which can promote their ability to apply the acquired professional knowledge and expertise; 4. Interactive teaching. The financial salon calls for on-site interaction between the expert and students or amongst students, which can optimize the learning effects; 5. The extra-curricular benefit. The learning model "Study + Salon + Survey + Scheme" can achieve a win-win result in both professional capabilities and personal qualities.

## 5 THE SCHEME OF IMPLEMENTING THE CASE-DRIVEN 4S TEACHING MODEL

The case-driven 4S teaching model is to be implemented by introducing, exchanging, examining, and answering the case, which can be basically described as follows:

### 5.1 *Step 1: Case introduction, financial study*

For most families in amateur financing, the basic objective of learning about personal finance is to establish the rational concept of finance, grasp basic knowledge of finance, and demonstrate basic expertise of finance. To help students basically understand the importance of personal finance and manage the key points of personal finance, this step is mainly conducted by teachers to guide the study, while students are proactively learning and thinking under the teacher's guidance, instead of cramming knowledge. They practice themselves in classroom communication and discussion of knowledge, ranging from active thinking, actively finding out and raising problems, and actively solving problems, to the active recognition. In each theme, teachers introduce some representative cases, explain the importance of finance, specify the key points of finance, or provide some clues of finance, which allows students to realize and think twice about their financial situation, and stimulates their learning. This is more obvious in our teaching materials. "Personal Insurance Finance", for instance, we have illustrated four examples, "Build the insurance pyramid before conducting the family finance", "Lausanne was hit into a car accident and Lausanne's parent had no alternative but went back to Tibetan", "Deep reflection ofYujuan Event" and "Consideration after the disaster", which gives incentive to the student's demand for the knowledge and expertise about personal insurance finance.

### 5.2 *Step 2: Case exchange, financial salon*

In this step, an exchange platform of collaboration and communication can be built up, which creates a learning atmosphere focusing on wealth, enjoying finance, and highlighting the importance of personal finance. In the two forms, "Student Finance Salon" and "Expert Finance Salon", it can change the student's financial concept, and improve the student's financial quality. In order to enhance the effect of the finance salon, the teacher generally assigns a task before the salon. For example, students are required to present the case study or raise the confusion and problem concerning personal finance, allowing them to be immersed into the salon atmosphere, and absorb the knowledge.

"Student Finance Salon" refers to a group that is spontaneously organized by several students, where they can exchange and help each other. Students can exchange their own financial experience and learning experience of the personal finance course; they can exchange ideas about personal finance knowledge and surveys; teacher can assess and exchange the finance scheme as prepared by students; they can plan and evaluate the finance based on real situations; they can organize and participate in finance simulation competitions, allowing them to put knowledge into use, etc. In this form, students can gain more knowledge, and practice the coordination and communication skills with each other, thereby promoting the exchange.

"Expert Finance Salon" refers to the professional lecture given by financial experts that features interaction, communication, and personalized design. Through this lecture, that combines knowledge with practice, students can understand the newest information about the financial market, and learn specialized financial knowledge; they can get much closer to all kinds of bank financial products; enjoy the professional financial instruction; and optimize their personal assets.

### 5.3 *Step 3: Case examining, financial survey*

As personal finance develops with time and practice, students are required to do more practical surveys and collect financial information. This is done by organizing the students to visit the practice base, or for groups of students to conduct surveys on the websites of financial institutions or branches. As a result, they will understand the situation of bank finance, know the channels of financing, know the types and characteristics of financial products, experience the operational flow of financial business, appreciate the financial culture, and strengthen their knowledge about the real financial market, thus justifying us to work out the financial scheme accordingly. As described in "Personal Insurance Finance", the teacher assigns a survey:

go to several companies offering life insurance, understand the situation of financial insurance, and design an appropriate scheme for financial insurance according to the family's condition. We can understand the operational effectiveness of the financial instrument and grasp the financial situation of the financial institution only through surveying, which also provides a reference and basis for us to work out the appropriate financial scheme.

## 5.4 *Step 4: Case answering, financial scheme*

The ultimate objective of personal finance is to understand the financial trend, understand financial demands, and tailor the financial plan depending on personal condition, in order to maximize the personal wealth and achieve the financial freedom, whether for common residents or financial professionals. By following the said three phases, learners can more or less understand the common methods and skills concerning personal finance, but how to design a financial scheme according to a given case, or how to plan for the future based on the available resources, is seen as the problem to be solved. In this phase, students play the role of finishing it, while teacher is responsible for guiding students to plan and finance through presenting the case study based on the given condition. "Personal Financial Overview", for instance, requires the students to make a financial analysis of Mr. Song's family's financial standing and design the comprehensive scheme of the family's financial plan based on the specified case, or "Personal Insurance Finance" requires the students to conduct insurance finance depending on the individualized insurance demand.

## 6 THE EFFECTS OF APPLYING THE CASE-DRIVEN 4S TEACHING MODEL

This model introduces a learning situation via an actual case, and conducts a practical survey that is, more or less, the same with the reality, and answer or probe into the case through financial planning and financial salon, thus improving the financial quality of learners. By doing so, it is much applicable because of the incorporation of theoretical teaching with practical teaching and teaching with social service.

First and foremost, it obtains the intended teaching objective of the course. The teaching objective of a personal financial course is to popularize the basic knowledge of personal finance and develop the basic skills of personal finance; as a result, the student's learning interest can be motivated. It also reinforces the theoretical and systematic knowledge base, develops analytical and problem-solving abilities, thereby allowing students to put their acquired knowledge to good use. The case-driven 4S teaching model combines network, market, and classroom, in other words, it incorporates the sky network, the ground network and the human network(Remote network education terminology). In this way, it helps students embrace

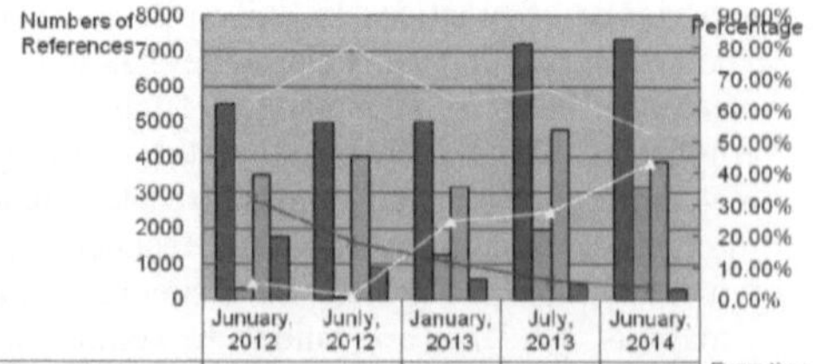


| | Junuary, 2012 | Junly, 2012 | January, 2013 | July, 2013 | Junuary, 2014 |
|---|---|---|---|---|---|
| Numbers of References | 5504 | 4976 | 5003 | 7176 | 7317 |
| >90(in people) | 288 | 63 | 1247 | 1983 | 3164 |
| 60-89(in people) | 3473 | 4009 | 3175 | 4761 | 3867 |
| <60(in people) | 1743 | 904 | 581 | 432 | 286 |
| >90(in percentage) | 5.20% | 1.30% | 24.90% | 27.70% | 43.30% |
| 60-89(in percentage) | 63.10% | 80.50% | 63.50% | 66.30% | 52.80% |
| <60(in percentage) | 31.70% | 18.20% | 11.60% | 6.00% | 3.90% |

Figure 2. The final exam scores distribution of HNRTU over two recent years.

the society or job gives more chances for students to practice, allows them to partake in brainstorming for analysing real problems, allows them to find out the appropriate method to solve real problems, and ensures they acquire a well-founded knowledge of financial skills. Meanwhile, the case-driven 4S teaching model combines theory with practice, guides the students to learn, discuss, survey, and plan while being involved in the case study, thereby attaining the course objective. According to the distribution of final exam marks for Personal Finance in two recent years (as shown in Figure 2), the number of students selecting Personal Finance is ever increasing. The percentage below 60 scores is ever decreasing, while the percentage above 90 scores is ever increasing. In January, 2014, the percentage above 90 scores reached 43.3%, accounting for approximately half of the students.

Secondly, it effectively improves the student's capability of lifelong study. Lifelong study capability is one of the necessary skills for survival in our current society, which mainly includes self-study capability and information quality capability. According to the statistics, as shown in Figure 3, financial majors or non-financial majors spent more hours in online learning for the past one and half a year. In this sense, the personal finance 4S teaching model can encourage students to conduct self-study and lifelong study aided by information technology. It is gradually becoming a habit.

By presenting the financial cases close to the life experience, the case-driven 4S teaching model guides the students to learn by themselves, drives them to learn and discuss further under the teaching organization consisting of teacher-student interaction, and organically combines learning objectives, working requirements, and survival skills, thus creating an open learning environment for students. The cases in the environment are collected by students themselves; some are familiar ones, while some are very close to the financial stories and examples in day-to-day life. By learning these case studies, students can understand and satisfy their own financial demands. These factors could inspire a craving for knowledge and learning, driving students to look into the case.

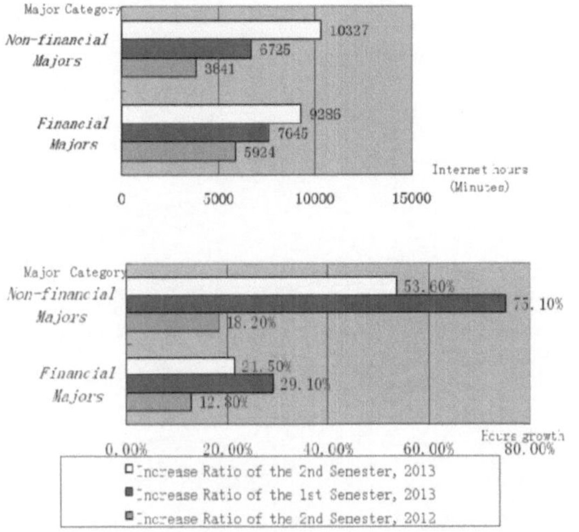


Figure 3. The changing trend of online learning hours before and after applying the teaching model.

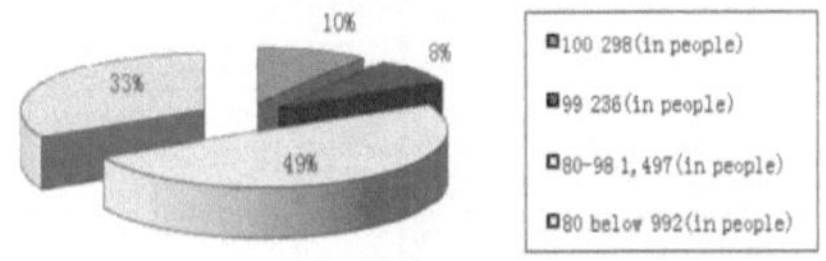


Figure 4. The scores distribution of HNRTU Personal Finance network knowledge in November, 2013.

Last but not the least, the model develops the student's knowledge about society. Personal Finance is defined by many schools as a general curriculum, so it means that personal finance capability has become a kind of general capability that everyone should demonstrate. The learning model "Study+Salon+Survey+Scheme" can be performed to train the student's basic financial capability and social communication capability, and this combination of training would be the key to developing the student's general social capability. In the second half of the 2013 semester, HRTU organized the Personal Finance Network Knowledge Competition and Case Design Match, which attracted 3,023 competitors in total. As shown in Figure 4, there were 534 competitors above 99 scores, accounting for 17.7%. According to the students, Personal Finance adopts the "case-driven 4S" teaching model, which provides the channels for students, such as financial surveys and expert salons. It encourages the students to focus on and grasp their financial environment, which develops their recognition and sensibility about the financial market; moreover, students can participate in the social financial consulting activities, thereby actualizing the financial interaction and communication between campus and workplace.

## REFERENCES


Joyce (US). Teaching model (The 7th Edition). China Light Industry Press, 2009

Li Qilong. Selected Works of Herbart Education Works. Zhejiang Education Publishing House, 2011

Zhou Li. The "4S" teaching model in the Concept of Practice-oriented. Journal of Hubei Radio and TV University, 2009 (2)

Wang Degang, Lian Zhiqiang. Promote the Course Teaching Reform by Adopting the Case-driven teaching model. Journal of Inner Mongolia Agricultural University (Social Science), 2013 (1)

Yi Zhiheng. The teaching model of 3D constructing the Personal Finance. Exam Weekly, 2011 (72)

Liu Jiao. The Application of Action-oriented Theory in the Course Reform of Personal Finance. Entrepreneur World, 2011 (5)

Wang Sihao. A Probe into the Teaching Reform of Investment and Fiance. China Urban Economy, 2011 (18)

Zhang Hongyan. The Discussion of Financial Teaching Reform as Targeted at the Practical Development. Pioneering with Science & Technology Monthly, 2011 (18)

Zhang Yanying. A Study of the Teaching Reform of Personal Finance as the Open Education General Course. Exam Weekly, 2012 (94)

Zhang Qiang, Huang Ming. Personal Finance. Hunan People's Publishing House, 2012

*Frontiers in Computer Education – Wang (ed.)*
*© 2015 Taylor & Francis Group, London, ISBN 978-1-138-02797-8*

# Practical teaching reform in the VHDL (Very High Speed Integrated Circuit Description Language) course based on the idea of CDIO (Concept, Design, Implementation, and Operation)

Jun Liu, YaDong Zhu, XiaoPing Yang, JinLi Liu & Zhen Wei
*Tianjin Key Laboratory for Film, Electronic, and Communication Devices, School of Electronics and Information Engineering, Tianjin University of Technology, Tianjin, China*

ABSTRACT: During the cultivation of electronic engineering talents for modern enterprises, practical teaching is an indispensable and important part. This paper analyses the question about the VHDL course in the college teaching system. The teaching hours, the new experimental items, and the ways of evaluation and examination of this course are redesigned in the CDIO framework. At the same time, this paper shows how to redefine the roles of the teachers. Several years of teaching practice show that it is efficient to inspire students' enthusiasm for learning and to improve their creative ability and engineering consciousness.

*Keywords*: CDIO; VHDL; Reform; Capacity building

## 1 INTRODUCTION

With the development of microelectronics technology and design techniques, it becomes more and more complex to design electronic circuits. Using VHDL (Very High Speed Integrated Circuit Description Language) can increase the development efficiency and shorten the development time. Now VHDL is widely used in industrial automation and control, intelligent instrumentation, digital signal processing, and communications fields with the IEEE's (Institute of Electrical and Electronic Engineers) help. It also plays an important role in modern industrial areas.

VHDL is a basic course for students of electronic information engineering, electronic science and technology, and integrated circuit design and integration of systems. Students can improve their practice, innovation, and teamwork abilities, and lay the foundation for the future of electronic design by learning from this course. This paper is a study on the teaching reform of the VHDL course under the guidance of CDIO, including its teaching content, the teacher's role, localization, and experiment teaching etc. This paper also presents a new training method to build the electronic engineering technology professionals' literacy.

## 2 CONCEPT OF CDIO

CDIO is the newest achievement of international engineering education reform in recent years, which is based on the framework of modern engineering education at MIT(Massachusetts Institute Of Technology), and advocated by dozens of the world's leading universities. CDIO is an initialism for Conceive-Design-Implement-Operate. It is a new educational framework. The students should learn actively by finishing several projects in the course. All the projects should not only limited in this course but also contain some related courses. All of these require the students connecting what they have learned to what they are learning actively in the projects. Specifically, it sets a specific project where the students learn the knowledge of the main areas of expertise in the completion of the project, during the process. So this model is also known as a project-based learning model.

## 3 PROBLEMS IN THE EXPERIMENT TEACHING PROCESS

A VHDL course, which has less hours but involving a wide range of knowledge, require the students have practical operations capability. Our teaching team, over years of theoretical and experimental teaching of VHDL, have found some problems:

(1) The course structure was unreasonable and the experimental projects were too simple. The course carried three credits. It would take 48 hours: 40 hours for lectures, and 8 hours for experiments. The course paid more attention to engineering science knowledge, whilst ignoring practical training and comprehensive ability.
(2) The roles of the teachers and students were unclear. The teachers centred on lecturing in class and lacked communication with students. This will make them lack of innovation. At the same time, the students experimented in a purely mechanical

manner and studied passively according to the experimental textbook. They could not correlate their learned knowledge and lacked ability to think independently.

(3) The students lacked creative consciousness and active knowledge expansion, so it was hard for them to build their own intellectual constructs. The students did not grasp the development trend of modern electronic technology which led to less interest and ignoring the cultivation of interest.

(4) Some students did not understand the use of this course. They choose this course only to get credits instead of mastering and using it. So the knowledge they had learned passively, was difficult to use in practical projects.

## 4 EXPERIMENTAL TEACHING REFORM BASED ON CDIO

In order to solve the existing problems in the experiment teaching process, our teaching team has carried on a further exploration. We have get the experience of the reformation of experimental teaching in the comprehensive practice.

### 4.1 *Arranging hours and content of the theory teaching based on CDIO*

VHDL never stands still. It is always changing and developing. It can be used at each stage of the digital system design. Therefore, it is important to arrange the teaching contents and hours logically. Practice has proved that this course should consist of 3 credits, including 32 hours for the theory teaching and 16 hours for the practice. Theory teaching focuses on a software development environment when using Quartus II, application design flow and method of depiction, basic structures, and basic function of sentences when using VHDL. So, this paper suggests arranging the theory teaching contents and hours according to Table 1.

According to the CDIO teaching idea, students are the principal part of learning. Teachers should build a good learning environment, develop learning resources, and serve students learning. So the teachers should prepare some design examples in which a knowledge point can be set by using multimedia courseware. An experiment is presented to enhance students' perceptual knowledge when needed. In this framework the students actively learn instead of passive learning passively. All can stimulate study interest and excavate their potential. When facing hard problems, students can easily grasp the key to solving these problems.

### 4.2 *Adjusting the experimental course contents based on CDIO*

The essence of CDIO is a process of "learning by doing." The students learn the main professional knowledge in a specific project. So the improved experimental course shows that the digital circuit

Table 1. Theory teaching contents and hours.

| No. | Theory teaching content | Suggested hours |
|---|---|---|
| 1 | Development of the FPGA (Field Programmable Gate Array technology), Overview of the VHDL, Difference between using VHDL and traditional design | 2 |
| 2 | Basic structure of VHDL | 2 |
| 3 | Method of depiction of architecture | 2 |
| 4 | VHDL description statement | 6 |
| 5 | Design of logic circuits and sequential circuits | 5 |
| 6 | Designing examples by using VHDL | 5 |
| 7 | FSM( Finite State Machine) design | 6 |
| 8 | Software development environment | 4 |

modules are designed in the projects. These projects and actual demands are closely linked. Each project requires students to learn the theory, design the programme, and solve the actual problems. The students can learn the characterization for each circuit and solve the specific problems in their application in the whole process. These cultivate the students' project consciousness. Therefore, this paper suggests arranging experimental teaching contents and hours according to Table 2.

The students can master usage of Quartus II, basic operators and data types of VHDL, basic structure and description methods of the program design, and concurrent statements and sequential statements of VHDL in required experiments, which lay a foundation to achieve elective experiments. In the elective Module 1 the students choose two projects to complete, which are same type of projects, and the students learn to design more complex electronic control systems in these projects. In elective Module 2, the students choose one project to complete which is more complex. In this project, the students learn to conceive the design, design the program, and operate software to implement the whole system. It can stimulate students' study interest, improve their practical skills, and educate their ability for creative thought.

According to the education idea of CDIO, the students do not work alone in the elective module. They can work in groups of two to three people. They actively complete experimental projects from selecting the project to implementing the project; they are active learners. The whole experimental project originates from practical engineering projects, and the projects are from simple to complicated and from independent to cooperative. These can stimulate interest as much as possible in learning in the opening circumstances.

### 4.3 *Adjustment of teachers' role and status based on CDIO*

Some scholars have pointed out that education concepts and methods of education must be changed in the CDIO mode. Metaphorically speaking, teachers must change their role from chef to nutritionist firstly,

Table 2. Experimental teaching contents and hours.

| No. | Experimental project | Suggested hours | The nature |
|---|---|---|---|
| 1 | Design of combinational logic circuits (encoder, decoder, etc.) | 2 | Required |
| 2 | Design of sequential logic circuits (counter, dividers, etc.) | 2 | Required |
| 3 | Design of Led digital display circuits | 2 | Required |
| 4 | Design of ADC (Analog to Digital Converter) controller | 3 | Elective 1 |
| 5 | Design of UART (Universal Asynchronous Receiver Transmitter) | 3 | Elective 1 |
| 6 | Design of stepping motor driver | 3 | Elective (1) |
| 7 | Design of VGA (Video Graphics Array) controller | 3 | Elective (1) |
| 8 | Design of frequency meter | 4 | Elective (2) |
| 9 | Design of DDS (Direct Digital Synthesizer) | 4 | Elective (2) |
| 10 | Design of electronic clock | 4 | Elective (2) |

the course must be changed from food to nourishment secondly, and finally, the school must also be changed from kitchen to nutrition room. All of these require the teachers to be designers and to avoid using the traditional cramming method of teaching. The teachers act as a guide to evaluate the whole learning process of the students. All the changes can help to build the students' creative consciousness. The students who can master more knowledge of related disciplines have the ability of independent thinking, ability of engineering, and ability of team cooperation. At the same time, this requires teachers to constantly improve their ability in areas such as engineering, designing the curriculum, and assessment.

### 4.4 *Adjustment of the examination method based on CDIO*

A set of reasonable examination methods can guarantee a teaching effect. Because of the characteristics of VHD, the traditional mode of examination of the course cannot effectively assess the actual operation and the students' ability for using knowledge. The examination method has three sections: usual-time appraisal (20%), theory examination (40%), and project appraisal (40%). Usual-time appraisal mainly includes four parts: attendance, discussions, attitude, and homework. The theory examination is used to examine how the students grasp the knowledge, and it adopts the form of a written examination. The project appraisal is judged by the teaching team. This is a comprehensive appraisal according to the selected experimental project including the conception, design, implementation, and operation of the project. This kind of appraisal pattern is established to evaluate the various abilities of the students. It can also form a diversiform learning style. As a result, the students who have solved concrete problems in teaching strengthen their responsibility, collective ideas, communication skills, and language abilities.

## 5 REFORMATION ACHIEVEMENTS BASED ON CDIO

In recent years, more and more students are interested in active learning. They actively participate in electronics clubs or unions. Some students join the national undergraduate or university undergraduate innovative experiments. Some were awarded second prize and third prize for work in the National Undergraduate Electronic Design Contest in 2013. Some were awarded two first prizes, a second third prize, and a third prize in Tianjin division. Others received awards and honours in an integrated circuit design contest in 2014.

## 6 CONCLUSION

In the teaching of the VHDL course, it is obvious to improve teaching quality and stimulate students' interest. It does not only develop students' comprehension and practical abilities, but also develops teachers' professional skills and ethics.

The employer put forward the consistent high praise. And they agree with this training method which meet a requirement of enterprise. Finally, we believe that relevant experience can help students to form training systems of engineering quality. It can also provide reference for related majors in Science and Engineering.

## ACKNOWLEDGMENTS

The research work was supported by the teaching reform and research project of Tianjin University of Technology, China (YB12-09), and an engagement fund from Tianjin University of Technology (LGYM201243).

## REFERENCES

Danhua, Cao. et al. 2012. Reform of Optoelectronic System Design Based on CDIO Model for Optoelectronic Undergraduates. *Research and Exploration in Laboratory* 31(4): 237–239.

Haimei, Tian. et al. 2011. Practice of comprehensive project based on CDIO education mode. *Experimental Technology and Management* 28(4): 130–132.

Hao Zhang. 2009. From "Chef" to "Nutritionist" — Discussion and Reference of CDIO Concept in Train Engineering

Management Course Student. *Higher Education Forum* 5: 62–64.
Hongwei, Liang. et al. 2011. Study on reform of experiment teaching and practical teaching for EDA Technology and Application. *Experimental Technology and Management* 28(1): 147–149.
Hui, Zhou. & Hua, Jiang. 2008. An Exploration on the Teaching Reformation of the HDL Course. *Higher Education Forum* 12(6): 184–186.
Jianzhong, Zha. 2008. Strategies for Education Reform: the relationships among "CDIO", Internationalization and Industry-Education Cooperation. *China University Teaching* 5: 16–19.
JunJie, Bai. et al. 2012. Teaching Practice and Exploration in VHDL. *Journal of Chongqing University of Science and Technology (Social Sciences Edition)* 1: 185–186, 188.
Liang, Kang. et al. 2012. Exploration and Application of CDIO Education Model in Students Technological Contests. *Research and Exploration in Laboratory* 31(2): 128–130, 147.
Miaosun, Cao. & Zhixing Liang. 2012. On the Transformation of Higher Engineering Professional Teachers' Role Based on the Concept of CDIO. *Research in Higher Education of Engineering* 1: 88–91.
Sheng, Chang. et al. 2009. Course-System Construction for Electronic Design Automation Technology. *Journal of Electrical & Electronic Education* 9: 38–40.
Shuowang, Wang. & Chengwen, Hong. 2009. CDIO: the Classic Mode of Engineering Education in MIT—An Unscrambling on the CDIO Syllabus. *Journal of Higher Education in Science & Technology* 28(4): 116–119.
Yaliang, Wang. et al. 2010. Development and practice of the experimental project based on the CDIO education mode. *Experimental Technology and Management* 27(2): 119–121.
Yuechang, Liu. et al. 2012. Engineering Experimental Teaching of Data Structure. *Research and Exploration in Laboratory* 31(8): 339–341.

*Frontiers in Computer Education – Wang (ed.)*
 *ISBN 978-1-138-02797-8*

# A research in the innovation of excellent engineers' training based on excellent consciousness cultivation

Xiao Liang
*School of Information Engineering, Wuhan University of Technology, Wuhan, China*

Kun Lang
*School of Management, Wuhan University of Technology, Wuhan, China*

ABSTRACT: The education Ministry proposed the "excellent engineers education training program", in order to adjust the personnel training structure and promote the education teaching reform. This paper first researches higher education, and explains the definition of excellence. Then, the new excellence consciousness is proposed. Moreover, the paper analyses the importance and necessity of developing excellence consciousness, and the relationship between excellent consciousness cultivation and excellent engineers training. Finally, the paper proposes the key path for excellent engineers training based on the cultivation of excellent consciousness.

*Keywords*: excellence, excellent consciousness; engineers training

## 1 INTRODUCTION

The Ministry of Education proposed *Excellent Engineers Education Training Program* in 2011, in order to cultivate a large number of excellence programmes for the industry, the world and the future, who have strong innovation ability, adaptability to the development of economic society and the professional abilities for the industry. This programme is an important guidance and demonstration for most activities, whose aims are providing more excellence, adjusting the personnel training structure and improving the talent training quality, promoting the education teaching reform and strengthening graduates' employment ability. Our government and universities all actively promote this programme's pilot implementation. The first batches of pilot training students have entered in September 2011.

However, the training project is still in the starting and exploration stage, whose training orientation, training standard, training mode, and training process are still not so clear in some part[1]. This paper proposes that excellence training's key and foothold in engineering practice ability, and start point should be the training of excellence consciousness. The consciousness cultivation must run through the whole process of excellence training and should be a motive power for the whole programme.

## 2 DEFINITION OF EXCELLENCE

In order to implement the program excellently, the focus is on understanding the definition of excellence. The word "excellence" is coming from JiangBiao Biography which is written by Yu Pu from the West Jin dynasty. Hu Zongzuo wrote the book *Friends* and said that 'the person, who enjoys more outstanding intelligence than his peers is Zhu Gege'. The Dictionary of Modern Chinese translates this word as outstanding and conspicuously superb.

In this programme, the definition of excellence does not have a unifying concept according to the current education status, and in addition there are lots of misunderstandings of this word. For example, in some universities, the one-sided focus on theoretical foundation and basic skills exists only towards the elite and the training is misplaced; besides, some colleges focus on the training of practice application talents, and unilaterally emphasize on practical capability. Even some teachers and professors believe that the excellence programme can only be used for a few members of the elites, otherwise it cannot highlight its pre-eminence.

In essence, however, education is a career, which can shape one's spirit and leads it towards healthy growth. Excellence in this programme stands for the human spirit and attitude, as well as one type of consciousness of improving the inner quality towards excellent personal quality.

Human's abilities are limited, but as human beings realizing their limits, they can surpass it and create something new. Only because of this excellence consciousness can people constantly realize their own excellence. Generally speaking, excellence, a relative concept based on the consciousness training, is not a privilege belonging to a few elites, but a life status for everyone.

## 3 CULTIVATION OF EXCELLENCE CONSCIOUSNESS

Current researches on consciousness are multiple, and its definitions are also different from different positions. ENCYCLOPAEDIA of CHINA (PSYCHOLOGY) defines consciousness as human's unique senior mental reflection form objective reality. Consciousness is awareness activities not only including feelings, perception, memory, imagination, and thinking, but also more advanced than them. It regulates human activities, and is the essence of being human. Therefore, the consciousness is the foundation of success[3].

Excellence consciousness is an inner request of excellence, whose manifestation forms are different in different environment. However, different forms are all aspiration of fully reflecting the life potential, embodying human goals, as well as wishes of constantly pursuing harmonious objects[2]. Persons with excellence consciousness exploit life potentials, and will always pursue the value and significance of life, improve life quality and intelligence and realize an individual life's excellence.

The cultivation of excellence consciousness is the goal of the education, especially for higher education. We should take different approaches and means in differentl stages of university education to awaken and nurture students' consciousness of excellence. In this way, we plant consciousness of excellence in the inner world of contemporary college students, and gradually precipitate it to the spirit and thought patterns of excellence. Finally, the pursuit of excellence should be a life status of consciousness Driven. This paper proposes that the cultivation of consciousness of excellence is the key purpose and foundation mission of implementing the excellence training program.

## 4 EXCELLENCE TRAINING BASED ON EXCELLENCE CONSCIOUSNESS CULTIVATION

Napoleon Hill (1883–1969), the earliest and the greatest scholar of modern Success Science, said that success comes from a person who has the consciousness of success; and failure roots in those who unconsciously produce losing awareness. Also, the key to one becoming excellent is whether one has consciousness of success, which means, excellence consciousness is excellence training's power source.

Based on the higher education personnel training, this paper first analyses the excellence consciousness training. Then it structures the excellence consciousness system and finally proposes the excellence training model based on excellence consciousness cultivation. Excellence consciousness is one type of self-awareness of striving for excellence.

Subdivided into the three aspects of thought, knowledge and ability, the excellence consciousness includes: self-consciousness, credit consciousness, and communication consciousness, responsibility

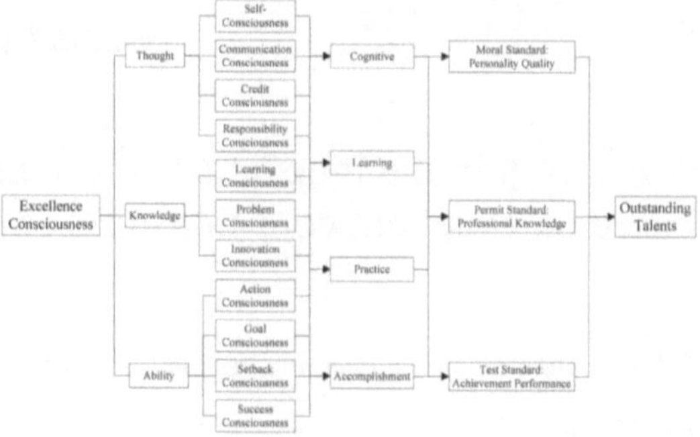


Figure 1. Excellence consciousness and excellent talent cultivating mode.

consciousness from a thought aspect, learning consciousness, problem consciousness, and innovation consciousness from a knowledge aspect; goal consciousness, action consciousness, setback consciousness and success consciousness from an ability aspect. Pelple with consciousness of excellence perfect their own personality traits, accumulate professional knowledge and create outstanding achievement from learning, cognition and practice. In the end, they grow up to outstanding talents which conform to ethical, entrance and inspection standards[4]. The training model is as follows.

Excellence consciousness is the key point of excellence training, which is usually not taken seriously. The current society is impatient, seeking a quick success; the college students follow suit. At the sometime, the colleges and universities' governors eagerly urge their students to follow the existing so-called right way, letting them stay in the passive situation, and overlook the foundation of excellence consciousness. However, the difference between the excellent and ordinary is the careless consciousness. In order to cultivate some excellence, consciousness plays a much more important role than other issues. Only by perception, learning, practice and self-cultivation, can individuals realize the excellence.

## 5 MEASURES AND METHODS OF EXCELLENCE TRAINING

Excellence Training Programme clearly proposes that the stratified cultivating different engineers talents is necessary to improve the quality of excellence, such as site engineers, design and development engineers and research engineers. This cultivating action needs to find joint training mechanism between universities and corporations, in order to strengthen practice activities.

As Ernst·Edward·Springer proposed that personality mind awakening is the core of education, first and foremost step of excellence training is the consciousness of becoming excellent. By improving students' awareness the programme can be much easier realised. The implementing subjects—universities and colleges are necessary for the consciousness cultivation.

Universities and colleges are well placed to design, organize, and reform the education models, in order to cultivate the excellence consciousness. The practical actions are as follows.

### 5.1 *Reinforce the educational foundation, enlighten rationality and morality*

Because of the exam-oriented education before college, developments of rationality and virtue are not unified, especially in some poor areas, or remote mountain areas, where this consciousness education has not begun. With the aim of cultivating excellence consciousness, the programme should first encourage students to shape their rationality and virtue, pursue the truth, beauty and goodness, and constantly transcend themselves through edification, initiation, guidance and elicitation forms.

Man is not born with reason and virtue, therefore practical life and education are the two key methods used to shaping them. Excellence training should abandon the traditional education model with knowledge education as a centre, textbooks education as a purpose, and put forward the new comprehensive education model with a combination of rationality and virtue, as well as a combination of practical experience and education, which guide students to adopt critical-reflective thinking and practical ability well.

### 5.2 *Construction of high-level teachers*

Education is an action among humans, especially among teachers. Excellence program implementation should be launched through two aspects. On the one hand, university education needs high level teaching team with practical experience. Full-time or part-time teachers should be better qualified with precious engineering experience, or working experience in some of well- known corporations.

On the other hand, teachers equipped with valuable professional abilities, would be better placed to implement the improved education model and education quality and thoroughly respect student's value option. As Socrates said that ignorance awakens personal self-knowledge, excellent teachers should first give up the thought that teachers are authorities on everything, and then, using the communications with students based on equal terms, guide them in the right direction in order to awaken their excellence consciousness. Patient and tolerant teachers appreciate the reasoning behind personal differences among students, and bravely admit ignorance and imperfection, which is favourable to students who are cultivating excellence consciousness, and pursuing knowledge and truth.

### 5.3 *Curriculum system highlighting the excellence consciousness*

Curriculum is the process of teachers teaching and students learning, which is the main carrier of talent training. It largely shows the orientation and key points of education and the knowledge, ability and quality that students are enjoying. A curriculum of the excellence education is set according to the requirements of excellence education and consciousness training.

On the one hand, practices and knowledge are two essential contents of the excellence education. It should consist of adaption to the new era, meeting career requirement, and improving practical abilities at important places. Also, we should pay attention to setting interdisciplinary curriculums, in order to cultivate interdisciplinary talents. On the other hand, the curriculum should be an effective means of students' realizing the mental excellence and improving their quality of life.. Courses include not only professional knowledge, but also consciousness training, such as e professional spirit, professional ethics, social responsibility, team work spirit and communication ability.

Course system setting should be scientific and rational, which is beneficial to awakening students' excellence consciousness, urging them to pursue the truth with strong interest. In the practical experiments, the new curriculum motivates students to study with positive, upward and striving spirit.

### 5.4 *Implementing personalized teaching to realize general excellence*

The purpose of the program is to train outstanding talents with consciousness of excellence. Excellence is the general character between training objectives and training objective in common[5]. While the generality is the same, a respect for individual development and adoption of the individual teaching model are necessary, because "Generality contained in all personality, no personality no generality". It is the encouragement of personality that considerably shows person's unique value and outstanding abilities.

The origin and destination of education are people, but there are differences and uniqueness between everyone. So, it is impossible for education to standardize production. The excellence program should more advocate the matters of personality. Only by cherishing an individual character can the stimulation of personal excellence consciousness be realized. Based on the consciousness, students bring full play to individual's potential and innovation, and achieve the excellence goal.

## 6 CONCLUSIONS

The excellence programme is not only training professional mechanical talents, but also cultivation of excellence consciousness. The current university students are lacking the consciousness of excellence rather than the professional knowledge, basic ability, brilliant brain or high IQ[6]. Universities' education model should be based on the awakening of excellence consciousness, combination of professional abilities and morals, and bound practical experience with theory innovation. Only in this way, can students be cultivated to become the outstanding youths in the new era. Excellence program helps students awaken the excellence potential and realize self- transcendence.

## AUTHOR INTRODUCTION

Liang Xiao (1984–); Male; Han; Pingshan County, Hebei Province; Information Engineering School of Wuhan University of Technology; teaching assistant; master; mostly engaged in the ideological and political education work and study.

Lang Kun (1984–); Female; Manchu; Qinghuangdao City, Hebei Province; Management School of Wuhan University of Technology; master; mostly engaged in the ideological and political education work and study.

**Supported by** the Teaching Research Project of Wuhan University of Technology (WUT: 2013053), the Special Project on Quality Education of Scientific Research Funds (WUT: 2013).

## REFERENCES

[1] Wang Bao-xi, Reflections on the Carrying Out of the "Plan for Producing Excellent Engineers" [J]. Journal of Higher Education Management, 2012(1): 15–19. (In Chinese).

[2] Wang Qiu-ge, On Education and Man s Excellence [D] 2007, Henan University. (In Chinese).

[3] He Shu-man, Consciousness- the basis of excellent success [J]. Modern Special Education, 2000(6): 45–47. (In Chinese).

[4] Nie Hui-ping, The Constitution of Talent Standard [J]. Commercial Research, 2006(18): 70–72. (In Chinese).

[5] Zhang Qing-jie, Excellent education excellent talents and excellent life, http://www.wutnews.net/news/news.aspx?id=55224 2011. (In Chinese).

[6] Long San-ping, Initial Analysis on Cultivation of Contemporary College Students' Success consciousness [J]. China University Students Career Guide, 2007(14): 68–69. (In Chinese).

*Frontiers in Computer Education – Wang (ed.)*
*© 2015 Taylor & Francis Group, London, ISBN 978-1-138-02797-8*

# A teaching model inquiry of innovative practice team based on the Conceive, Design, Implement, Operate (CDIO)

ShangFu Hao, BaiLi Sun & YuanQiang Wang
*Information Science and Engineering College, Hebei North University, Hebei, Zhangjiakou*

ABSTRACT: In this paper, taking Hebei North University innovative practice team as an example, a new teaching model is set forth which is based on the concept of CDIO engineering education, a student-centred, and actual project-object. Practice shows that this kind of teaching model can improve students' scientific literacy, exercise students' exchanges and cooperation capacity and shows that this is one of the effective ways to cultivate innovative, complex and applied talents.

*Keywords*: CDIO; iInnovative practice team; teaching model

## 1 INTRODUCTION

At present, China's higher education in professional settings, training programs, teaching ideas, teaching staff and other aspects has a serious disconnect phenomenon from social needs and enterprise products. In March 2014, the Vice Minister of Education Lu Xin said that China will revolutionize higher education adjustment, that more than 600 local colleges will gradually be turned to vocational and technical colleges, providing modern vocational education that focuses on training engineers, senior technicians, high-quality labourers, etc. [1]. The general curriculum modules in colleges and universities emphasize the basic theory of learning, relatively lacking professional course modules and engineering practice modules. During traditional teaching, teacher is as the main body, materials are for the centre, weakening the position of students, ignoring teamwork capacity and innovative practice ability.

CDIO stands for Conceive, Design, Implement, Operate which is proposed by Professor Edward Crawley in MIT [2]. The CDIO engineering education concept mostly takes a product, production process and a product life cycle as a carrier, through the process of product demand analysis, principle design, prototype implementation until the late running to firm students' foundation of scientific knowledge, develop their lifelong learning, teamwork and system running capabilities, and cultivate students' comprehensive quality ability to construct, design, implement and run a kind of product in enterprise and social environment [3].

## 2 THE TEACHING MODEL INQUIRY OF AN INNOVATIVE PRACTICE TEAM BASED ON CDIO

An innovative practice team means that under the guidance of teachers, students participate in project development. According to subject direction, teachers absorb students who are interested joining the research group, assigning students to develop modules of a subject, and students implement project in accordance with the specifications so as to achieve the purpose of joint research of teachers and students. With the platform of province key discipline of medical informatics, Information Science and Engineering College of Hebei North University has built more innovative practice teams. Through practice and exploration, the graduates from innovative practice teams gained valuable skills in technical theoretical knowledge, personal problem-solving skills, teamwork and project system capacity that practice the engineering education concept of the CDIO.

### 2.1 *Engineering practice and specialized knowledge learning*

The university professional training programmes are usually step by step and traditional, which cannot meet the needs of research team.

First, the teachers start teaching straight after their graduation without the experience of the whole product development process, lacking the knowledge of social and enterprise skills, and the ability of engineering practice training, although possessing in-depth knowledge on a single discipline. The CDIO concept makes higher demands on teachers, in a sense that teachers need to improve the ability to integrate theory with practice, involved in the related enterprises for engineering practice, and be able to cooperate with businesses in related fields. For nearly three years, the teachers in the team of Information Science and Engineering College have been involved in dozens of large domestic and foreign software companies and electronic technology companies for study and practice (eg. U.S. Google IT companies, Oracle (China) Software System Ltd., GTA IT Co., Ltd., ZhongLian Information Industry, Taiwan Chun Original Technology Co., etc.) which have laid the practical foundation for guiding the team building.

Second, students of the team are in different grades, different professions and with different interests which cannot fully meet the needs of professional theoretical knowledge on a subject. According to the CDIO education programmes and subject contents, teachers assign the related learning contents in accordance with students' interest orientation and professional characteristics. During the learning process, teachers communicate with students regularly, answer their questions, guide students through a clear research area, subjects' contents and the corresponding development prospects, that make students participate in various research activities in their spare time.

In order to make use of professional knowledge and practice skills effectively in order to achieve the purpose of innovation, a united, harmonious and cooperative atmosphere must be built within the team, giving full play to the students' individual interests, potential abilities, and ambitions. With this kind of atmosphere, each member will integrate himself into the team in order to develop their interpersonal skills and the ability to adapt to different environments.

### 2.2 *Strengthening team engineering practice conditions*

Research laboratories at universities are important places for cultivating innovative, complex and applied talents, and also the incubators bases for excellent engineers and scientific research achievements. The CDIO concept points out that a single experimental curriculum teaching is difficult to cultivate students' innovative ability, so it is necessary to integrate related laboratories and build new research practice labs in order to develop the concept of engineering education. With the support of the college, the research team of the Information Science and Engineering College has established the Internet of Things research lab, a virtual hospital lab, a mobile internet lab and a large data lab, etc. Using these large laboratory equipment, students can combine theory and practice better, knowing trends and the latest developments of the project. Meanwhile students' professional knowledge will be expanded as well.

Scientific research is the process for team members to gradually improve professional knowledge and engineering education. While implementing these distinctive scientific experiments, students' abilities of identifying and solving problems are developed gradually, and the product development specification becomes habitual. Through the practice in the research laboratory, undergraduates' research literacy of truth-seeking, realistic, enterprising collaboration and dedication is cultivated gradually, laying the foundation for becoming an excellent engineer with corporate standards.

### 2.3 *The team carries out comprehensive innovative practice in accordance with "tutorial system"*

Professor Zha Jianzhong introduced the CDIO model to the "learning by doing" study approach, and proposed that "learning by doing" method will be one of education reform strategies to cope with the industrial development demands for innovative engineering talents under the economic globalization nowadays [4]. Here the concept of "learning by doing" and the project-based learning model are introduced into the innovative practice team, advocating the student-centred to develop the engineering practice ability-target teaching mode. Referencing the CDIO syllabus, "we will provide students with a rich design and implementation of the projects, so that students grasp the theory knowledge better in case of practice and active learning." [5] Within the existing basic knowledge and term limit, through the actual project driving, students' ability of identifying problems, solving problems and knowledge application is trained. Meanwhile, by project cooperation and exchange, students' personal qualities are also improved which makes them develop all-round in engineering practice.

Currently, there are medical information technology teams, the "Internet of Things" team, a virtual hospital team, a mobile internet team and a large data team, etc. in our college. Each team is composed of backbone teachers, young teachers, senior students and junior students forming an echeloned team. The backbone teachers directionally guide young teachers, and both of them guide team students to conduct the research practice, choose the right projects, and counselling students in order to determine the practical and innovative practices.

The following example is for the Internet of Things team to illustrate the innovative practice team's work. Around the project design, the necessary knowledge and skills, being oriented towards engineering practice, the two lines of hardware development and software development are combined organically in a team, guiding students to learn and master the practical ability to form an effective whole. After three years of practice, a practical knowledge system and engineering skill training scheme are formed in the team as shown in Figure 1.

For the hardware direction, it is carried out around hardware development of Internet of Things and information transmission. The specific line is shown in Figure 2.

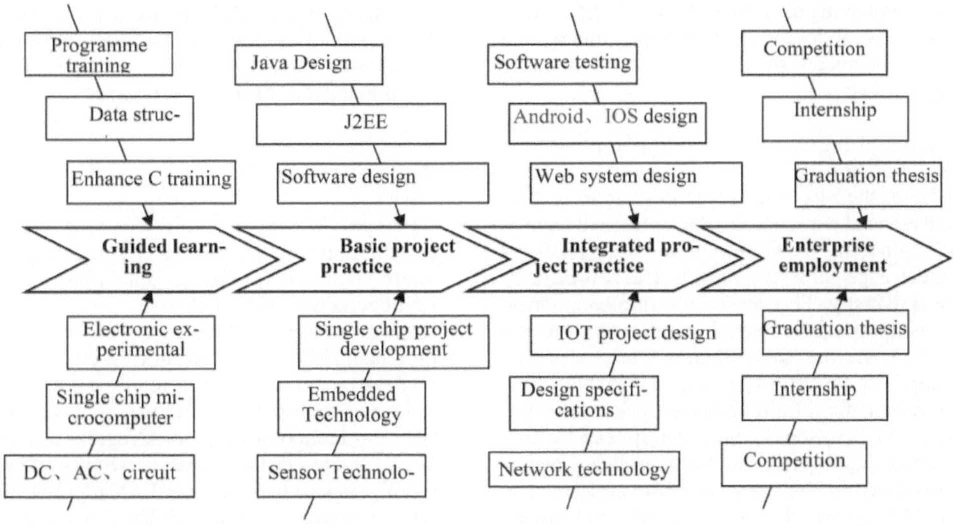


Figure 1. The diagram of practical knowledge system and engineering skill training.

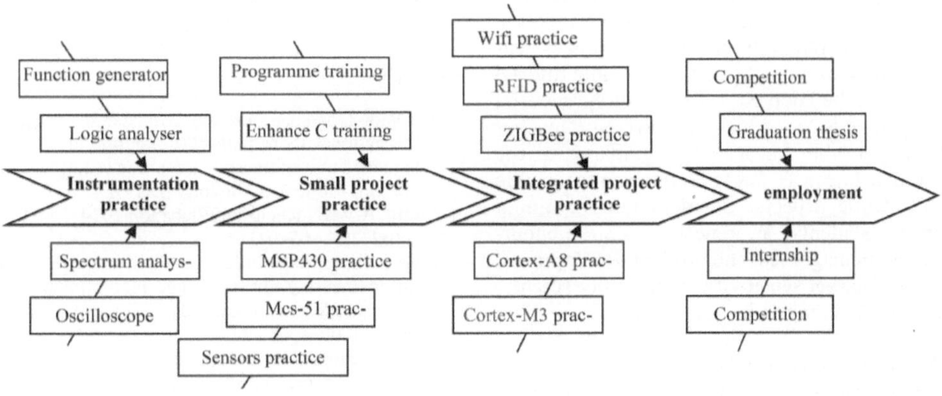


Figure 2. The diagram of hardware development knowledge system and engineering skills training based on IOT.

### 2.4 *Team emphasis of students' science and technology competition*

Science and technology competition provide a platform to display the science and technology work that is developed by a teacher-student team, and also being an opportunity to test the team's achievements in scientific research. It encourages students to develop thinking methods of identifying problems, analysing problems correctly, and dealing with problems as well. Science and technology competition also cultivates students' interpersonal skills of cooperating with instructors and classmates, gradually achieving the goal of solving scientific problems independently [6–7]. Our college actively encouraged teams to participate in domestic and foreign high-level scientific research competitions. The teams prepared research works carefully, took part in many competitions that were students-centred, and have gained excellent results. In some influential technology contests, such as China Software Cup, the Students Software Design Competition and Bi Sheng Cup, the National Electronic Innovative Design Contest, the team got two first-prizes; in the Internet of Things innovative applications designed Competition of National Colleges, Hong Kong, Macao and Taiwan Students Computer Application Competition, and Chinese Students challenge of Google Android application development, the team got three second-prizes; in smart home APP application design contest, the team got the third prize. In recent years, the college teams take scientific competitions into talents training programs, making great efforts to cultivate students' innovation, creative enthusiasm, and inspire their creativity continually.

### 2.5 *Existing problems*

After nearly three years of practice, the innovative practice team of Information Science and Engineering College has made some achievements, having a certain understanding of the concept of CDIO engineering education, and having also set a suitable CDIO teaching practice scheme for the team. However, there were still some problems during the practice, mainly in the following aspects:

### 2.5.1 *Teachers' engineering practice ability cultivation*

Most of the teachers have certain achievements in their respective areas of expertise, but lack practical experience and engineering experience in social and business environment, especially for the young teachers directly from the university. The key to the implementation of CDIO is the need for related engineering practice capabilities. Therefore, strengthening teachers' practical training is our top priority. One of it is that teachers take part in short-term training in enterprise; the other is that new teachers should enter enterprises for one year's engineering experience, and then begin teaching at university. In this way, teachers can know the throughout life cycle of the product which will broaden their horizons of engineering studies [9].

### 2.5.2 *Undergraduate business tutor appointment*

In our country, doctoral, master's academic evaluation and dissertation accreditation are all carried out by academic staff and entrepreneurs, so that double tutorial system may also be adopted to foster entrepreneurial engineering talents, in which a university teacher would be responsible for academic guidance of the project, and a business tutor would be responsible for non-academic aspects. Business tutors' joining can promote students' awareness of entrepreneurship, business management and product development, enhance their market sensitivity, opportunity seizing, and interpersonal communication abilities.

### 2.5.3 *Research project credit system*

There are credits for some practical aspects in college's training programs such as compulsory experiments, curriculum design, etc.; while no credits in other practical ones such as various scientific and technology competitions, academic papers, patents, and so on. If there are credits for these technological activities, students' research enthusiasm will be incentive greatly; furthermore if there is also a minimum credit requirement for research, an impact on all students' scientific literacy and innovation will be produced.

## 3 CONCLUSION

After several years of research and practice, the innovative practice team has built a suitable CDIO engineering education outline, which is student-centred and the actual project-object, expanding the practice teaching contents, improving students' scientific literacy, arousing their enthusiasm and fully inspiring students' activity. Through the practical research training, the team's knowledge of architecture has also been perfected, the team's innovation has been improved, and the students' research interests, personal skills and sense of teamwork have all been improved.


## ACKNOWLEDGMENT

- Corresponding author: Xiao Zhang School of Information Science and Engineering Hebei North University, Hebei, Zhangjiakou, China, xz1965cn@yahoo.com.cn
- Higher Education Research Project in Hebei Province, 2011: "The Study and Practice ofTeacher-Student Teaching Research Team" (GJXH2011-25)
- "12th five-year-plan" in Rural areas of National Science and Technology Support Program Sub-project, 2012: "Integrated Demonstration Study of three-tier Health Service in Rural Area" (2012BAJ18B08)
- Zhangjiakou Science & Technology and Earthquake Bureau Project, 2014:"Fixed Assets Management System Development Based on Internet of Things Technology (1321019B).


## REFERENCES


[1] Zhang Guangkai. Deputy Minister of education said the national more than 600 undergraduate course colleges and universities will transition vocational education account for 85% of the total [EB/OL]

[2] http://www.guancha.cn/Education/2014_05_10_228396.shtml. 2014-05-10

[3] Gu peihua, Shen minfen, Lu xiaohua, etc. Re-recognized the engineering education: The international CDIO cultivating mode and method [M]. Beijing: Higher Education Press, 2009

[4] Wang gang. The CDIO engineering education mode of reading and thinking [J]. China Higher Education Research, 2009(5): 86–87

[5] Zha Jianzhong. On CDIO Model under "Learning by Doing" Strategy [J]. Researches in Higher Education of Engineering, 2008, (3): 1–6

[6] Gu Peihua, ShenMinfen, etc. From CDIO to EIP-CDIO: A Probe into the Mode of Talent Cultivation in Shantou University [J]. Researches in Higher Education of Engineering, 2008, (1): 12–16

[7] Sun Cheng. Exploration and Practice of Innovation Education in Science and Technology Competition [J]. Journal of Guangdong University of Technology (Social Sciences Edition), 2008 (7): 148–149

[8] Feng Lin, Zhang Wei. Undergraduates' Innovation Experimental Program under the View of"Quality Project" [J]. Research and exploration in laboratory, 2008, (6): 27–28

[9] Zhang Yuecong. A Research on Cultivating University Students Quality of Innovation in Extra-curricular Activities of Science and Technology [J]. Higher Education Forum, 2009 (4): 72–74

[10] Duan qingru, Kan Lianbao, Wu guozhong, etc. CDIO education mode sinicization research review [J].Heilongjiang Researches on Higher Education, 2012(10): 33–37

*Frontiers in Computer Education – Wang (ed.)*
*© 2015 Taylor & Francis Group, London, ISBN 978-1-138-02797-8*

# Entrepreneurship education research in college and university based on tacit knowledge theory

Na Chen
*Dean's Office, Shijiazhuang University of Economics, Shijiazhuang, China*

XiaoQin Cai
*Center for Scholarship Dissemination, Shijiazhuang University of Economics, Shijiazhuang, China*

ABSTRACT: This thesis states the tacit knowledge theory and relationship between the tacit knowledge theory and entrepreneurship education based on the working experience of the author in university and some related document literature. Then it goes on to discuss the problems of China's entrepreneurship education research in colleges and universities in the present stage based on tacit knowledge theory. And it expresses the author's opinions on the development of China's entrepreneurship education research in colleges and universities in terms of the idea and goal of entrepreneurship education, teaching method, curriculum system, practice platform, and social support mechanism, etc.

*Keywords*: Tacit knowledge theory; College and university; Entrepreneurship education

## 1 INTRODUCTION

From the beginning of the 21st century, China has begun to implement the entrepreneurship education. At the present stage, the entrepreneurship education has been carried out in some considerable scale and achieved preliminary results. The entrepreneurship education, a subject with strong practical characteristic, aims to cultivate the students' entrepreneurial ability. The entrepreneurship education has become an important concept and component of China's higher education. How to carry out the entrepreneurship education and improve college students' entrepreneurial ability has been a hot topic researched extensively among scholars. Some scholars have reached a consensus in their researches based on tacit knowledge theory. The exploration of tacit knowledge theory and the research on it has to some extent promoted the development of the entrepreneurship education in colleges and universities. However, it is not easy to put the theory into practice. The entrepreneurship education research in college and university based on the tacit knowledge theory need a process to be established and implemented. There are still some problems in the entrepreneurship education research in college and university.

This thesis attached emphasis to research on problems of China's entrepreneurship education in colleges and universities in the present stage based on tacit knowledge theory. And it presents the author's opinions on the development of China's entrepreneurship education research in colleges and universities in terms of the concept and goal of entrepreneurship education, teaching method, curriculum system, practice platform, and social support mechanism, etc.

## 2 TACIT KNOWLEDGE THEORY

In 1958, Michael Polanyi, the British physicist and philosopher, in his "Personal Knowledge", states for the first time the tacit knowledge theory. Polanyi believed the human knowledge could fall into two categories: the first category is called explicit knowledge, which could be described by written words, chart and mathematical formula and could be combined, stored, looked for again and converted into related practice by different mechanism. It is characterized by normalization and systematization. It is easy to communicate and share with people that explicit knowledge. Another category, which is called tacit knowledge, is rooted in individual and his mental model. By contrast with the former, it is difficult to be code and communicate with. Technical, restricted by circumstances, it belongs to personal experience which is hard to share with others. Highly personal and hard to standardized are characteristics of tacit knowledge.

Compared with the explicit knowledge, it is difficult to clearly express or to formulate the tacit knowledge. By some ways, it can be put into words. It is a highly personal knowledge which is individual and private. It has not been or never could be described via formulation, but could only be combined by unsystematic fragmentary knowledge unconcerned by people, which could be understood in an integral manner. To acquire the tacit knowledge, compared with the explicit one, more practice is needed for the subject of knowledge, which can be interpreted as an experience that demands person in the scene or an understanding process. To grasp the skill well and to realize the deep connotation of the tacit knowledge, it is necessary for the subject of knowledge to practice and practice again. Meanwhile, in the process of acquiring tacit knowledge, some uncertain or diversified factors may restrict the subject of knowledge. The mutual conversion between tacit knowledge and explicit one could only be accomplished under some certain condition.

## 3 RELATIONSHIP BETWEEN TACIT KNOWLEDGE THEORY AND ENTREPRENEURSHIP EDUCATION

The research on tacit knowledge theory has opened up a new perspective of entrepreneurship education, and promotes the development of entrepreneurship education. To do a deep research on tacit knowledge theory and look for new method of entrepreneurship tacit knowledge are of great importance for the development of entrepreneurship education research in college and university.

The past psychological theory believed that, knowledge and ability belong to two categories which interact with each other. While, Robert Sternberg, the American psychologist believed there was an overlap between knowledge and ability. This overlap is just the tacit knowledge. Therefore, the cultivation of entrepreneurial spirit and entrepreneurial ability to a large degree comes from the generation and acquisition of entrepreneurship tacit knowledge. A considerable part of entrepreneurial abilities belong to the category of tacit knowledge, such as to grasp the business opportunity, to identify entrepreneurial project, to form the business groups, to grasp the entrepreneurial inspiration, to integrate the resources and to coordinate the relationships, etc. The main goal of entrepreneurship education research in college and university is to cultivate the entrepreneurial abilities of college students, rather than to view the survival and growth of pioneering enterprise as the only target.

The theory that the knowledge type decides the course type and teaching method has been recognized extensively by educators. There are lots of differences between the explicit knowledge and tacit knowledge in terms of educational aim, teaching content, teaching method, evaluation standard, etc. Therefore, during the implementation of the entrepreneurship education, the traditional teaching method which gives priority to lecture in class has met with great challenge. The tacitness, individuality, situationality, integrity and cultural feature of tacit knowledge demand the subject of knowledge to take part in the comprehensive practice activity under certain situation as a supporting platform whether in the stage of acquiring knowledge or in combing and converting stages. In this way, the individual can comprehend and transform the world. Thus, entrepreneurship education in college and university should be implemented on the basis of practical knowledge rather than on the basis of theoretical knowledge, which is a starting point to ponder over comprehensively the entrepreneurship education system in college and university.

## 4 PROBLEMS FOR THE ENTREPRENEURSHIP EDUCATION FOR COLLEGE STUDENTS BASED ON TACIT KNOWLEDGE

### 4.1 *The tacit knowledge theory has not infiltrated into the concept of entrepreneurship education*

Influenced by the traditional teaching method, and restricted by the traditional teaching resources, entrepreneurship education in colleges and universities usually put emphasis on the teaching of explicit knowledge rather than tacit knowledge. Under the influence of such education concept, most of colleges and universities set up the training system of entrepreneurship education according to the transferring pattern of explicit knowledge, and treat entrepreneurial knowledge and ability as a kind of explicit knowledge. The acquisition and transformation of tacit knowledge has not been given in-depth understanding. The entrepreneurship education has not been well combined with discipline construction and talent training scheme, and has not conformed to the college students' growing pattern. Besides, some one-sided ideas have derived from this entrepreneurship education concept, such as, entrepreneurship education in colleges and universities are nothing but entrepreneur education or entrepreneurship education aims to cultivate the students into business bosses, etc.

### 4.2 *The teaching method based on tacit knowledge has not been fully implemented*

According to the features of tacit knowledge, the explicit knowledge acquired by the students from the receiving-type teaching is dispersed, which requires the subject of knowledge to integrate with the tacit knowledge acquired from experience-teaching and inquiry learning approach. Thus, in entrepreneurship education, the experience-teaching and inquiry learning approach which call for subject of knowledge to take part positively are good for the college students to acquire tacit knowledge for entrepreneurship. Many colleges and universities have paid attention to the importance of the subject of knowledge's participation in entrepreneurship education. However, the extent and depth of such participation have not been achieved well enough. The main teaching method has still been receiving-type approach. The subject of knowledge has not been fully engaged in teaching process. In short, the above mentioned factors lead to an insufficient acquisition of the tacit knowledge of entrepreneurship education of college students. The college students will not be guaranteed to be able to transfer the knowledge into abilities in experience-teaching environment, which harms the effect of entrepreneurship education.

### 4.3 *The practice link which reflect the internalization and transformation of tacit knowledge is weak*

Due to the fact that tacit knowledge is understandings and experiences acquired by subject of knowledge via participating in and doing practice in specific situations, practice is the primary channel for college students to gain the entrepreneurial ability. In recent years, some colleges and universities have set up several entrepreneurship training base by cooperating with enterprises, trying to create a real market scene offering tacit knowledge of starting a business. By doing that, the college students' entrepreneurial ability could be transferred and improved. At present, most of the college students lack opportunities of entrepreneurial practice. But owing to the limitation to the scale of school training resources and practice base outside school, most of the entrepreneurship education in colleges and universities has been carried out in campus, which allows only a small number of college students to participate in the training. More often than not, only a very few people are starting an undertaking, while the most are just waiting and seeing.

### 4.4 *Multidisciplinary entrepreneurship education curriculum system which reflect tacit knowledge has not been implemented*

The integrity, diversity and complexity of tacit knowledge of entrepreneurship decide the multidisciplinary feature of entrepreneurship education curriculum system. In recent years, some colleges and universities have come to realize the importance of entrepreneurship education curriculum with multidisciplinary and open features. Diverse and interdisciplinary curriculum resources and teachers have been introduced to educate and train the students' entrepreneurial ability in a comprehensive way. In general, the entrepreneurship education curriculum system in most colleges and universities still features a strong disciplinary independence. The entrepreneurship education curriculum system has not reflected the integration of entrepreneurship education with professional education, scientific education with humanistic education. The teachers of entrepreneurship education are not diverse. A separation between entrepreneurship education and general education, professional education has greatly influenced the students' acquiring of tacit knowledge of entrepreneurship education.

## 5 THE DIRECTION OF DEVELOPMENT OF CHINA'S ENTREPRENEURSHIP EDUCATION IN COLLEGES AND UNIVERSITIES BASED ON TACIT KNOWLEDGE

### 5.1 *To ensure the concept and goal of entrepreneurship education based on tacit knowledge*

Colleges and universities should notice the features and follow the rule of tacit knowledge in its formation and break the traditional training model of entrepreneurship education, integrating tacit knowledge into the concept of entrepreneurship education. And a scientific and comprehensive theory is needed to instruct entrepreneurship education. An innovation mechanism including system preparation and resource integration is in urgent need. The traditional teaching methods featuring transmitting of explicit knowledge have to be changed to advocate the participation of subject of knowledge. By this way, combined with an interdisciplinary and open curriculum system, a practice platform, plus various resources, a training system aiming at cultivate college students' innovative spirit; innovative consciousness and innovative ability could be implemented to accord with the acquiring and transferring pattern of entrepreneurship education tacit knowledge.

### 5.2 *To adopt the teaching method based on tacit knowledge*

In order to improve the college students' entrepreneurial ability based on tacit knowledge, students' active participation and firsthand experience should be attached emphasis to in the entrepreneurship education. More real or simulated scenes should be created for students and more experience-teaching models should be adopted. Besides that, some other teaching methods or approaches should also be taken into consideration. Such as case teaching method or master-to-apprentice mentoring system can both be adopted in classroom teaching.

Case teaching is an effective way to transferring tacit knowledge of entrepreneurship education. The selected business cases explained by teachers in simple languages can reproduce the concrete problem situation in entrepreneurial process, which enables the student to more easily grasp necessary knowledge to start a business and to comprehend the entrepreneurial skills from this. Master-to-apprentice mentoring system can be also applied to learn tacit knowledge of entrepreneurship education. Some successful entrepreneurs will be employed and invited to act as mentors for college students who can understand the entrepreneurial skills through the interaction between the masters and apprentices, gain all-round instruction in entrepreneurial activity process. Those employed entrepreneurs can both teach entrepreneurial knowledge and share experience of solving problems in starting a business.

### 5.3 *To establish an open and interdisciplinary entrepreneurship education curriculum system*

Entrepreneurship education should be linked with individuality education. College students' individuality, interest and learning characteristics should be respected. An multivariate, layered education model should be set up, which could be achieved by implementing general entrepreneurship education in junior grades and carrying out elite education in senior grades, and setting up a correspondingly open and multidimensional entrepreneurship education curriculum system. The interdisciplinary feature of entrepreneurship education curriculum meets the requirement of acquisition of tacit knowledge, which can help students to grasp the multidisciplinary theory and knowledge needed for entrepreneurship and even achieve mastery. Therefore, entrepreneurship education in colleges and universities should be based on interdisciplinary system, integrate various aspects of knowledge, organize teaching content revolving around the problems in the entrepreneurial process, and introduce multidimensional, interdisciplinary curriculum resource and teachers.

### 5.4 *To build entrepreneurial practice platform with multi-subject participation*

Entrepreneurial practice and scene simulation are main ways to acquire tacit knowledge of entrepreneurship education. Owing to the limitation to entrepreneurial resources and fund, in the primary stage of entrepreneurship education, colleges and universities can concentrate all their efforts on building incubator part in campus to breed, hatch and support the college students' entrepreneurial team. Having brought the mature entrepreneurship project into market, colleges and universities could make a further step by building a larger entrepreneurial platform united with enterprises, public welfare institutions and government. Some colleges and universities with better condition may make full use of their facilities to build Science and Technology Industrial Park in campus to offer support to the students' entrepreneurial team in their project selection, fund raising, market promotion and human capital, etc. Those schools will accumulate experience and resources step by step, search for cooperating institutes outside school and build entrepreneurial practice platform system with multi-subject participation to offer all-round, multivariate breeding service to students' entrepreneurial team.

### 5.5 *To form a mass-participation entrepreneurial supporting mechanism for college students*

As the tacit knowledge of entrepreneurship mainly comes from specific experience, only classroom activities and limited social practices can hardly guarantee the acquisition of it. It is the common responsibility of colleges and universities and the whole society to cultivate undergraduates equipped with rich entrepreneurial practical knowledge and abilities. Their success can promote employment and economic growth; create common wealth for the whole society. Thus, it is necessary to form a benefit sharing mechanism for entrepreneurship education in colleges and universities which is led by government and participated by the whole society. And it is important to construct an entrepreneurial environment encouraging entrepreneurship and tolerating failures, with government, colleges and universities, enterprises, and social welfare institutions, etc. participating in the activities to cultivate entrepreneurial abilities of students. A social service system can thus be constructed to give support to college students' innovative undertaking.

## 6 CONCLUSION

Faced with new opportunities and challenges of social development, the major task of China's entrepreneurship education in colleges and universities is to do research on the generation and transformation patterns of entrepreneurship tacit knowledge and to form entrepreneurship education concept and goal based on tacit knowledge, to build an entrepreneurship education system improving the college students' entrepreneurial abilities.

The relationship between tacit knowledge theory and entrepreneurship education is a problem demanding much more deep research of scholars. Due to the author's limited knowledge and experience, the statement on the relationship between the two has not covered all the points. The thesis presents the author's opinions based on the working practice in university and some reference documentation. Readers' valuable advice are welcomed and hoped.

## REFERENCES

Chai Xudong. 2011. University Entrepreneurship Education from the Perspective of Tacit Knowledge, *Research in Higher Education of Engineering* (1): 81–86.

Ding Guifeng & Wei Wenjie. 2013. A Research on Undergraduates' Entrepreneurship Tacit Knowledge Learning. *Global Education Outlook*(9): 87–93.

Jiang Wenliang. 2011. The Construction of a Tacit-knowledge-based System of Carve-out Education, *Journal of Yangzhou University (Higher Education Study Edition)* (6): 60–63.

Yi Kaigang & Sun Jingxu. 2013. Entrepreneurship Education in Universities Based on Tacit Knowledge Mining: Change of concept and Countermeasures, *Heilongjiang Researches on Higher Education* (11): 81–84.

Zhang Suwei. 2011. Plan to Explore the Implementation of Entrepreneurship Education for College Students in Hebei Local Colleges and Universities in View of Tacit Knowledge, *China Market* (48): 187–196.

*Frontiers in Computer Education – Wang (ed.)*


# Look at the existing problems of cultivating talents in Hebei Province for the use of foreign language—The necessity of adopting a new foreign language education policy

Limin Liu & Jinfen Yue
*College of Foreign Languages, Agricultural University of Hebei, Baoding, China*

ABSTRACT: In order to obtain details of college graduates in Hebei Province in the foreign language use and provide a data support in personnel training, the research group launched a survey of questionnaires among graduates and employers. The questionnaire results show the use of a foreign language differs in occupations and employers have different requirements for graduates' language ability. In view of this situation, it is necessary to consider the problems of educating language talents, and discuss the necessity and urgency of adopting a new foreign policy in teaching college English.

Foundation item:2013 year Hebei province, the development of Social Science Research (201303194)

## 1 INTRODUCTION

In order to get first-hand material of information on college students' use of foreign languages (mostly English) after graduation, a research group of five teachers and ten students from Agricultural University of Hebei conducted a questionnaire survey in the summer of 2013 in the two cities of Shijiazhuang and Baoding, among ninety five graduates who had finished college studies within the last three years. The questionnaire results showed that these graduates are quite different in using English, mostly connected to their respective professional directions, gender, and occupations. Whereas the overall situation is satisfactory, most employers have stronger requirements for English abilities from the recruited members. The Ministry of Education in 2010 promulgated the national long-term education reform and development plan (2010–2020), in which it clearly put forward the idea "to cultivate a large number of language talents with profound international vision, familiar with international rules, and active in international affairs and competition" [1]. According to the questionnaire, it is necessary to modify and revise the current foreign language education policy, in order to meet the requirements of cultivating talents in the changing society.

## 2 A STUDENT QUESTIONNAIRE

The student questionnaire includes options and self-statements. Ten questions are given to cover a wide range of fields, relating to respondents' gender, age, graduation school, time spent on learning English, evaluation of foreign language abilities, and use of foreign languages at work. Besides, the students are also required to state their views on foreign language use and their regrets in the learning process. Finally, they give suggestions to college students on how to learn a foreign language.

### 2.1 *The questionnaire results roughly fall into three categories*

#### 2.1.1 *Questionnaire results in five institutions*

In the five institutions surveyed, mostly schools, government organizations and so on, 25 students (100%) in order to get a title or to go abroad, or to visit or travel, will continue to use English (learning). When talking about regrets in learning English, 70% of the respondents choose to "master basic listening, reading, writing, and translation skills, while the actual use of English in real life is rare". As an advice to college students, 90% graduates think that "students should practice English more and learn the Western culture at school".

#### 2.1.2 *Questionnaire results obtained at eleven enterprises and companies*

In the survey of eleven enterprises or companies, mostly computer, clothing, medicine, real estate, automobile and logistics, the foreign language use has obvious practical features. Among the fifty respondents, there is a great relationship between the foreign language use and the nature of their work. Thirty two of them engaged in design, drug development, circulation, services, trade and other fields will use English 100%. Ten of them are engaged in assembling, packing, transportation, production, the ratio of using English is 20%. Eight of the administrative personnel will actively learn and use the language. When the fifty respondents are required to state their "regrets" in English learning, 80% of them think they only grasped the basic skills in learning, speaking, reading and writing, but rarely used them in reality. In addition, almost all of them expressed the wish for cultural training, customs, and beliefs of the foreign language. In the advice column, 90% of the respondents suggest "more learning of the spoken English and the Western culture at school".

#### 2.1.3 *Questionnaire results of twenty individuals and unemployed graduates*

According to the survey of twenty individuals and unemployed graduates, their foreign language use follow their career development, and next study for a further promotion. Interest in English learning starts from the very beginning at school, with two divisions of either continuing or quitting totally. Of the ten unemployed graduates surveyed, nine believe that it is necessary to learn English in order to find a job or pursue a postgraduate study. Of the 1ten self-employed graduates, six have vague ideas of English learning and the English they learned at school seems forgotten. Of them, four will not learn English unless traveling abroad or visiting.

### 2.2 *Characteristics of a foreign language use*

#### 2.2.1 *The nature of occupation determines the efficiency and extension of the foreign language use*

Graduates engaged in foreign businesses, such as the overseas sales department of Baoding Great Wall Motors, use English at work. Their abilities and knowledge become rich and extensive with their working experience. Graduates in organizations or institutions, such as teaching, secretarial, administrative and other fields, use English in a small range with low efficiency, but experiences like studying abroad or competing in a work situation can improve their initiative of learning English. Among the unemployed graduates and postgraduates, the foreign language use differs from man to man. English remains in use mostly as a tool and it seems to be of little value for intercultural communication.

#### 2.2.2 *From a "book type" to "practical"*

The questionnaire results show that companies engaged in foreign trade use English extensively, mostly for foreign trade services (reception, correspondence, negotiation, etc.), foreign language materials, studying abroad, report writing etc. Take Great Wall Motors and North China Pharmaceutical Factory as an example, among the twenty three graduates engaged in overseas trade, six graduates in North China Pharmaceutical have once gone abroad to participate in training or meetings, while fifteen graduates in Baoding Great Wall Motors are mostly engaged in automobile export and new product development. These twenty one graduates found difficulties in adapting to a foreign language situation, especially in oral communication and data translation, as well as grammatical structure of language data. When they are required to comment on "regrets", they all mentioned the cultural shock, namely confusion and embarrassment in intercultural communications. The knowledge they learned in English books becomes real and meaningful with their deeper involvement with a foreign culture. After using English for a while, they become very skilful and confident in language communication and reading accuracy.

#### 2.2.3 *(FL+professional) compound talents*

Compound talents (FL+professional) are essential workforce resources for China's international development. Among the ninety five respondents, self-employed and postgraduates excluded from the group, there are fifteen foreign language majors and sixty non-majors. Of the fifteen majors, eleven in national institutions, three in foreign trade correspondence and one full-time English interpreter. Of the sixty non-majors, fourteen were in institutions, twenty nine engaged in foreign trade correspondence, sales, logistics, product research and development, ten in packaging and transportation, and seven as the company's administrative class. Obviously, foreign language talents no longer equate "foreign language majors", and compound talents of foreign language majors and "professional + language" for non- foreign language majors mode, and are more in line with China's internationalization development. It should also be noticed that of the ninety five graduates, those frequently using English account for 42%, those occasionally using English account for 40%, and less than 20% of the graduates never use English.

## 3 AN EMPLOYER QUESTIONNAIRE

### 3.1 *Questionnaire design*

| questions | options | | | |
|---|---|---|---|---|
| 1. What's your requirement for foreign language use? | zero (0%) | normal (5%) | high (78%) | advanced (17%) |
| 2. Your evaluation on your staff's language ability? | Hard to say (5%) | Just-so-so (64%) | Good (28%) | Perfect (3%) |
| 3. At what level do you think your staff need improving? | International and intercultural consciousness (28%) | Listening, speaking communicative abilities, (66%) | Language skills in literature and writing (0%) | Oral interpretation and translation (6%) |
| 4. What's the major consideration in recruiting staff? | Basic professional knowledge (11%) | Professional ability foreign language ability (78%) | Coordination abilities (8%) | Political quality and professional knowledge (3%) |
| 5. What's your requirement for communicative ability? | Good at interpersonal communication (0%) | Good at international communication (0%) | Normal abilities in communication (5%) | Communication both nationwide and international (95%) |
| state your expectations on graduates in a foreign language. | Having international consciousness, fluency in a foreign language application, abilities of cross culture communication and knowledge, confidence and courage to accept challenges, innovation, patriotism and unremitting self-improvement spirit | | | |

### 3.2 *Analysis of questionnaire results*

The employers surveyed include government organizations, national institutions, big provincial companies, as well as foreign investment enterprises.

The above questionnaire indicates that these employers surveyed have different requirements for graduates of foreign language use. Companies with foreign businesses require more of English level graduates, while schools and the government require a relatively small number.

#### 3.2.1 *Great expectations of high-quality talents*

The employers surveyed had great expectations for graduates of a foreign language use. 78% of the employers required that graduates should obtain a high level of the English language ability, 17% required an advanced level of the English language ability, including listening, speaking, reading, writing and translation, etc. With the development of China's internationalization, many companies open international businesses, thus needing high-quality English talents.

#### 3.2.2 *General satisfaction with the English language abilities*

The employers surveyed felt satisfied with graduates' language abilities. 64% of the employers felt satisfied, while only 3% felt perfectly satisfied. On the one hand, China's college English language education has achieved success to some degree, on the other hand, there are still deficiencies in cultivating foreign language talents, such as excessive emphasis on knowledge and ignorance of practical abilities.

#### 3.2.3 *More specific demands for the English language abilities*

Requirements for graduates of the English language abilities tend to be more specific. 66% the employers required that graduates should have listening and speaking abilities, as well as the intercultural communicative competence. 78% of the respondents required that graduates have a strong professional ability and foreign language abilities in the recruitment. 95% of the employers required that graduates have higher domestic and international communication abilities. A compound type of talents with a professional + English and the international consciousness continues to sell well in the market, which also shows the cultivation of English talents in our country should take the market demand as a criterion, timely adjusting the direction of personnel training.

## 4 TWO QUESTIONNAIRES INTEGRATED TO MANIFEST PROBLEMS

### 4.1 *Reform out-dated foreign language education policy and emphasize culture in cultivation*

The formulation of foreign language education policy is restricted by two aspects of factors, the internal being the inherent law of foreign language education while the external being globalization and national development needs [2].

College English Teaching Requirements has been implemented for ten years, and with China's internationalization, some problems begin to appear. Among the ninety five graduates, those who often use English account for 42%. When asked about regrets and advice, the "cultural confusion" is most often mentioned. For example, graduates experienced embarrassment in dealing with foreigners, and knew little about the taboos, etiquette, customs etc. Teachers often put too much emphasis on basic language skills, and ignore the penetration of cultural knowledge, the result of which is that students are skilful in the language use but ignorant in the cultural background. The coexistence of language and culture has been for so long ignored, leading to the loss in intercultural knowledge in college English education [3].

### 4.2 *Lay solid foundation for talent cultivation and formulate global education policy*

With the rapid development of the Internet, the Chinese multinational enterprises and the strong appeal for China's participation in the international stage, China is in great need of a foreign language education policy, which is also the premise and guarantee to further promote the internationalization of China. In this regard, it is worthwhile to draw lessons from the USA, Japan, South Korea and other countries. America promulgated a language policy including a foreign language ability standard and the education funding system, which carries multiple educational goals to meet a deep political, economic target [4]. Japan, in implementing the foreign policy, gradually has shown "pragmatism" tendency. Students can study English according to their own needs, interest and ability. The aim is to cultivate "Japanese who can use English" [5]. The South Korean government has always seen the national English level as an important factor to participate in international competition and promoting the economic development. The former president Lee Myung-bak in his inaugural address proposed "to greatly enhance the school English education", emphasizing that "English ability is the key to encouraging international investments to Korean enterprises." [6] The internationalization of China makes it urgent to formulate policies to keep pace with the times. Based on curriculum requirements, a new policy is aimed at developing an international awareness and cultivating compound language talents.

### 4.3 *Combine schools, students and markets to cultivate English talents*

Cultivation of English talents is an effective guarantee of China's international development and an important scale to measure the national quality. "A national language popularization degree often reflects the international consciousness of its citizens, the degree of internationalization, and the development prospect of the future …[7]".

To cultivate talents, we must effectively combine schools, students and market demands. Demand-production-supply will make a sound circle to fulfill a national goal of talents cultivation. For example, in 2013, a Hebei pharmaceutical company specified requirements for intensive English in its notice of staff recruiting in the fields of chemical engineering and technology. North China Pharmaceutical Group and Peking University, Chinese Academy of Sciences, Meiji Seika, America Chiron Company, etc. have close cooperation in the establishment of Microbial Engineering Research Center, national drug antibody drug development of State Key, Laboratory of National Research and Development, and a postdoctoral scientific research workstation. Workers in Baoding Great Wall Motors, in the course of transition from the pickup to SUV, translated into English and studied "68 details of fine management in Toyota Business School". Requirements for English talents grow with internationalization.

### 4.4 *Guarantee the future development*

China's growing internationalization provides a solid soil for cultivating English talents, and has produced unexpected results. For example, Confucius Colleges have attracted more than 100 million overseas students today, compared with 30 million ten years ago. Chinese Bridge, a competition show for foreign students in China to compete in Chinese language, gets more attention today. Chinese multinational companies have been built overseas. Chinese Teachers Working Abroad greatly promotes exchanges between China and the world. Formulation of foreign education policy should adapt to the changed situations and further enhance and revise the old, rigid stereotypes. It is crucial for teachers to build confidence in their students and teach them the values of autonomous learning. A harmonious atmosphere should be built to encourage people to learn, use, and speak English happily.

## 5 CONCLUSION

Hebei province is located in the Beijing-Tianjin-Hebei Economic Zone. Its graduates render services mostly in this area. Talent cultivation cannot do without international foreign policy guidance, so universities and employing units should pay great attention to this fact, in order to adapt to China's current situation. Meanwhile, the Ministry of Education, as the ultimate authority on foreign education policy formulation and issues, should widely collect information on reforms and, in conjunction with experts, scholars, outstanding teachers and institutions, work at the formulation of a new education policy, effectively implementing people-oriented philosophies, and contributing to a rounded scientific development, with the aim to improve the development of the economy, society, and human beings.

## REFERENCES


[1] Central government. Plan of National Medium and Long-term Educational Reform and Development (2010–2020) [Z]. Beijing: Xinhua (Xu Yan, 2010.06).

[2] Meng Zhen. Considerations on the Foreign Language Education Policy [J]. Journal of Zhenjiang College, 2012 (4), p29: 3–5.

[3] Wang Wei. Reflections on Foreign Language Education Policy in China, [J], Journal of Hubei Radio and TV University, 2009, (5): pp. 117–118.

[4] Yang Dexiang. The Orientation and Planning of Foreign Language Education, [J]. Transportation University of Lanzhou (Social Sciences Edition) 2005, (2): pp. 113–116.

[5] Zhang Zhenai, Wang Kefei. Study on South Korean International Language Education [J]. Foreign Language Teaching and Research, 2010, (09): p346: 18–25.

[6] Zhuang Zhixiang, Discussion on Training of New International Foreign Language Talents [J]. Language World 2012, (2): p42: 4, 15.

[7] Meng Zhen. The Research Status and Development of Foreign Language Education Policy [J]. Journal of National University of the Inner Mongolia (Social Science Edition), 2012 (1): pp. 7–11.

*Frontiers in Computer Education – Wang (ed.)*
*© 2015 Taylor & Francis Group, London, ISBN 978-1-138-02797-8*

# The revision of college English curriculum requirements and talk about cultivation of college foreign language talent again—taking agricultural university of Hebei as an example

Limin Liu
*College of Foreign Languages, Agricultural University of Hebei, Baoding, China*

ABSTRACT: *College English Curriculum Requirements* has been issued for more than ten years until now. Under its guidance, university English teaching has cultivated a large amount of foreign language talent. The Agricultural University of Hebei has been strictly abiding by its provisions. With the internationalization of China and its increasing demand for foreign language talent, there is a need to revise the *Curriculum Requirements*. It can be used to correct certain teaching notions and guide the work in management and teaching.

Foundation item: 2013 year Hebei province, the development of Social Science Research (201303194)

## 1 INTRODUCTION

Since the expansion of college enrolment in 1999, in order to regulate college English teaching, the State Ministry of Education issued the college English curriculum requirements in 2004 (on trial), put forward three requirements of cultivating foreign language talents, that is, the general requirements, higher requirements, and even higher requirements [1]. With the progress of our education and higher education reform, the Ministry of Education in October 2007 formally promulgated the university English teaching requirements (hereinafter referred to as the requirements), clearly put forward the university English teaching goal as training the students "comprehensive ability of using English, especially listening and speaking ability, so that they can use English to communicate effectively in the future study, work and social contacts, and enhance their ability of autonomous learning, improve the comprehensive cultural literacy in order to adapt to the need of China's social development and international exchanges [2]."

Until today, curriculum requirements has been implemented for nearly ten years (from the start of trial), University English teaching has made huge achievements, but there are some problems. The Agricultural University of Hebei has been paying attention to English teaching of the students, the ongoing classroom practice and teaching reform, in order to cultivate the national and social needs of the talents of foreign languages.

## 2 MAKE CLEAR THE CONCEPTUAL ISSUES OF FOREIGN TALENTS AND IMPLEMENT THE CURRICULUM REQUIREMENTS

### 2.1 *Foreign language talents under college English education*

Traditionally, a talent is a brilliant man of wide learning. The Ministry of education in the Outline of the National Medium and Long Term Talent Development Plan (2010–2020) pointed out that the talent is "to have certain professional knowledge or special skills, creative work and make a contribution to the society. Talent is the ability of workers and high quality of human resources. Talent is the first resource in economic and social development of China [3]." When it refers to the foreign language talents, certain English professional knowledge or special skills are required.

Here, it needs to clarify the three kinds of foreign language talents. Only with a clear concept, will a definite object in view be set as to the talent cultivation.

#### 2.1.1 *Narrow sense of foreign language talents*

The narrow sense of foreign language talents. It refers to specialties in languages and literature, or professional translators. They are mostly engaged in a study of foreign languages, of Western literature or literary translation, of interpretation and other aspects of such works. They are the elite of foreign language talents, truly representative of a nation's foreign language level. Such figures, a modern foreign language teaching and research authority figures are, for example, Yang Xianyi, Dai Weidong, Hu Wenzhong, etc.

#### 2.1.2 *Broad sense of foreign language talents*

The broad sense of foreign language talent is the combination of modern English major and market demands. Personnel training should be meet market requirements. This kind of foreign language talents, mastering a foreign language, receiving education or training on foreign trade, secretarial specialty, tourism, etc., are engaged in language related occupations after graduation. They are among the most popular compound applied talents of foreign languages. Because of the specialty of targeted fields, compared with the characteristics of narrow foreign language talents, mostly engaged in academies and research, they are more widely distributed in society. These talents are the most vigorously trained versatile foreign language personnel at present in our country. This kind of talents are mostly engaged in foreign trade, translation and other social work, whereas in the academic circle they are seldom found.

#### 2.1.3 *Mass type of foreign language talents*

The object of teaching English at university are all college students. Their pre-school education gave them a certain foreign language foundation and university English education only added to training them into useful foreign language talents in our society. They use English skilfully in study, work and life, and contribute to our society with their talents. One of the guiding people of modern Enlightenment, shipping expert Yan Fu, China's supreme merit award winner, expert in "the two bombs and one satellite", and a scientist Tsien Hsueshen, are examples worth having in mind. They are not engaged in foreign languages, but have made great contributions by using a foreign language. This is what our country cultivates to render social services nowadays.

### 2.2 *English education at university should conscientiously implement the curriculum requirements*

#### 2.2.1 *Releasing goal*

Curriculum requirements was released with the intention to fully implement college English teaching reform and to meet the needs of talents cultivation in the state and society in the new era. The nature of college English teaching is summarized as "guided by foreign language teaching theories, and with English language knowledge and application of skills, intercultural communication and learning strategy as the main content, uniting many kinds of teaching models and methods into an integrated teaching system."

The Agricultural University of Hebei, as a provincial agriculture and forestry university, recruits large numbers of students every year. How to do well in college English teaching, is not only related to the overall development of the university, but will affect the overall quality of college students in Hebei province. This year's enrolment in school has reached 12230 students. In cultivation of these students, we should seriously carry out the curriculum requirements in English teaching, and fully implement the reform. With the teaching theory as the instruction, it is necessary to meet current national and social demands for foreign languages talents, which is also an obligation of every English teacher.

#### 2.2.2 *Three levels of requirements*

According to the curriculum requirements, English teaching at the university stage is divided into three levels, namely, the general requirements, higher requirements, and even higher requirements. This is the standard that non-English major undergraduates should reach after studying and practicing English at university. The comprehensive ability of three requirements includes: 1. listening comprehension; 2. oral expression; 3. reading comprehension; 4. writing ability; 5. translation skills; 6. recommended vocabulary. College English teaching material is written under the six specific capacity requirements.

The foreign language teaching material that the Agricultural University of Hebei uses is *New Horizons* (designed and written closely to the spirit of the curriculum requirement, whose language skills, language application, and the language culture embodies the basic requirements of the foreign language course, not only as an instrument of learning language, but also culture.) It provides a learning platform [4] for all the college students in learning English. University English teaching of Agricultural University of Hebei can be divided into three levels: undergraduate class two, class three and technical level, according to the principle of "teach students in accordance with their aptitude", which focus on teaching in a different culture and different direction, respectively, to the three kinds of talent requirements. The English class teaching of undergraduates class two is divided into two different levels, A and B. The two levels are determined at the beginning of the new semester after a mass English exam, and restructured after the semester ends. The purpose of construction and reconstruction is to adapt to the students' changing situation and to encourage and guide the students to learn initiatively.

#### 2.2.3 *Curricula of Agricultural University of Hebei*

The curriculum requirements put a particular emphasis on "giving full consideration to the cultivation of listening and speaking abilities", and giving students enough time and credit. Moreover, it highlights that the university English curriculum should "give full consideration to students' cultural quality cultivation and international cultural knowledge." This requirement takes into consideration the fact that the Chinese internationalization process has greatly accelerated and that

China hugely focuses on the internationalization of foreign language talents.

English teaching curricula of the Agricultural University of Hebei include the *New Horizons* series, namely, the reading and writing courses, an extensive reading course, comprehensive training, and an audiovisual tutorial. In order to improve students' listening and speaking abilities, in the year 2013, the extensive listening and vocabulary handbooks are added to the original teaching curricula. What is more, the school radio station also broadcasts listening programmes at certain times. Frankly speaking, a lot more work has been done to build a better course system in the cultivation of students' listening and speaking abilities. However, it needs to be pointed out that students' listening and speaking abilities will not naturally improve by doing some textbook homework, and they also needs teachers' guidance in the classroom. As to the cultivation of students' international knowledge and intercultural quality, our school also needs to make further research and improvement in the curriculum system, teaching requirements, and training of teachers.

#### 2.2.4 *Learning assessment and evaluation*

The Curriculum requirements clarify students' learning assessments, which includes a formative assessment and a summative assessment. Of the two, the formative assessment is introduced with more details and it claims that it "is conducive to students' autonomous learning process and effectively monitored". At the same time, the formative assessment can "use classroom activities and record of extracurricular activities, online autonomous learning record and other forms for the pro-curatorial, evaluation and supervision in students' learning process in order to achieve effective learning."

English evaluation in our college is a combination of a formative assessment and a summative assessment, a comprehensive evaluation system of teachers, classroom, and network. In university English final exam, a course score accounts for 20%, and other comprehensive ability test scores account for 80%, which, as a summative assessment, examine students' listening, reading, writing, and translation abilities. The course score usually comprises four results, namely, class attendance, students answering questions and positive attitude, quizzes, and homework completion. In order to prevent teachers from giving points that are too high or too low, each class usually makes a distribution ratio, such as the 90–95 accounts for 10%, 80–89 accounts for 40%, 70–79 accounts for 40%, 60–69 accounts for 10%, and specific adjustment is made according to specific circumstances, such as 0 attendance usually deserves 0, and so on. Internet testing (iTest) and a new platform of English teaching and learning system have been introduced since the year 2013. It not only helps students form autonomous learning but also helps teachers in the formative assessment to give students a scientific and a reasonable performance result.

## 3 AFTER TEN YEARS OF IMPLEMENTATION OF THE CURRICULUM REQUIREMENTS, ACHIEVEMENTS AND PROBLEMS COEXIST AND SOME EDUCATIONAL IDEAS STILL NEED ADJUSTING

### 3.1 *Ideas of education need to keep pace with the times*

Not only the school teaching management, but also the teachers and the students in a classroom all need to reject of the stale, narrow-minded beliefs and establish the scientific concepts of teaching and learning, that is "respect for talent, respect for individuality, and give full scope to the talents".

#### 3.1.1 *School administrators*

As a school administrator, one should be full of confidence to cultivate talents, to understand his/her students, and to work out the training plan for their development. Jiang Zemin emphasized on the importance of training talents, saying, "the competition of today's world and the future, fundamentally speaking, is the competition of talents. [5]" Recent years have witnessed a downfall in students' overall quality due to an expansion in enrolment, thus some schools become pessimistic in cultivating talents, and lack confidence in their own students. They show neglect in the teaching management and neither go to discover the problem, nor try to solve it. Also, in some schools, the phenomenon that one student gains more favours than another for no reason, greatly undermines the teaching quality, thus fairness or equality is hard to maintain.

#### 3.1.2 *Teachers, directors of class teaching*

As directors of class teaching, teachers should be good at discovering different personalities of students, and design respective teaching plans according to their achievements, personalization and their interest in English, avoiding "drawing water with a bamboo basket in vain" and "cannot see the wood from the trees", the two typical defects in modern education [6]. In fact, foreign language teachers at university English teaching have greater say in determining the success of the whole education. As the directors and participants of teaching, they can grasp the learning dynamics and mentality of students on time, and their teaching philosophy and methods will directly affect the cultivating effect.

#### 3.1.3 *College students as the training objects*

Contemporary college students, the training objects, have a common problem, impetuousness. On the one hand, a society or family influence affects them, on the other hand they lack a purpose for learning. The key of talent cultivation is to inspire the soul and self-confidence. University English teaching should instil in students the philosophy of life, "you are beautiful just the way you are", and cultivate their autonomous learning motivation as well as confidence in learning English.

### 3.2 *To update the teaching material and curriculum system*

Research on cultivation of foreign language talents has been going on for many years and have made success in the teaching reform. However, no matter how successful the teaching system and curriculum setting are, they need upgrading according to the market and social demands. The school should discard those outdated and inappropriate teaching materials and reform considerably based on its own specific circumstances. Blindly indulged in authoritative textbooks or successful curriculum, and totally forget its own reality, it can only cultivate students without social demands, or no use at all with only a name.

### 3.3 *To further implement education reform*

Foreign language teaching and research should be focused on the needs of the country and the personnel training. The purpose of the study is to find out and solve the problems, and ultimately to guide the teaching practice. At present, formalism and serious plagiarism exist in the foreign language study. Repeating word for word what others say, scratching the surface when serious study is needed, lack of innovation and visionary perspectives, and without hard-working, only to name a few. University English teaching reform will be divorced from reality and fall into the rigid ideas, if the education theories are to be separated from the teaching practice.

## 4 CONCLUSION

The curriculum requirements have been issued for more than ten years and achieved a very high performance. It is the guidance and compass of university English teaching and reform. In the development of Chinese internationalization and society today, universities need to revise the requirements, and appreciate their basic spirit, further promoting the reform of university English teaching, changing the teaching concept, and founding on cultivation of outstanding foreign language talents. Teachers, working on the frontiers of education, should put the cultivation of talent as their focus of work, and constantly improve their sense of responsibility and professional level, and give their support to the teaching cause in China.

## REFERENCES


Department of higher education, University English curriculum requirements [Z]. Shanghai: Shanghai foreign language education press, 2004.01.

Department of higher education, University English curriculum requirements [Z]. Shanghai: Shanghai foreign language education press, 2007.09.

Jiang Zemin, Jiang Zemin, anthology volume third [M]. Central Document Editing Committee of the people's Publishing House 2006.08:43.

The Central Government, the outline of the national medium and long term talent development plan (20102020 years) [Z]. Beijing: Xinhua (Introduction, 2010.06).

Zheng Shutang, a new vision of university English (reading & writing) [M]. Beijing: foreign language teaching and Research Press, 2008.03 (5).

Zheng Shutang, a university English teaching method research – Research on the current situation of university English teaching foreign languages circle four [J]. 1997.03 (5):23–26.

*Frontiers in Computer Education – Wang (ed.)*
*© 2015 Taylor & Francis Group, London, ISBN 978-1-138-02797-8*

# A discussion based on Problem Based Learning (PBL) in the teaching reform of the course in microcomputer principles and applications

XiYang Liu & XianMei Chen
*Hainan University, College of Information Science and Technology, Haikou, Hainan, China*

FuHua Ma
*Guangxi University for Nationalities, College of Information Science and Engineering, Nanning, Guangxi, China*

ABSTRACT: The Problem Based Learning (PBL) teaching method is a teaching method of a problem based learning. According to the PBL concept, in the teaching of the microcomputer principle and application course, a teacher designs a problem, encouraging students to ask questions, guiding the students' autonomous learning, discussions, and cooperation in groups. Thus, the students can improve their learning enthusiasm and coordination skills, and good teaching results can be achieved.

## 1 INTRODUCTION

In the information age, with the expanding knowledge and the rapid development of science and technology, socialization is more and more complex. This situation makes students learn more and become more knowledgeable, and many students began to question what to learn and how to learn it. Without correct guidance in study, students tend to lose interest and motivation. The PBL, a teaching method of problem-based-learning, is a good type of method in guiding students' learning. Recently, we used the PBL teaching method in the course of Microcomputer Principle and Application interface technology. The teaching practice revealed that teaching under the PBL mode in the course of Microcomputer Principle and Application has more advantages in order to improve current college students' learning interest, and to cultivate the students' thinking about control systems and their ability of applying these to solve actual problem. This method played a significant role in improving the teaching quality and its effects.

## 2 THE PRINCIPLE OF PBL TEACHING METHOD AND THE FEASIBILITY TO APPLY PBL IN THE COURSE OF MICROCOMPUTER PRINCIPLE AND APPLICATION

Problem Based Learning (PBL) is a problem-oriented teaching method and a student-cantered education way based on the real world.[1] In the late 1960s, Barrows, an American neurology professor pioneered this method at McMaster University in Canada. It was mostly used in medical education, but now it has become a popular teaching method around the world.

The PBL teaching mode emphasizes that the study is put in the complex and meaningful problem situation, and the problem is the starting point for a study. In the PBL mode, students are encouraged to ask questions which have no fixed solutions. After the students' autonomous learning, group discussions and mutual cooperation, students can analyse problems, set up hypotheses, simulate and do experiments, and solve problems. Finally, they can learn the scientific knowledge behind a problem. This process can make students realize that they have already known what they need to learn and try their best to solve a problem; they discover and use all kinds of information resources through a variety of ways in trying to solve practical problems. Meanwhile they can take the initiative to obtain opportunities in which to apply knowledge, which can stimulate the students' learning interest and enthusiasm, cultivate students' ability to master and apply knowledge.

Microcomputer Principle and Application course is a professional basic course in many colleges and universities currently, which occupies an important position in the specialty construction and course system.[2] Microcomputer interface technology is a hardware and a software application technology accompanied by the overlapping and mutual basis, which is the important technical foundation for students to learn subsequent courses, do graduation design and future work. The course content is made up of the theory and practice, which can be tedious, boring and abstract. The course has a high basic, advanced, cutting-edge, and comprehensive requirements. It also requires that students have strong practical abilities. In traditional teacher-cantered teaching mode, students are not clear about the course objectives. Therefore, they think that it has no practical use for their future study and employment.

At the same time, the teachers are usually feeling useless about this. So the learning of this course is based on the dilemma that 'students find learning hard, while teachers find teaching difficult'. The traditional teaching process ignored the students' cognitive main role, which gives a priority to teaching, and it is difficult to stimulate students' learning enthusiasm because of the lack of interaction. How to achieve the teaching goal of the Microcomputer Principle and Application course with abstract and complex content, broad knowledge and strong practicality in limited teaching time? The PBL teaching mode is especially suitable for the teaching of the course.

## 3 PBL MODE OF MICROCOMPUTER PRINCIPLE AND APPLICATION COURSE

The PBL teaching mode and the characteristics of the courses of Microcomputer Principle and Application put forward a new challenge to teachers. It not only requires teachers to have deep understanding of the concept of PBL and master skills, but also requires teachers to have a high level of professional knowledge and skills. Thus, teachers can blend the PBL teaching mode in the microcomputer principle and application class gradually and effectively. For the PBL concept, teachers present some project problems to students through typical instances of microcomputer interface technology, and guide students to find solutions in the process of finishing the project of microcomputer interface technology. Students study by themselves and cooperate with each other, as well as simulate and do experiments, in order to have a further consolidation of the knowledge about microcomputer principle and application, to master the implementation steps, to know each job specification and the matters of every step , to lay a solid foundation for the study of subsequent course and the graduation design, etc.

### 3.1 *Design of a problem*

According to the teaching target which is set, the problem should be complex, open, real and explorative;[3] both students' knowledge and skills should be considered, having a moderate difficulty and conforming to the general requirements of the syllabus.

### 3.2 *Organizing learning team*

Allow students are free to form several study teams, each team of six to eight people. They can solve the problem through mutual consultation and cooperation among team members.

### 3.3 *Guiding the team to determine the learning goals*

A teacher should guide every team to have an elementary discussion, which should include relevant factual information, a concept, knowledge hidden in questions, the groups' initial ideas for solving problems, etc. The purpose of the discussion is to promote each team to establish overall learning objectives and learning programs by team members' mutual agreement, simultaneously establishing learning goals and learning programs for each team member.[4]

### 3.4 *Independent learning*

In the guidance of learning goals, each member separately explores and tries to solve the problem. At this stage, the teacher first must provide relevant references and materials, including the Internet and the laboratory. According to these resources, students will actively try to get information, start to make hardware solutions, choose the corresponding electronic components, and formulate the corresponding circuit principle diagram.

### 3.5 *Reporting and counselling*

After each team self-exploration, teachers should organize team members to report the progress of each team, the existing problems, and the solution to the problems. Then, teachers give answers to students' questions, helping students have a better understanding of relevant knowledge, giving the necessary inspiration and guidance for possible solutions to the existing problems. According to the teachers' opinions on the solution, each team should make the appropriate adjustments, modifications and refinements at the same time.

### 3.6 *Simulation and experiment*

Students should verify the feasibility of the program through debugging the hardware system and software system by Proteus. Meanwhile, if we have enough good condition,we can organize the students in the lab to do it, or organize the student to weld the PCB, debug the system, and test the feasibility, which also can improve the practical ability of students.

### 3.7 *Discussion and summary*

Each team displays and shares their collected information and solution, then teachers and all teams analyse and evaluate each team's ability from the collected information, solutions, process requirements, advanced technology and coordinated cooperation among team members in solving the problems. All teams discuss together and exchange views, sum up the harvest and experience in the learning process, and think over the problems that appear in the process of solving problem.

## 4 IMPLEMENTATION OF THE PROGRAM OF PBL

Recently, when we taught a microcomputer Interface technique of the Microcomputer Principle and Application course, students had been in the state of being tired. In order to mobilize their learning enthusiasm, we implemented the PBL teaching mode. Thus, according to the actual situation of the students, we gave them questions as follows: Design a square wave generator within the scope of the knowledge using 8086/8088 as microprocessors (To reduce the difficulty, the output frequency of the square wave generator is unspecified). The course knowledge related to this issue including the Hardware system and Software Systems are as follows.

### 4.1 *Knowledge of the hardware system*

The Knowledge of hardware system are as follows:

(1) 8086/8088: Selection of the maximum and minimum operating mode, function of the pins, a physical address, a logical address, offset addresses, memory, and the interruption technology.
(2) 8255A: Working Principle, the distribution of port addresses, the content of work control word, C port reset / set control word.
(3) 8253: Working principle, the distribution of port addresses, selection of the working modes (six operating modes), the content of work control word; the Equation [2] about the initial count value is below:

$$N = \frac{f_{clk}}{f_{out}} \quad (I)$$

where $f_{clk}$ = frequency of the input clock pulse signal, $f_{out}$ = frequency of the output signal.
(4) Integrated operational amplifier: Non-inverting input, inverting input, virtual short, virtual off, negative feedback.
(5) DAC0832: Conversion principle, conversion accuracy, the choices of work modes (single buffering, double buffering, direct access). The Equation II between analog voltage output and the corresponding digit is below:

$$D_x = (U_x - U_2) * \frac{0FFH}{U_1 - U_2} \quad (II)$$

where $U_1$ = the upper limit of the voltage, $U_2$ = the lower limit of the voltage, $U_x$ = the output voltage.
(6) Oscilloscope.

### 4.2 *Knowledge of software systems*

The Knowledge of software systems are as follows: Instruction (including prescriptive statements and indicative statements), the format of assembler, programming methods of assembly language, 8255A's initial programming; 8253's initial programming, Proteus simulation software, emu8086 assembler software.

### 4.3 *Implementation*

Then the students were divided into several teams, each team of six to eight people. After guiding the teams to establish learning goals and learning programs, we allowed each team member to study independently. A week later, we organized the teams to report what they had done, then counselled and answered their existing problems, and guided them to make the appropriate changes and adjustments on solutions to the problems. After another week, we checked and accepted their achievements.

From the feedback, we found that the students developed two solutions: (1) Using 8086/8088 and 8253, and the working model 3 of 8253 and giving it an initial set value, and outputting a certain frequency square wave from the output terminal of 8253. Some students used the two timers / counters of 8253. One of the square wave output of the counter is another counter's clock pulse signal, which is used in order to improve the role of frequency division. (2) Using 8086/8088, integrated operational amplifier and DAC0832. Some students added 8255A between 8086/8088 and DAC0832. An analog current signal out of DAC0832 is converted into a voltage signal by integrated operational amplifier. According to the different settings of digital quantity, its output voltage is correspondingly high or low. The same delay subroutine is called to meet the requirements of the square wave. The output voltage signal is unipolar or bipolar square wave from a virtual oscilloscope on Proteus, a simulation software.

When talking about the learning feelings and experiences, the students all agreed that the PBL teaching mode improved everyone's interest and motivation in learning, helped further understanding on computer architecture, assembly language and interface technology, and had more direct perceptions of control principles of microprocessor and the workflow of some chips, and had preliminary understanding in comprehensive application of hardware and software systems. We all believed that the PBL teaching mode played an important role in cultivating the ability of solving problems, cooperation and coordination, and the training of overall quality.

### 4.4 *Existing questions*

However, there are some problems in the specific implementation of the PBL teaching method, as follows:

(1) Some students showed a perfunctory attitude to the task because of the lack of self-study time and discussion time.
(2) Student's ability in communication and cooperation is obviously inadequate. Some students cannot adapt to the changes in teaching methods, and some even had resentments. The quality of teaching was affected.
(3) The PBL teaching method emphasizes that students play the main role in learning. For the students with poor self-discipline, it is hard to learn things.
(4) The PBL teaching method makes higher demands for teachers, who need to be mentally prepared, as well as possessing the necessary skills. In addition to this, inadequate preparation can affect their teaching effectiveness.

(5) The PBL teaching method requires that teachers' guiding behaviour must be "timely", "moderate", and a "proper law", or it will also affect the teaching effectiveness.
(6) The PBL teaching method requires improved teaching conditions. Lack of learning resources in the library, computer network resources, teaching software, and lack of a hardware laboratory will have a direct impact on students' learning attitudes and learning results.

## ACKNOWLEDGMENTS

This work was funded by the education research project funds of Hainan University, ID: hdjy1330.

## REFERENCES

[1] Chun Hong Wang. 2012. An exploration Based on PBL on university computer foundation teaching model. *The Science Education Article Collects*: 47–48.
[2] Yan Wang, JunHui Zheng, FaSheng Yi. 2009. An exploration of case teaching of microcomputer principle and interface technology" course. 23. *The Computer Education*: 136–138.
[3] Hmelo CE. 1998. Problem-based learning: Effects on the early acquisition of cognitive skill in medicine. 2. *The Journal of the Learning Science*: 173–208.
[4] Xi Qiu Chen. 2007. PBL teaching method in the application of the network engineering training course. 26. *Education and Vocation*: 153–154.

*Frontiers in Computer Education – Wang (ed.)*
 *ISBN 978-1-138-02797-8*

# Research into the teaching reform of the C# course based on One Case in the End

Xiaojing Wang
*School of Information Science and Technology, Jiujiang University, Jiujiang, Jiangxi Province, China*

ABSTRACT: In view of the lack of coherence and integrity caused by the phenomenon of poor teaching results in the Object Oriented Programming C# course teaching, it is proposed to use the methods of the One Case in the End in the process of teaching C#. An analysis of the teaching process and teaching effect evaluation methods needs to be made as well as the key problems that need attention, and the process of teaching. Finally, carrying out the practical applications has obtained good teaching results.

*Keywords*: One Case in the End; robot; case teaching; C#; programming

## 1 INTRODUCTION

### 1.1 *Introduction of One Case in the End*

Constructivist learning theory holds that students obtain knowledge is by way of construction of meaning, in certain situations, with the help of others, using the necessary learning materials. That is to say, the learner to acquire knowledge quantity does not depend on his memory or reciting ability, but the ability of decision by itself according to the experience construction the relevant knowledge of the significance [1]. The teaching method of "One Case in the End" is based on this theory arising.

One Case in the End, also called One Pass, refers to at the beginning of a course, the teacher chooses a typical case or materials, and combination all knowledge points of course to the case throughout., help the students construct their own knowledge system and gain new knowledge in the learning process, improve the multi ability and scientific literacy [2]. Here's Case, can be either one example or one material can also be a person, a thing, a topic.

### 1.2 *Introduction of C# course*

"Object Oriented Programming C#" is the result of the work of a Microsoft Corporation researcher Anders Hejlsberg , it is an object oriented language, and it derived from C and C++ programming languages, has the characteristics of security and stability, is simple and elegant, and almost all of the research latest achievements of software development and software engineering: object-oriented, type safe, component technology, automatic memory management, cross platform exception handling code, version control, code security management and so on. According to the TIOBE ranking in September 2014, data show that C# ranked fifth in nearly 100 types of programming languages, the market prospects are very optimistic.

Conforming to the trends of technology development and career needs, C# language has become an important branch of programming languages at colleges and universities. As a program design course, case teaching method is the most important teaching method. However, the coherence and integrity of the case is often ignored by teachers, only using a large number of cases, the teaching effect is not good. At the end of the course, students cannot macro grasp the course content. A lots of students almost cannot develop a project by using curriculum knowledge, that is to say that it cannot arrive at the teaching target.

The "One Case in the End" teaching method is applied in some humanities teaching, and has obtained the ideal teaching effect. R shows that One Case in the End teaching thought of using the invariant situations to focus the attention of students, the contact of all knowledge can help student to bulid a good knowledge construct, which avoid students cannot do what to do. Moreover, students to construct the intrinsic link between the old and new knowledge of the use of comparative, reasoning, inductive method, and put the old knowledge to new knowledge, to achieve positive transfer of knowledge [3]. However, there are few applications used to implement One Case in the End teaching method in science and engineering courses. The author found that the results of the above study are also suitable for programming courses. With object oriented programming C# as an example, the objective of the course is solving practical problems through programming, this process is usually integrated application of course content. Thus, organizing teaching with the one case method fits the teaching goal of the course.

Table 1. The task of teachers and students in C# teaching.

| | Teacher | Student |
|---|---|---|
| BC | Compiling a series of virtual robot cases | Preview knowledge |
| IC | Using Case combined knowledge points strictly | Understand digestion |
| AC | Activating case by homework, professional competitions and other forms | Internalization of migration |

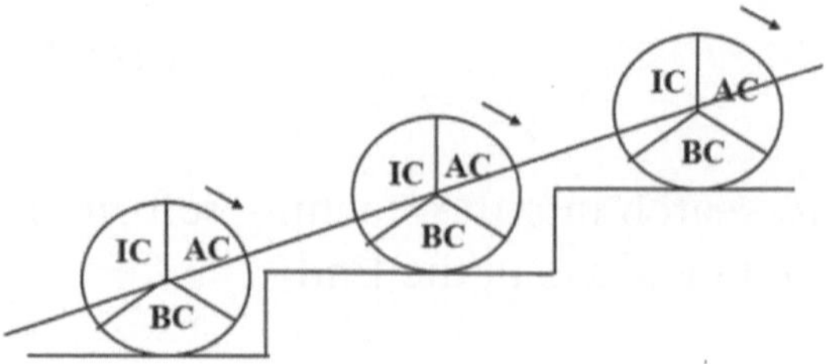

Figure 1. The schematic diagram of C# teaching process.

## 2 APPLICATION PROCESS OF THE ONE CASE IN THE END IN THE C# COURSE TEACHING

### 2.1 *Case selection*

The core problem of the One Case in the End is the selection of cases. When selecting cases, teachers should grasp the principle of interesting, practical, comprehensive, applicability [4]. Among them, the interest can captivate students' attention, and enhance the atmosphere of the classroom. The practicality is determined by the characteristics of the C# curriculum, the curriculum itself is used to solve the practical problems of production and management, so a course case also be practical. Comprehensive is to enable students to complete course knowledge construct through one case. Applicability is to choose a suitable case for students to accept, should not be too difficult and too complex, and can use some simple practical projects.

In the process of teaching in the past, the author used a chat robot case, and found that students are very fond of this case situation. The "robot" is a science and technology project of popular; The "robot" case is a universal case which can stimulate enthusiasm and increase interest in learning. At the same time, this case combined with embedded technology can achieve real robot versatility, which is practical and comprehensive. Of course, teachers can choose other cases , for example, students learning life is closely related topics, that the Grant Management Information System, or the Dormitory Management Information System and so on.

### 2.2 *Design of teaching process*

The teaching process is divided into several cycles according to time. Each cycle consists of three phases which are: before class, in class, and after class. Each cycle's end to end are linked together, to form a complete teaching process. The task of teachers and students in each cycle is shown in Table 1. The whole teaching process is shown in Figure 1. The BC, IC and AC respectively represent the time before class, in class and after class.

(1) Before class, a teacher compiles a series of virtual robot cases with C# language. The teacher needs to prepare various versions of the virtual robot and robot functional fragments based on the content main-line of Object Oriented Programming C#. The various versions of the robot include a process-oriented robot, an object-oriented robots, robots who can chat, a hungry robot, a robot who will handle the data, and a robot with memory, etc. Robot functional fragments include greetings, choices, repeated processing capabilities, build robot family members, and so on.

The robot case combines a course knowledge perfectly. The main knowledge points of C# course including C# program structure control statements, class, object, constructors, method, overloading, inheritance, polymorphism, abstract class, delegates, form controls, ADO.NET database access technology, multithreading, etc. Teachers should design a series of cases that integrate knowledge points closely, with different forms of virtual robots, different functions, and different personalities. And the case has become an organic whole, at the end of the course, the virtual robots are manufactured successfully.

Designing robot database.Designing robot language database, food database, user database, combined with the Connection object, Command object, DataReader object, DataAdapter object, and Dataset object, implement virtual robot based on database.

(2) In class, the teacher uses case by combine with knowledge points strictly. Case should be strictly matching and knowledge. Build a robot class, creates a robot object, initializes a robot parameters (build baby robot), defines a robot methods (functions), and creates a robot future generations, build a robot family, defines robot interface , design and implement a robot database and so on.

Throughout the Object Oriented Programming C# course teaching process, teachers uses different forms of virtual robots as one case. Almost all the knowledge points are explained using one case. Students happy learning, and gets a systematic learning experience, At the same time so that students interested in robotics technology products, students explore the spirit of innovation ability and scientific attitude, inspire students' curiosity and creativity, enhance analyse and solve practical problems.(3) After class, students digest cases using the following forms, such as experimental, homework, work, explanation, competition, and so on.

Here, the after-class has two meanings, one of which is after the end of each class, and the second is after the course is over. After each class, students must review the content of what they have learned, and then expand

Table 2. C# course content based on the One Case in the End teaching method.

| C# teaching plan | Robot Case morphology and function |
|---|---|
| Grammar foundation | Process-oriented chat robot |
| Class | Robot class |
| Object | Robot object |
| Constructor | Baby robot |
| Method | Robot with various acts |
| Overloading | Robot with different personalities |
| Inheritance | Robot's descendants and family |
| Abstract class | Aspirations of robot's grandparent |
| Delegate, events | Object-oriented chat robot |
| Windows Forms Controls | Robot's appearance and shape |
| ADO.NET | Robot of processed data<br>Robot has memory<br>Robot's food library and language library |
| Integrated Project | Play go bang virtual robots<br>Business services virtual robot |

it to a class case in the form of homework, which has to be consolidated. Allow students to use their imagination to realize such robots play Rubik's cube, and robot play go bang, and so on. The students digest knowledge in interesting games. The students apply old knowledge to new knowledge consciously to achieve positive transfer of knowledge in a coherent situation. After the end of the course, all programs that students learned about the robots are assembled by students to complete a sizable virtual robot which has many functions, and then students consolidate the gains by showing and explaining the demonstration program, or join in contests. Thus, the project was born, and at the same time, students can easily construct a C# knowledge system.

The correlation between teaching content and robot case shown in Table 2 is based on the One Case in the End teaching mode in C#.

The following two programs are chatting robot with C# language, were used in the form of process-oriented and object-oriented form to write. After comparing the two examples, we can see that one case in the context of teaching may remain unchanged, so that students focus on absorbing new knowledge.

```
/********* Process-oriented robot: *********/
Console.WriteLine("Hello, I'm Robot! ");
int fullLevel = 5;
   while (true)
{
   string str = Console.ReadLine();
   if (fullLevel == 1)
   {
      Console.WriteLine("I'm hungry! Have
something to eat?");
      int food = int.Parse(Console.ReadLine());
      if (food <= 0)
   {
         Console.WriteLine(" En, I'm angry! ");
         Console.ReadKey();
         return;
      }
      if (food > 10)
      {
         Console .WriteLine ("I'm full! ");
         Console.ReadKey();
         return;
      }
      fullLevel += food;
      }
   if (fullLevel == 0)
   {
      Console.WriteLine(" Goodbye next life!");
      break;
   }
   if (str.Contains("what") || str.Contains("name"))
      Console.WriteLine("I'm Robot! ");
   else if (str.StartsWith("how old") && str.Contains
("you"))
      Console.WriteLine("Tow!");
   else if (str.StartsWith("your") && str.Contains
("girlfriend"))
      Console.WriteLine("In the future there will
be.");
   else if (str.Contains("weather"))
      Console.WriteLine("Today is sunny.");
   else if (str.StartsWith("88") || str.Contains("Bye"))
   {
      Console.WriteLine(" Goodbye!");
      break;
   }
   else
      Console.WriteLine("Are you from Mars? ");
   fullLevel–;
}

/********* Object-oriented robot: *********/
class Robot
{
   public string Name { get; set; }
   private int FullLevel { get; set; }
   public void SayHello()
   {
      Console.WriteLine("I'm {0}", Name);
   }
   public void Eat(int foodCount)
   {
      if (foodCount <= 0)
      {
         Console.WriteLine("En, I'm angry!");
         Console.ReadKey();
         throw new Exception(); //Exit
      }
      if (foodCount > 10)
      {
         Console.WriteLine("I'm full£¡");
         return;
      }
         FullLevel = FullLevel + foodCount;
   }
   public void Speak(string str)
```

```
    {
      if (FullLevel <= 0)
      {
         Console.WriteLine("I'm hungry!
Have something to eat?");
         string no = Console.ReadLine();
         int n;
         if (int.TryParse(no, out n))
            this.Eat(n);
         return;
      }
      if (str.Contains("what") || str.Contains("name"))
         this.SayHello();
      else if (str.StartsWith("how old") && str.
Contains("you"))
         Console.WriteLine("Tow!");
      else if (str.StartsWith("your") && str.Contains
("girlfriend "))
         Console.WriteLine("In the future there will
be.");
      else if (str.Contains("weather "))
         Console.WriteLine(" Today is sunny. ");
      else if (str.StartsWith("88") || str.Contains("Bye
"))
      {
         Console.WriteLine(" Goodbye!");
         Console.ReadKey();
         throw new Exception(); //Exit
      }
      else
         Console.WriteLine("Are you from Mars? ");
      FullLevel–;
    }
}
```

### 2.3 *Teaching effect examination system*

C# course is both a theoretical and a practical course, evaluation of teaching effectiveness in two ways: written examination and software works. The purpose of the written examination is that students master the basic concepts, the basic theory, and the basic skills. The purpose of the software works assessment is that testing students' practical ability and ability to solve problems. Both are indispensable.

As a routine examination form, written within the stipulated time, the student independent, without the aid of any help to complete the course under examination. If they do not understand completely course theory, is more difficult to pass the examination. Of course, teachers should grasp the difficulty of examination papers. By analysing examination papers, teachers can find shortcomings and deficiencies in teaching, thus the teaching can be improved.

The software works reflect the students' design capabilities, programming capabilities, and even professional capacity, and is directly reflected on the C# course teaching objectives. Teachers should provide some spare topics for students to choose from, usually works done by another student. Works need to be completed over a longer period of time, usually outside the classroom. To test the quality of the work, students must pass a brief reply process to ensure the authenticity of the work.Finally, teachers give students a comprehensive assessment of the result, the result should be a comprehensive written exam, work achievements, classroom performance, and experimental achievements. Teachers should summarize of teaching methods of "One Case in the End" in practice improve themselves.

## 3 THE KEY PROBLEMS THAT NEED ATTENTION

### 3.1 *Moderate degree of difficulty of teaching cases*

If the case is too easy, they will put the simple question to complicate when in use advanced techniques (such as threads, commission, events). Too complex a case will make the students lose confidence. Therefore, the process of this case should be brief, and used to cover all knowledge points.

### 3.2 *Compare One Case in the End with one theory and more cases*

This method of One Case in the End is superior, but there are also disadvantages, that will cause illusion to students: C# can only solve this problem of the case. The reason is that always limited in one example. Therefore, teachers must make up for this deficiency by increase examples. Especially for beginners, it is necessary for a knowledge point to use multiple small typical examples to explain. In addition, when they are in a single situation for a long time, students appeared fatigue, and thus lose interest. Therefore, it is advantageous to add a small example. Teachers should have strong technical reserves.

This method of One Case in the end demands that teachers have the ability of finding, analysing and solving teaching problems in a short time when all program segments are transferred into one project, and each knowledge are linked intrinsically to curriculum teaching case, each program must have relevance.

## 4 CONCLUSION

The author has used the teaching method of the One Case in the End in the Object-Oriented Programming C# curriculum teaching for the last three years, and it had an ideal teaching results. This method solves the continuity, integrity and systemic problems of curriculum cases, while training students in the feel of the project, enhancing their interests in learning, and ultimately enabling students to easily make sense of the construct C# knowledge system, multi-angle, depth and systematically mining the course content, and strengthening their professional skills.

## ACKNOWLEDGMENT

This research is supported by 2014 Jiujiang University Teaching Research and Reform Project Grand

No.XJJG10. Thanks to organizers for all the work and efficient communication.

## REFERENCES

W.D. Yang, N. Jia. Constructivist Learning Theory Review [J]. Journal of Socialist Theory Guide. 2011(5): 77–80

Z.X. Xiao, Thinking of "One Case in the End" Case Teaching Method in the "Microeconomics" [J]. Journal of Hubei University of Economics (Humanities and Social Sciences). 2010(9): 166–167

H.F. Yang, "One Case Pass": Teaching Meaning and Classroom Implementation [J]. Education Digest. 2013(1): 59–62

Yu Dong-mei. The Exploration of a Multi- element Case Teaching Method in the Program Design Course [J]. Journal of Chinese People's armed Police Force Academy. 2013, 29(11): 77–79

*Frontiers in Computer Education – Wang (ed.)*
 *ISBN 978-1-138-02797-8*

# An investigation and analysis into the quality of medical graduates in Zhejiang

Chuan-Hong Zeng
*Zhejiang Chinese Medical University, Hangzhou, Zhejiang*

ABSTRACT: In order to understand the quality of medical graduates' training situation in Zhejiang Province, this paper conducts surveys and analyses, and puts forward some suggestions to improve the quality of medical graduates.

*Keywords*: Quality of training, graduates, employers

## 1 SURVEY AND DESIGN

The research group conducted interviews and questionnaires on 23 hospitals in Zhejiang, to understand the quality of medical graduates and the requirements of the employers, to know the views and suggestions of employers and graduates, and to improve the quality of personnel training, teaching, and job satisfaction. The 557 copies of survey questionnaires were distributed, of which 551 valid copies were recovered, at the effective rate of 98.92%.

## 2 SURVEY RESULTS

### 2.1 *The overall evaluation on the performance of graduates*

Employers objectively evaluated all aspects of the quality of graduates using the 12 evaluating indicators on the performance of graduates. Medical graduates get the same recognition by employers with excellent moral character, solid professional knowledge and skills. However, medical graduates' research and innovation capacity, relationships, practices and organizational management are not so good.

### 2.2 *Matching professional settings and socio-economic needs*

From the survey by the employers, medical specialty settings and the curriculum needs are adapt to socio-economic needs in Zhejiang. The research group received the same conclusion in the survey on the graduates, that most graduates believe the specialty curriculum fits social needs, and 90.54% graduates have learned them by working with their professional counterparts.

### 2.3 *The state of talent cultivation*

The survey results show that the top three main problems in quality training is the ability of analysing and

Table 1. Employers' statistical evaluation of the performance of graduates.

| Project | Excellent | Good | General | Poor | Very Poor |
|---|---|---|---|---|---|
| Moral character | 76.47% | 23.53% | 0 | 0 | 0 |
| Professionalism | 52.94% | 41.18% | 5.88% | 0 | 0 |
| Basics | 35.29% | 64.71% | 0 | 0 | 0 |
| Expertise | 41.18% | 58.82% | 0 | 0 | 0 |
| Practical ability | 29.41% | 58.82% | 11.76% | 0 | 0 |
| Research and innovation capability | 17.65% | 58.82% | 23.53% | 0 | 0 |
| Relationships | 38.10% | 42.86% | 19.05% | 0 | 0 |
| Organization and management capacity | 29.41% | 58.82% | 11.76% | 0 | 0 |
| Teamwork | 50% | 43.75% | 6.25% | 0 | 0 |
| Foreign language proficiency | 23.53% | 64.71% | 11.76% | 0 | 0 |
| Computer Skills | 17.65% | 76.47% | 5.88% | 0 | 0 |
| Comprehensive quality | 23.53% | 70.59% | 5.88% | 0 | 0 |

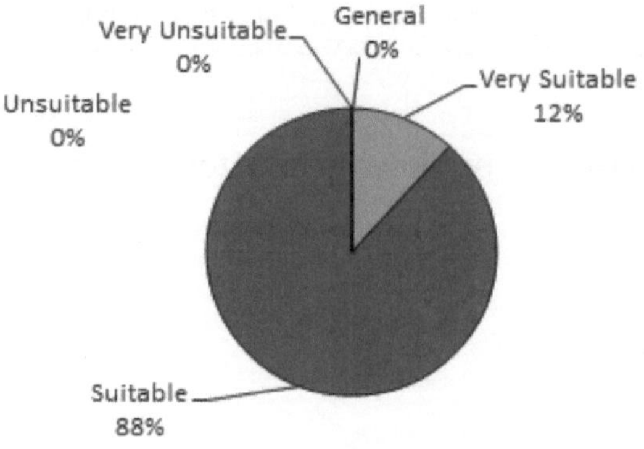


Figure 1. Matching on professional setting and socio-economic needs.

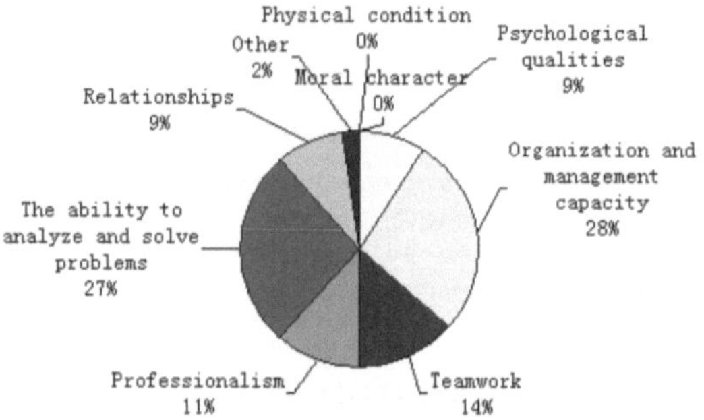


Figure 2. The problems of quality training.

solving problems (27.27%), organizational management capacity (27.27%) and the ability of teamwork (13.64%).

### 2.4 *Reform of personnel training*

In terms of medical talents training and teaching reform, the survey found that there are three suggestions significantly higher than other proposals. They are strengthening the teaching internship trainee and social practice areas, strengthening the cultivation of application ability (25.53%); developing interpersonal communication skills and coordination capacity (23.40%); strengthening the research and innovation capacity-building (17.02%).

## 3 DISCUSSION AND RECOMMENDATIONS

From the survey of employers and graduates in Zhejiang, medical graduates have the solid foundation of the theory and professional knowledge and high professionalism, etc. But in terms of research and innovation capacity, relationships, practices and organizational management, medical graduates are not recognized. Based on the above issues, the research group put forward the following recommendations.

### 3.1 *Improving the practical teaching system, reforming teaching mode*

Because medicine is a very practical subject, medical colleges should focus on strengthening students' hands-on approach and practical ability. Teaching programs should fully reflect a practical teaching system. Curriculum development should increase the proportion of practice teaching. On clinical teaching, most medical colleges have currently adopted bedside teaching. Medical colleges should establish the university clinical skills centre (or simulated hospital), unified management, should reform assessment methods, and increase practice assessment efforts.

### 3.2 *Constructing the system of extracurricular scientific and technological activities*

Medical colleges should develop the system of extracurricular scientific and technological activities, set up scientific research training courses and special academic lectures, hold academic forums, academic festivals, academic essay competitions, organize students to seminars and other academic activities of research projects, and open laboratories which are equipped with professional instructors. Awards and publicity has made outstanding achievements of students and instructors in technological innovations.

### 3.3 *Strengthening the teaching staff*

The quality of teachers determines the quality of students. In the investigation forum of graduates, the research group repeatedly learned that students expressed praise and gratitude to good, serious, and responsible teachers. Medical colleges should strengthen the teaching staff planning, creating a number of teachers with advanced educational philosophy, a high level of professionalism, and a strong sense of responsibility; they should plan and systematically train a number of teachers who are adapted to medical education and research.

### 3.4 *Well-organizing student activities*

Students' comprehensive ability depends both on their own quest for knowledge, but also on the atmosphere and can affect the learning environment. Medical colleges should give full play to the role of the second, third class, well-organized and colourful, healthy and progressive student activities, in the creation of "patient communication" course, based on the introduction of "medical humanities", "interpersonal" and other elective courses or seminars.

### 3.5 *Helping students develop a scientific college planning*

From the results of this survey, many graduates reflected on the lack of learning goals at college; they did not know what to learn and how to learn it during their time at medical school. Schools should help students under their guidance to develop overall goals as well as targeted, detailed, specific, and individualized milestones. Universities should offer 'College study guides', electives, and lectures, advertised on the website.

## REFERENCES

Chen Liqun, Lin Dajing Thinking to build a harmonious environment for the development of colleges [J]. Chinese Journal of Medical Education, 2006, (1): 9–11.

Wu Qidi. Vigorously promoting the teaching evaluation work, and effectively improving the quality of higher education teaching [J]. Higher Education Press, 2005: 13–21.

Zhu Jiayong, Qi Ping. Localization of medical institutions and regional economic development compatibly [J]. China Higher Medical Education, 2007, 18 (2): 29–31.

*Frontiers in Computer Education – Wang (ed.)*
*© 2015 Taylor & Francis Group, London, ISBN 978-1-138-02797-8*

# Educational quality improvement in the 'Principles of Communications' course, based on bilingual teaching and research teaching mode

Lei Yu, Li Zhou, Jiali Luo & Yuhan Zhou
*Department of Electronic Information Engineering, Nanchang Hangkong University, Jiangxi Province, China*

ABSTRACT: A research teaching mode centred on problem solving and the combination of teaching and research is presented based on the characteristics of the course Principles of Communications and bilingual teaching. A complete research teaching system is established through revision of the syllabus, development of the interactive platform and design for the practice session. The results of the practice show that not only are the students' academic grades improved, but their enthusiasm for the course are also stimulated, and their research capabilities are enhanced as well. Thus, a solid foundation has been laid for the cultivation of innovative inter-disciplinary talents who will meet the needs of the information technology development.

*Keywords*: Research Teaching Mode, Bilingual Teaching, Principles of Communications

## 1 INTRODUCTION[1]

Principles of Communications is an essential specialized basic course regarded as a vital connection between the previous basic courses and the subsequent specialized courses. With the development of science and technology, and the deepening of international communication, more and more communication engineering professionals are required to be equipped with abundant foreign language skills in our country in the information age. In 2002, Professor Zhou Li at Nanchang Hangkong University took the lead in introducing the original teaching material from the University of New South Wales in the curriculum construction of the bilingual course – Principles of Communications[1]. Soon afterwards, Southeast University[2], Jiangxi University of Science and Technology[3], and Nanjing University of Posts and Telecommunications[4] successively began to open similar bilingual courses.

Dominated by the traditional teaching ideas, most universities merely focus on the inculcation of theoretical knowledge instead of strengthening students' research capabilities during the teaching process of Principles of Communications[5]. Research teaching, on the other hand, pays more attention to improving students' scientific literacy when imparting knowledge. Rich in content and involving all aspects in communication process of signal, channel, modulation, demodulation, and synchronization, Principles of Communications provides vast development space for research-based courses[6]. Therefore, it is necessary to implement the course reform and the research teaching method.

## 2 RESEARCH METHOD

Firstly, on the basis of the characteristics of bilingual teaching and Principles of Communications, we revise the syllabus available for research teaching. Secondly, an interactive platform is developed to support it. Finally, during the teaching process, a design for teaching and practice is conducted based on a problem-solving design. In addition, website and demonstration software are further applied to assist it in order to achieve better research results.

Research method is illustrated in Figure 1.

## 3 THE ESTABLISHMENT OF A BILINGUAL TEACHING MODE FOR PRINCIPLES OF COMMUNICATIONS BASED ON ORIGINAL TEACHING MATERIALS

Students at colleges of engineering in China mostly use Chinese textbooks when professional courses begin in their junior and senior years. In the meantime, they are required to take a professional English course due to the implementation of the continuous line in foreign language teaching policy. They are expected to learn all

[1]This paper is supported by a grant from the Research of Teaching Reform in Colleges and Universities in Jiangxi Province (No: JXJG-14-8-19) and the Teaching Reform Project in Nanchang Hangkong University (No: JY1435).

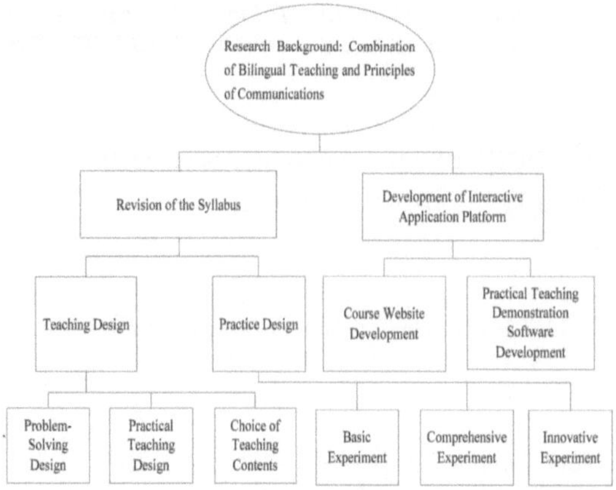


Figure 1. Research method for research teaching mode.

professional courses in English within dozens of hours. Obviously, it is almost impossible for teachers to drive home the concepts, or explain the problems in such a short time. Thus, this makes it difficult for students to have a thorough understanding of their professionals. Hence, we can see that the curriculum is both time consuming and fruitless. Bilingual teaching in engineering courses, on the other hand, not only improves teaching efficiency, but also makes it possible for students to learn more systematically and deepen their understanding. In addition, bilingual teaching stimulates students' initiative, so that they are more likely to acquire as much knowledge as they can in a short time, and they are also trained in independent reading and self-teaching skills. The original edition of Principles of Communications, which is informative, new in content and practicality, is not only beneficial to the improvement of students' English proficiency, but also guides students to use English dictionaries which enable them to increase their reading quantity in English documents, enrich their knowledge, and provide a thorough grounding in quick obtainment of first-hand information on advanced technology in the related field.

## 4 THE ESTABLISHMENT OF A RESEARCH TEACHING MODE BASED ON BILINGUAL TEACHING

Traditional expository teaching method is generally adopted in Principles of Communications. Most teachers only focus on the inculcation of theoretical knowledge during the teaching process. Although it maximizes the amount of knowledge transmission, students' lack of active thinking and questioning opportunities makes the knowledge unrenewable after being transmitted by the teacher. This is the teaching method that puts students in a passive position. It is contrary to the original purpose of teaching, and students' scientific literacy development is therefore neglected.

Table 1. Performance of bilingual teaching mode.

| Grade | Number of Classes | Bilingual Teaching | Pass Rate of CET-6 |
|---|---|---|---|
| 2002–2004 | 3 | No | 21.76% |
| 2005–2011 | 15 | Yes | 53.29% |

Principles of Communications is an important professional course served as a link between some related professional courses. On the one hand, knowledge acquired in the previous basic courses such as Signal and System and High Frequency Circuit, is essential for learning Principles of Communications. On the other hand, knowledge learned in Principles of Communications is also indispensable for the subsequent courses such as Mobile Communication and Fiber Communication. Rich in content and involving all aspects of the communication process, it provides a vast development space for research-based courses. Therefore, due to the introduction of research teaching mode in Principles of Communications, the traditional teaching method will be changed and students will have greater opportunities not only to develop their innovative spirit and practical abilities, but also to improve their scientific literacy and attitude, which may lead to the improvement of their capabilities to analyse and solve practical problems by using their knowledge.

The research teaching mode of Principles of Communications is established based on bilingual teaching, aiming to improve teachers' teaching method and develop students' abilities. Case teaching and heuristic teaching methods are adopted in the classrooms to stimulate students' enthusiasm for the course. We also design an English course website based on Zend Framework and develop the practical teaching demonstration software to enhance students' professional skills and research capabilities.

## 5 EDUCATIONAL RESULTS

We succeeded in teaching the method by means of teaching and developing the capabilities of problem solving when the research teaching mode was established and carried out on the basis of the policy: 'To combine teaching with scientific research, to promote research with teaching and to enrich teaching content with research results'.

The research teaching mode reform of Principles of Communications (bilingual) has taken place in six classes at our faculty from 2009 to 2011. After collecting and analysing the information of Principles of Communication' for ten years, we can find that:

1. Students' English level is greatly improved by bilingual teaching. (The number of students that passed CET-6 has greatly increased. See Table 1)

Table 2. Performance of research teaching mode.

| Grade | Number of Classes | Research Teaching | Pass Rate | Excellent Rate | Evaluation of Teaching quality |
|---|---|---|---|---|---|
| 2002–2008 | 12 | No | 62.54% | 12.95% | 91.3 |
| 2009–2011 | 6 | Yes | 98.61% | 39.02% | 97.6 |

2. Students' academic grades (including pass rate and excellent rate) are prominently improved through the reform of research teaching mode. Students' enthusiasm for the course is stimulated and the evaluation of teaching quality is also improved. (See Table 2)

## 6 CONCLUSION

We have explored a research teaching mode suitable for bilingual teaching through research and practice. Our teaching purpose is achieved through the following methods: 1. Revising the syllabus and multimedia works for research-based Principles and Communications (bilingual). 2. Setting experiments for research-based Principles and Communications (bilingual) and compiling the related experiment instruction books. 3. Developing the practical teaching software and English course website. Hence, not only do we stimulate students' enthusiasm for learning and enhance their research capabilities, but also lay a solid foundation for cultivation of innovative inter-disciplinary talents who will meet the needs of the information technology development.

## REFERENCES


Dong Heng, Cao Shike. Research of 'Principles of Communications' Teaching Reform – from Bilingual to English Teaching Information Technology. 2011. 11:51.

Shen Huifang, Xie Bin. Research and Exploration of 'Principles of Communications' [J]. China Science and Technology Information. 2011(5): 1–2.

Song Tiecheng. Exploration and Practice of 'Principles of Communications' Bilingual Teaching [J]. Journal of Peking University (Philosophy & Social Sciences). 2007: 95–98.

Wang Shugen, Jiang Xin. Summary of the Historical development Stage on National Bilingual Teaching [J]. Proceedings of National Conference on Surveying and Mapping Subject Education. 2007: 385–389.

Wang Shiyuan, Duan Shukai, Wang Lidan. Practice and Discussion on Undergraduates Bilingual Teaching [J]. Journal of Southwest Agricultural University (Social Sciences Edition). 2010. 8(1): 192–195.

Wu Ping. Summary of Bilingual Teaching in Five Years [J]. China University Teaching. 2007(1): 37–44.

*Frontiers in Computer Education – Wang (ed.)*


# Application research of virtual experiment technology in the power electronic technology course teaching

Xin-yu Yang, Tian-lin Niu, Wei-wei Li, Lin Zhang & Qiang Zhang
*College of Air and Missile Defense, Air Force Engineering University, Xi'an, Shaanxi, China*

ABSTRACT: This paper studies the application of the virtual experiment in the *power electronic technology* course theory teaching, on the basis of analysis of the course teaching content, curriculum reform direction about the power electronic technology and the background of the electrical engineering requirements. The paper uses different means of virtual simulation experiment, creating favourable conditions for the smooth running of the course teaching and meeting the needs of the students' individualized learning through the optimization of the teaching content. After teaching practice of many years, the practice that leads the power electronics course teaching reform by virtual experiment teaching has made a significant impact on training students to analyse and solve problems, to develop practical and innovation ability. This teaching practice also promotes knowledge, and the ability and quality of the students' coordinated development.

*Keywords*: virtual experiment; power electronics; teaching reform; practice teaching

## 1 INTRODUCTION

Electrical engineering has strong engineering properties. The professional demand of society in general can be divided into three categories-research (including application research), application, and professional skills (also known as professional skills), according to the development of science and technology in the electrical engineering field in China. The training demand of electrical engineering undergraduate talents should mainly belong to application, according to the nature functions of professional technical university. Therefore, the basic theory and the training of practical and comprehensive ability should be paid attention to, especially the training of application ability. So it is important that status of theoretical analysis in the major and leads to mathematical tools applied in the study of electrical engineering are much more. Meanwhile, as the main and basic means, the experiment research is very important in engineering major. Without the basic test conditions, it is difficult for the students to grasp the basic knowledge of this major. So simulation has become a kind of widely used method for the specialty with the development of computer technology. Based on above analysis, experimental research, simulation, equivalent and analogy are important scientific methods for the specialty.

Power electronic technology is an interdisciplinary technology that powers electronic devices that are used for power transformation and control. It is the integrated interdisciplinary of many subjects that include power transformation technology, modern electronic technology, power semiconductor devices, and automatic control technology. Then the all are applied in the field of power electronics technology. Since the electric circuit structure and the waveform drawings of the teaching are more and complicated, so it is difficult for the students to study. For the curriculum teaching process, there are a large number of waveform analyses in the theory teaching content. Traditional approach applies to a large number of charts and draws voltage and current waveform figure of converter process. The analysis shows that the process lacks the authenticity and the teaching efficiency is low. If the students can observe the voltage and current waveform after rectifier from the oscilloscope by combining the experiment demonstration, it will play a good role in promoting on the students' in-depth understanding to the teaching content. But as a result of the teaching content update quickly, the laboratory construction fund shortage and other objective conditions, the teachers cannot be in the class presentation one by one and it is impossible to allow the students to verify all circuit by experiment one by one. Thus it cannot be completely synchronized in practical teaching for power electronics teaching content. How to strengthen the effect of experiment teaching, practical teaching and teaching mode that combines innovation theory and practice, and how to meet the demand of the current theoretical teaching, how to maximize the realization of the professional training target-engineering and technology talents, they are the targets that are worthy of continuous efforts in the teaching reform.

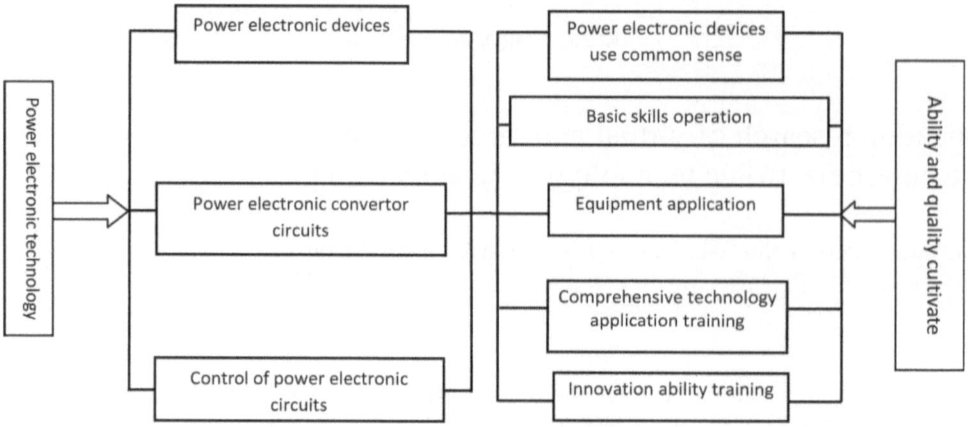


Figure 1. Course content system block diagram of power electronic technology.

With the rapid development of virtual computer technology, building virtual experiments with the use of electronic technology simulation software and a computer has greatly improved the level of power electronics practice teaching course. Virtual experiment is a new experimental method that constructs an experimental simulation environment using a computer and achieves the effect and purpose of experiment through the establishment of circuit, and the analysis of data and the circuit function. There are many types of specific application software, for example Multisim is one of the common tools for electronic technology design and training, which can be used to test and analyse control circuits of power electronics. MATLAB/SIMULINK is mostly used to rapidly finish the simulation and calculation of the power electronic circuit topology structure. In particular, the power system toolkit modules after the version of MATLAB5.2, provide the ideal tool for the research and simulation of power electronic technology; it makes the experiment circuit that originally must be jointed with real power device to finish by 'soft experiment' with simulation research on the computer. Since virtual experiment does not almost need any investment, it also does not have the risk of burning components and endangering the personal safety in reality, it is an ideal means to strengthen the practical teaching effect.

This paper explores the virtual simulation software using that MATLAB is its main tool. It constructs a virtual experiment, which is useful in the classroom teaching, in order to realize the specific idea and method with the combination of theory and practice, and then it can raise the theory teaching effect of the power electronics course.

## 2 DIRECTION OF COURSE REFORM AND DEVELOPMENT

### 2.1 *Course content characteristics analysis*

As the basis of electrical engineering, especially in the comprehensive engineering technology, the application of power electronic technology course is main target. So the teaching content is divided into three parts: power electronics devices, power electronics converter circuit, and the control technology of power electronic circuit. According to the subject development and the needs to meet the military training transformation, we update and adjust course content and build a new type of power electronics course content system, which is shown in Figure 1.

Power electronic technology is based on power electronic devices, power electronic devices control is a means of implementation, and various circuit topology structures are the specific application. This kind of structure determines that the course has the following features: (1) Discipline overlapping, wide range and good practicality; (2) Quick development and advanced content. It is hard to meet the requirements of teaching for the traditional experimental method and it is necessary to pay attention to the innovation of practice teaching content synchronization; (3) Strong applicability. Course content does not only combine closely with the daily life, but with the new power equipment especially, it is more widely used in the new weapons system. This requires certain comprehensive design ability and analytical ability of the students, so that they master the comprehensive utilization ability of the new weapon system power equipment, and improve the capacity of the first position.

### 2.2 *Course teaching reform target*

These characteristics of the course determine that the teaching reform should be aimed at making the students understand the technical movements of the power electronic technology development, the new application field, and the application in weapon system power through the course of learning. In accordance with the requirements of the curriculum reform teaching goal, it must be emphasized that equal attention must be paid to the theory and practice; the principle of theory, as the service of application in theory teaching, and the comprehensive application of power electronic technology in weapon equipment should be strengthened appropriately. This method requires that the students must be good at outward learning on the basis of the basic circuit, but is difficult to bear expanding demands of the practice which rely only on the fundamental experiment method. At the same time, the power electronic technology is developing rapidly, all kinds of new types of power electronic devices, the circuit topology and control technology appear constantly, and power electronic technology applications are increasingly becoming widespread. This poses a challenge for the course practice teaching of the power electronic technology. Therefore the study on virtual experiment teaching is of great significance in cultivating innovative talents for electrical engineering in the power electronic technology course.

## 3 THE ANALYSIS OF APPLICATION OF VIRTUAL TECHNOLOGY IN POWER ELECTRONICS TEACHING REFORM

### 3.1 *Physical and virtual combination to enhance the cognitive level of the senses*

In the knowledge system of the power electronic technology course, the component part is the foundation and the premise of other contents. The main teaching involves structures of various types of devices, work principles, characteristic parameters, using method and the drive and protection etc. The purpose of teaching is to lead students to master the use method of all kinds of power electronic devices proficiently and the problems that should be paid attention in use. For the primary sensory of devices, the teaching method in a real way can be used. However, to really improve their abilities to use the devices, there will be very large limitations and low efficiency if using visual illustrations in theory lessons. Properly combine with the simulation test and make every module necessary for the master electrical components add to the SIMULINK model editor window, you can quickly set up simulation model to realize the calculation simulation of power electronics devices, to master the deep cognition, such as device working principle, performance parameters and so on and to improve the students' basic use skills to the components. Certainly, the disadvantage of the simulation software is that the elements in circuit are all symbols and very different from the physical test. We can increase immersion of the virtual experiment software into the multimedia auxiliary teaching software on the basis of the existing, realize the complementary advantages through the design of multimedia electronic technology experiment system software and enhance intuition and operability of the virtual experiment.

The main part of the power electronic technology course is the five types of circuits which has a basic transformer circuit, a rectifier circuit, a chopping circuit, ac power control and ac frequency conversion circuit, and an inverter circuit. The teaching purposes

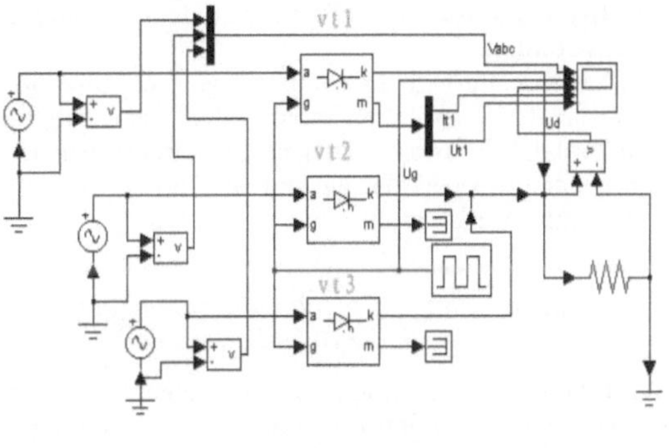


Figure 2. Rectifier circuit diagram.

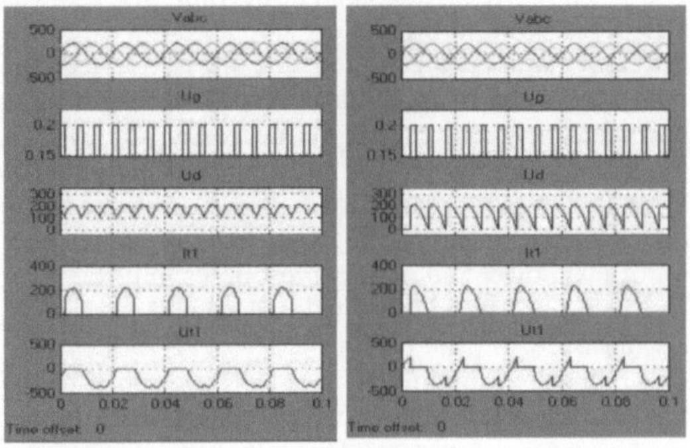


Figure 3. Simulation waveform when $\alpha = 0°$ and $\alpha = 60°$.

of this part require students to have four types of ability: drawing, analysis, calculation, and selection. Drawing means drawing main circuit diagram; analysis means analysing the working principle of the circuit and drawing out the circuit work waveform combined with the circuit diagram; calculation means calculating some important electric parameters, such as the average of the rectifier circuit output voltage, inverter circuit outputs voltage harmonic, and withstands voltage of the power switching device, etc. Selection means choosing the appropriate power switch devices which mostly involves in rated voltage and rated current of a power switch device. For four kinds of ability of the trainees, the organization way that regards the physical experiment as the forerunner and the virtual experiment as the subject can be used. To conduct the virtual experiment, we first need to draw an experiment circuit on the software platform, thus drawing ability can be promoted. The experiment components can easily change in the software testing platform, especially as it can lower the difficulty and inexperienced risk of exploratory experiment; through analysis of the simulation software and computer waveform display function, guiding students to analyse the circuit and select the component intuitively become easy. As shown in Figures 2 and 3, it is the contrast results figure that common cathode three-phase half wave controlled rectifier circuit and operation simulation in the Simulink environment simulation circuit. It has intuitive demonstration effect for students to master and use circuit.

Therefore, through the comprehensive combination of physical and virtual technology the sensory cognitive level of the basic components and circuit structure can be strengthened and the students' ability to learn can be improved overall.

### 3.2 *Guided with the content optimization, actively cooperated with the virtual experiment*

With the new development of power electronic technology, the course information is being continually increased. But the formulation of new wide professional teaching plan, the teaching period be compressed further, but the content become much more. Therefore, how to effectively impart knowledge for different majors in the limited time and how to lead the bold innovation of teaching content, teaching means, and teaching methods are the necessity of developing a discipline situation. Tuning the course content to form a set of rapid change for the transformation of the physical experiment is of certain difficulty to spend a lot of manpower and material resources, thus keeping the continuity through virtual experiment is of great significance.

Aiming at the optimization of teaching contents, the half-controlled devices physical experiment content can be appropriately cut and the full-controlled devices and power integrated devices of experiment content can be strengthened in the virtual experiment teaching. Compared to the traditional phase-controlling technique, new computer control technology and PWM control in independently virtual experiment can be properly increased; the chopper technology and inverter part can be increased in the virtual demonstration experiment class.

On the basis of the innovation of the teaching means and methods, we should implement the idea of quality education and innovative education into the classroom. The course can adopt the teaching experiment method of the heuristic type, a contrast type research, etc. The most basic approach is to give the virtual experiment a full play in the class, and increase the experimental efficiency and saturation course. In addition, make use of virtual simulation technology to enhance the teaching of images and promote the students' perceptions. Pay attention to summary and contrast through the experiment teaching, such as in Figure 3, the different output voltage, current waveform and other knowledge points under the different load parameters about the full-controlled bridge. The virtual simulation experiments have the advantages of intuitiveness and accuracy. A certain number of virtual simulation practices should be done in extracurricular studies to deepen and consolidate the theoretical knowledge.

### 3.3 *Flexible application of virtual teaching methods to improve theory teaching effect*

In view of different content, a virtual experiment can be divided into different types. To enhance the experimental effect, different ways of virtual experiment should be applied flexibly in the power electronic teaching practice.

The learned knowledge in view of power electronic components is suitable for a large number of virtual demonstration experiments. An experiment is presented to enhance the real sensitivity of the students to the knowledge they have learned and inspire the initiative and enthusiasm of the students' thinking. The experiment is designed by the teachers carefully, it can guide the students to observe, analyse, induce, and summarize the relevant laws according to the experiment phenomenon. Two ways that include video and animation demonstration can be mostly adopted under the condition of the network. Video presentation makes the whole experimental process made by the teachers filmed completely. It repeats the experiment scene by storage and transmission of the computer network to make the students watch the experiments. Animation showing is made by Flash, Author ware and so on, especially in science and engineering classes. It can show the experiment process accurately, simulate the actual process and phenomenon that the teachers find hard to describe with language, and highlight the key points and the difficulties. In the virtual demonstration experiments, the interactivity provided is limited, and the experimental condition lacks authenticity, so attention should be paid to strengthening the teachers' guidance and the summary.

In the learning process of extensible knowledge of the power electronics course, in addition to the use of hardware experiment, a large number of simulation experiments can be made. Simulation experiments simulate the experimental equipment, such as the instrument, the facility, the component and so on, and the experimental conditions with software in a computer, then display them using the text, graphics, or animation. A virtual laboratory can regard MATLAB Simulink and management software' as a platform, the students in the lab interact with the experiment software after they login into the virtual laboratory. They can make full use of the abundant experimental equipment in the virtual equipment library, design the experimental process, build experimental devices, operate experiment, finish statistical analysis of the experimental results, etc. according to the needs of the experiment. Finally, they write the experiment report and the whole experimental process is finished on the computer. For example, the students are free to use all kinds of resistors, capacitors, transistors, integrated circuits, and other virtual devices of the software to finish the circuit that is required in the lab, and then connect to a virtual switch and power, as well as the signal generator, the measuring instruments, the oscilloscope, and other virtual instruments to conduct the experiments and observe and record the experimental data and results. The simulation experiment operated by software should pay special attention to standardizing the experimental skills of the students and make them develop good training habits.

Anyhow, we should formulate the corresponding specification and apply the ways of virtual experiment flexibly according to different teaching contents, so that the students can summarise the knowledge they have learned more conveniently and more intuitively, and improve the ability of analysis and application in the basic principle of power electronics technology.

### 3.4 *Develop the advantages of virtual experiment and promote the comprehensive ability*

The teaching purpose of this course is to make the user have certain experiment ability, accurately and skilfully use the knowledge in order to analyse and solve relatively simple and practical problems, lay foundations for the study of theory and principle, and improve students' innovation consciousness and innovation ability at the same time.

A target is regarded as traction, to guide students to learn from reception of exploration and independent cooperation through the virtual simulation experiment, to encourage students to think independently, to stimulate the initiative of learning, and improve the ability of independent thinking. Accordingly, to carry out a comprehensive design of the virtual power electronics experiment and to reconstruct the curriculum in several comprehensive design type teaching experiment projects, such as an application of the DC speed regulating device, the intermediate frequency induction heating power supply, and switching the power supply and inverter. Implementation of the project bring in responsibility system during the process, in the form of project development, the students can be divided into group and select their own topics. A complete power electronics experiment system includes power electronic devices, a variable flow circuit, and a control circuit, which involves the power electronic system design, installation, debugging, and other working processes. These tasks can be refined according to labour, to ensure that each goes smoothly and exercises the students' team consciousness and coordinated ability.

Through the combination of comprehensive virtual experiment design, theoretical teaching, and real experiment, teaching effect of experiment process is greatly enriched, the students' ability of understanding of the theory, practical ability, and scientific research is improved, the consciousness of cultivating innovation is achieved, and the rigorous scientific attitude and scientific research methods are cultivated. Also, the students' ability to analyse the problems by using computer tools and solve the actual problem of power electronic circuits is improved.

## 4 CONCLUSIONS

Virtual experiment teaching can cultivate the students' scientific attitude and innovation spirit, arouse subjective initiatives of the students, make the students master the basic theory knowledge, and promote the students' cognitive development. Based on virtual experiment teaching, developing theory teaching of the power electronics course and using the advanced virtual simulation technology lead to electrical engineering teaching go smoothly. It is easy to carry out according to the students' aptitude and meets the needs of individualized learning. After several years' of teaching practice, combining the development trend with characteristics of power electronics, as well as weapons and equipment by means of virtual experiment teaching, focusing on the optimization of the teaching contents, breakthrough in the reform of teaching methods, and adopting flexible virtual experiment methods, have had a significant effect on training students to analyse and solve problems, improving their practical ability and innovation ability, and promoting the coordinated development of the students' knowledge, ability, and quality.

## REFERENCES


Du Baoqiang, Ye Huiying. Design and implementation of virtual experiment teaching system for circuits cheater [J]. Foreign Electronic Measurement Technology, 2005.7

Ge Xiaohan. An exploration into the reform in power electronics teaching [J]. Journal of Jincheng Institute of Technology, 2012.4

Tang Jie, Lin Li, Liu Jiafang, Li Haina. The exploration and practice of the research-based teaching for power electronics technology course [J]. China Modern Educational Equipment, 2011.03

Wu Chunhua. Application study of virtual experiment teaching in modern distance education [D]. Central South University, 2006

Zhang Yu, Cai Lijuan, Han Chunling. Preliminary study on the teaching reform of electrical and electronic teaching under the new curriculum concept [J]. Education and Vocation, 2012.02 (5)

Zhang Fengrui, Zhu Qinghui. Exploration and practice of virtual experimental teaching model based of information technology [J]. Journal of Nanyang Institute of Technology, 2009.07

*Frontiers in Computer Education – Wang (ed.)*
 *ISBN 978-1-138-02797-8*

# An analysis of the construction of university career guidance system under the new employment situation

Donghe Pang, Zhixiang Tong, Hongli Zhang, Xiaohui Meng, Changbin Sun & Zihang Fu
*School of Computer Science and Technology, Harbin Institute of Technology, China*

ABSTRACT: With a higher number of students enrolling at colleges and the slowing down in global economic growth, the original employment model has been unable to adapt to these changes. This paper begins with the building of a career guidance system at Harbin Institute of Technology and then discusses the strengthening construction of the career guidance team, reforming a career guidance curriculum, and building employment information systems for the purpose of improving the reform of the graduate career guidance mode.

*Keywords*: career guidance, university student, Harbin Institute of Technology

## 1 INTRODUCTION

With the continuous expansion scale of computer science education, the number of university students increased dramatically in recent years. However, the employment rate and the job quality of graduates whose major is in computer science are not ideal. According to *2014 Global Employment Trends Report* by the International Labour Organization in January 2014, global economic recovery is very slow, and the growth in the labour market is slow globally as well. In 2013, global unemployment reached 202 million people. In 2014, the growth of the global employment rate remains weak and unemployment continues to rise, especially among young people and many potential workers. The global employment situation is not so optimistic, while the situation of student employment is even less encouraging. There is a gap between the requirements of companies and student's professional ability. In addition, there is a deviation between the employment psychology of graduates and the expectation of companies. Building a scientific and reasonable career guidance system, in order to help students understand society, have clear career goals, make career plans, enhance their overall qualities, find suitable jobs, and integrate into society, is particularly important.

The School of Computer Science at Harbin Institute of Technology (HIT) attaches great importance to the career guidance system. Scientific, standardized, and systematic career guidance is the objective need for students' growth and the cultivation of their talents. From the point of view of the university, this is to train professionals to adapt to social needs, eventually achieving the education policies of the Party, and the target requirements for talent cultivation. From the point of view of society, it is the requirement of science and education strategies and the need for enhancing society and the economy.

We should establish the concept of comprehensive personal development through a career awareness programme, career planning, and a career exploration process, which will influence students throughout their life.

## 2 TOTAL INVOLVEMENT AND ESTABLISHMENT OF A THREE-DIMENSIONAL CAREER GUIDANCE TEAM

The career guidance system at the computer college of Harbin Institute of Technology, whose career guidance personnel is made up of career guidance teachers, instructors, class advisers, and professional teachers, who all together cooperatively integrate aspects of work such as recruitment, teaching, management, career guidance, and so on. In addition, they integrate ideological education into a career guidance system, reaching the aim of a comprehensive education. Different types of career guidance play different roles from different perspectives, to guarantee the consistent focus and concern of the leadership on a college student's career guidance, as well as active coordination and participation of related functional departments and teachers. In this system, since entering school, every aspect and every stage of career planning of students have corresponding instructors to provide a service to help and guide to them, and when recruiting companies select graduates, they can gain authorized and reliable information provided by career personnel from different departments of school.

## 3 REINFORCING ABILITY CULTIVATION AND IMPROVING CAREER COMPETENCE

Helping students to gain more qualities and abilities needed by enterprises is the basis for improving career competence, as well as the key to improving career quality. There have been students with great school achievements, but with bad ability in programming practical applications, therefore when gaining employment they would reveal their shortcomings and be confronted with frustration many times. This situation has enlightened the teachers who are responsible for students' career prospects, which shows that there is a certain distance between the original emphasis on cultivation through research, and a solid theoretical basis with enterprise emphasis on practical operational ability. After all, there are more than 80% of graduates having to take careers each year, and 50% of undergraduates entering into a career. In the meantime, when emphasizing theory, the question on how to improve the practical ability and innovative ability of students is also a subject that concerns colleges. Colleges have a deep understanding that making a good career work does not just mean sending students to enterprises, but also generates feedback about the needs of enterprises.

Under the guidance of research topics "11th five-year plan" – "tracking survey of graduate talents", of higher level education academy in Heilongjiang Province, the college surveys the demand for talents of some important employers, such as Baidu, Huawei and Langchao etc., in forms of questionnaire investigation, communication and interviews. Then we summarize the real requirement of companies for the students' ability, and report it to the leaders and career guidance committee of college in time. In the meantime, when the college emphasizes theoretical cultivation, it also seizes the opportunity to cultivate the practical ability of students, and emphasizes experimental work, with setting of 96 class hours for software design and development practice, and 48 class hours for computer design and practice, as well as 48 class hours for embedded system design and practice courses, as to the requirements of enterprises for embedded technology and talents. The college also strives to develop an annual innovation project plan, a research centre student academic club, effectively improving the practical ability and engineering development ability of students, and improving the core career competence of students.

## 4 STRENGTHENING THE COOPERATION BETWEEN COLLEGES AND ENTERPRISES AND GETTING THROUGH THE GREEN CHANNEL TO EMPLOYMENT

Students lack clear understanding of industries and enterprises, knowing little about their demand for human resources when finding jobs. Besides, companies only enter colleges when recruiting, so they have no opportunity to find out more about students. This leads to a universal problem. In order to address this problem, the department of computer science established an HR base, student clubs, and high quality courses together with enterprises, thus building up an enterprise-oriented platform for students, in order to help them stand out while looking for jobs in outstanding enterprises.

The department encourages students to join in Education Ministry-Microsoft's key laboratory for language and sound, HIT-IBM research laboratory of China, Harbin Institute of Technology-Tencent joint lab and other personnel training bases. The department held 6 Microsoft summer school programmes and other brand activities, setting up a broad platform for the majority of students to get in touch with the latest technology and industry applications. The CS department, Xinchao, Sina, Baidu, Sogou and such large IT companies established student clubs, and by organizing related technical training and skills contests, they invited enterprise personnel to carry out lectures about technology and culture. They built up a bridge of understanding between students and enterprises. During recent three years, the department of computer science of HIT has nearly 200 students recommended for a direct interview by the clubs. The clubs have become the buffer pool for students to "warm-up" to their ideal companies and the "pool of talent students" for enterprises to attract outstanding students. In addition, Department of Computer Science in Harbin Institute of Technology also sets up excellent courses with enterprises, bringing in the latest technology and typical cases into the classroom. Because most banks use IBM servers, many undergraduates who learned IBM server related courses are preferred by many companies.

## 5 CONSTRUCTING PERFECT SYSTEM OF CAREER GUIDENCE FOR LEADING STUDENTS TO FIND A CLEAR EMPLOYMENT DIRECTION

Career planning and employment guidance committees are set up in each year to constantly improve the whole process of an employment guidance system. In the situation where junior students are unclear about professional knowledge and deep understanding of the industry situation, the department held "masters in small classes activities in depth, and invited well-known experts and scholars to join classes and be face-to-face with students to have in-depth discussion, and to get close with students to talk about the history of computer science, career planning, and academic and industry prospects. A number of senior professors from our department actively promote Computers Yesterday, Today, and Tomorrow" lecture courses, helping younger students to solve the problems such as what is a computer? What is its use? What do we learn from them and put in future use?" They also address a series of other problems.

To deal with the confusion about career direction and job-searching skills of senior students, the school actively carry out Approach to Workplace activities, inviting well-paid graduates, employers, and HR staff to give lectures or workshops to share career advice and job experience, and guide students in order to wisely find their jobs.

The school continues to use innovative training methods. It makes full use of various resources to provide a variety of forms, multi-angle training, guidance, and training for students. These include courses, forums, seminars, employment networks, counselling, school-enterprise cooperation in training, experiential training, and so on. Among them, experiential training, has been attracting the attention of a growing number of colleges and universities. It references the company's human resources training to take a new, efficient model, in order to make the training object acquire new knowledge, skills, and attitudes through personal intervention practice methods. It lets students learn teamwork, communication skills, and decision-making abilities in the activities by themselves, and then make it a habit of thinking and behaviour. So that students benefit endlessly in employment and future work.

## 6 PERFECT SERVICE SYSTEM, HELPING STUDENT TO ACHIEVE THEIR DREAMS

The School of Computer Science and Technology constantly sums up experience in a working environment, and builds and improves the student employment service system. Through the preparation of the School of Computer Science Employment Guidance Manual, the school opened a variety of information platforms of all types such as QQ group, Fetion Groups, and public mailboxes to solve various problems that students encounter in the course of employment.

The college formed a weekly calendar on employment to help students to organize employers, and on the day of recruitment, we remind each student to participate in recruiting by Fetion. In the quarter of employment cases, the employment information system of the school has withstood various challenges. It costs less than half an hour for releasing employment information to each student. The system received high praise from employers and students. The high-quality service offered by teachers in our school is highly trusted and affirmed by the students. Before the signing a contract, most students discuss the conditions and treatments with their teachers who are in charge of career guidance, especially regarding the choice of multiple offers, where a teacher can give a student an adequate guidance and help.

As there are many students from poor families have trouble in employment, the college could use help of the committee of the Boat of Hope. Based on friendly service, the committee focuses on developing students' skills and ability. It regularly organizes network architecture and web design training sessions to enhance students' professional skills, and insists on carrying out the story-sharing sessions and lectures, known as "I dare say" to continuously improve their communication skills. During recent years, students from poor families in our school not only achieve a 100% employment rate, but also can find high-quality jobs.

## REFERENCES


Cai Xin. Research on Political Education in Employment Guidance of Universities in new situation [D]. Shanxi Normal University, 2012.

Chen Wanqing. 2012 Global Employment Situation Is Even More Severe. Report, 2012.

Hao Zhenwei. Thinking and Exploration of Construction of University Career Guidance procession, Education and Teaching forum, 2014.

Sun Hongqi. Research on Construction of Career Guidance Curriculum System of Universities in New Strategies [J]. Chinese Adult Education, 2011.

Sun Liang. Research on Innovation of Universities Career Guidance System [D]. Anhui University, 2011.

Yang Qian. On the Construction and Perfection of University Student Career Guidance System. Education and Teaching forum, 2014.

Zhu Yanhua. On the innovation of employment guidance mode universities under the new situation [J]. Theory Circle, 2006.

*Frontiers in Computer Education – Wang (ed.)*
*© 2015 Taylor & Francis Group, London, ISBN 978-1-138-02797-8*

# Decomposition and correlation design of a micro-video course content based on knowledge points

M. Wang, H.D. Chen & Y.S. Xue
*Hainan Normal University, Haikou, China*

D. Kang
*Hainan College of Economics and Business, Haikou, China*

ABSTRACT: The division of a micro-video course content is needed in order to think about decomposition and correlation design of the course content. This paper analyses the principle of dividing a micro-video course content, and defines the connotation and classification of knowledge points in a micro-video course. Based on the analysis of instructional objectives, the decomposition of a micro-video course content can be implemented by the methods of classification analysis, interpretative structural modelling, and hierarchy analysis. In addition, based on analysing the type and relations of knowledge points, ontology metadata modelling of knowledge points can make the knowledge points correlate with each other, ensuring the integrity of the micro-video course content, and the reusability of the instructional micro-video with a knowledge point-based design.

## 1 INTRODUCTION

As Khan Academy, MOOC (Massive open online course), and other online course resources have been paid attention to all over the world, the related research on online courses is becoming an important issue. One common characteristic of these course resources is that the main carrier is instructional micro-video. In the research, these kind of courses will be called micro-video courses, which are the total of instructional contents and instructional activities in a certain subject or theme, represented by a micro-teaching video. It includes instructional contents and learning environments with specific teaching objectives and strategies. The main carrier of instructional content is the instructional micro-video which is a kind of instructional and educational micro resource.

Therefore, the micro-video course content is composed of several instructional micro-videos based on instructional objectives and instructional strategies, according to some sequential structures. To ensure the reusability and sharing of course resources, the division of micro-video course content needs to consider the decomposition and correlation of the course content, which means we need to consider how to divide the course content into specific knowledge points, according to instructional objectives. Those knowledge points are correlated with each other, thus the course content cannot only be presented by instructional micro-videos, but must also ensure the integrity of the course content and the reusability of the instructional micro-videos. How to design the decomposition and correlation of a micro-video course content will be discussed and explored in the research.

## 2 THE PRINCIPLE OF DECOMPOSING COURSE CONTENT

The division method and granularity of course content have played a decisive role in the application of knowledge points in the curriculum knowledge system. Whether the decomposition of micro-video course content is scientific and reasonable will be directly related to the practice and application of a micro-video course. The design of the micro-video course content is based on the concept of knowledge point design from the perspective of the micro-level design, therefore, the decomposition of the micro-video course content needs to pay attention to the following principles.

### 2.1 *Loose coupling of course structure content*

Although the micro-video course is composed of a number of instructional micro-videos with intact and independent meaning, which does not mean that the micro-video courses do not have a course structure, the knowledge structure, or that there is no correlation between the videos. The system and structure of the course content do not have fixed standards, and what is more, there will be difference among subjects. However, the loose coupling of a course structure needs to be considered in order to achieve the reconstruction and sharing of the course, and construct dynamic, open, sharing, and personalized courses, which means that the course content is relatively independent. However, the content needs be correlated in some reasonable structure relationship in order to ensure modularity and relevance of the course content.

### 2.2 *The teaching content is formed in several knowledge points*

The micro-video course is a new course form that meets the needs of fragmental learning and micro-learning. In order to conform to the characteristic of fragmental learning style and meet the needs of learning context, instructional micro-videos include a micro and a short knowledge content, which means that the course content needs to be decomposed in a certain model and method. Therefore, in the demand of keeping course systematized and content miniaturized, the decomposition of a micro-video course content is mainly based on- knowledge points design in the research.

### 2.3 *The reusability of knowledge points*

From the perspective of a resource construction, the content design of micro-video courses should consider the characteristics of reusability. In the domain knowledge ontology base, the Reusable Knowledge Point (RPK) has constructed a reusable knowledge object with finer granularity, which has made the knowledge sharing on the knowledge point of domain knowledge ontology, and can effectively improve the efficiency of a course design and development. The knowledge point is a type of an independent and reusable course component. Furthermore, the curriculum always consists of learning objectives and several knowledge points which are from the domain knowledge ontology base.

## 3 THE DECOMPOSITION AND CORRELATION DESIGN OF THE COURSE CONTENT BASED ON A KNOWLEDGE POINT

The decomposition design of a micro-video course content is to divide and design knowledge points, definite teaching sub-objectives based on analysing teaching objectives and learners' characteristics, and then determine the content of knowledge point according to teaching sub-objectives, at last the course content can be divided into a series of knowledge units and knowledge points.

The size of a knowledge point varies from different instructional designers. The structure of curriculum is based on the knowledge point which is the smallest knowledge unit. So we shall analyse that how to carry on the division of knowledge point and the decomposition strategy based on knowledge points below.

### 3.1 *The connotation and classification of a knowledge point*

#### 3.1.1 *The connotation of a knowledge point*

The knowledge point is a general, abstract, and old concept, which means a different content from the different perspective. A knowledge point is a collection of knowledge with relatively intact logical sense from the aspect of content. In this research, the knowledge point refers to the basic unit for transmitting instructional information, and it is a set of knowledge that have relatively intact meaning, and can carry out the relatively independent evaluation. The knowledge point is the micro-videos' smallest learning unit; the size of knowledge point is related to the course structure and content.

#### 3.1.2 *The classification of knowledge points in micro-video courses*

The knowledge point in a micro-video course has a top-down and hierarchical structure. According to the content of knowledge granularity, knowledge points contain a meta-knowledge point and a compound knowledge point. The meta = knowledge point is that the content is indivisible, and the compound knowledge point is composed of two or more than two meta-knowledge points, as shown in Figure 1. The micro-video course can contain one or more knowledge modules, and a knowledge module (knowledge unit) is composed of a number of knowledge points.

The size of knowledge granularity affects the reusability of knowledge points and the length of the instructional micro-video. In general, a knowledge point with smaller granularity contains less knowledge capacity, and a shorter duration of the instructional micro-video. For example, the micro-video course Laundry Tips from the website named several minutes net (http://www.jifenzhong.com/lesson/272) is a knowledge module which consists of nine instructional micro-videos; each instructional micro-video explains one knowledge point. For example, those knowledge points respectively explain how to wash towels, how to clean a bra and panties, how to wash a silk scarf, how to wash shirt collars, and so on. The length of each instructional micro-video is generally not more than 2 minutes.

### 3.2 *Course content decomposition based on teaching objectives*

In order to ensure the effectiveness and integrity of a course content, the course content which is divided into knowledge points, needs to analyse teaching objectives, therefore the decomposition of micro-video course content mainly adopts the following methods.

#### 3.2.1 *Classified analysis*

Classified analysis is where we first need to determine the main categories of the course content, conclude knowledge content into several subjects or modules according to the category, and then decide on the structure and range of the teaching content. It is suitable for the content of courses without rigorous logic levels or programmes, for example, the course case Creative Handicraft in Kindergartens. This research mostly used classified analysis, which classified the handicraft material, such as jars, cotton, reusable bags, cups, paper, and so on, and then analysed and designed creative handicraft according to the material used.

3.2.2 *Interpretative structural modelling*

Interpretative structural modelling is an effective method to analyse and reveal complex structures, which can decompose complex, disorder relationship of each element in the system into clear, multi-level, hierarchical structure. Many courses have adopted this method at Khan Academy.

The processes of using interpretative structural modelling method to analyse a course content are as follows: First, extracting the core concept or the theory of course content, and then carrying on the teaching around these keywords and concepts. Second, finding the relationship among the core keywords, form a knowledge network by using a connection between two knowledge points, and then form a relatively intact course knowledge circle.

Because the structure relationship between these courses content is more complex, in order to promote learners to have the overall framework and macro understanding of course content, we can create relevance between knowledge points by the knowledge map. For example, micro-video courses "trigonometry" at Khan Academy is about the edges and corners' relationship between a plane triangle and a spherical triangle, and it includes 39 instructional micro-videos. Therefore this micro-video course is divided into a number of core concepts, including basic trigonometry, trigonometric, trigonometric equation, trigonometric application, etc., and then the relationships between those concepts is established.

3.2.3 *Analytical hierarchy*

Analytical hierarchy is a type of a content analysis method and an inverse analysis process, which is used to reveal the subordinate skills required by teaching objectives. If the course has a strict knowledge structure and a clear hierarchy content, this method can be adopted. The level of knowledge points is related to the difficulty level of the knowledge points. If the knowledge point is at the bottom, the difficulty level is lower. If the knowledge point is at the top, the difficulty level is higher. For example, a micro-video course Preparatory Arithmetic and Algebra at Khan Academy is a preparatory course for algebra, and both of them have a clear knowledge hierarchy in the whole algebra course structure. Furthermore, there is also a clear hierarchy on knowledge points of multiplication and division in the Preparatory Arithmetic and Algebra course.

3.3 *Knowledge modelling for promoting knowledge correlation*

The correlation and aggregation of the knowledge content is the important guarantee of realizing the integrity and structuralization of a course content and the correlation and systematization of learns' thinking. Knowledge modelling is one of important ways to realize the correlation and aggregation of the knowledge content based on the semantics. There is a description on correlation for knowledge content through ontology modelling for micro-video knowledge points as follows.

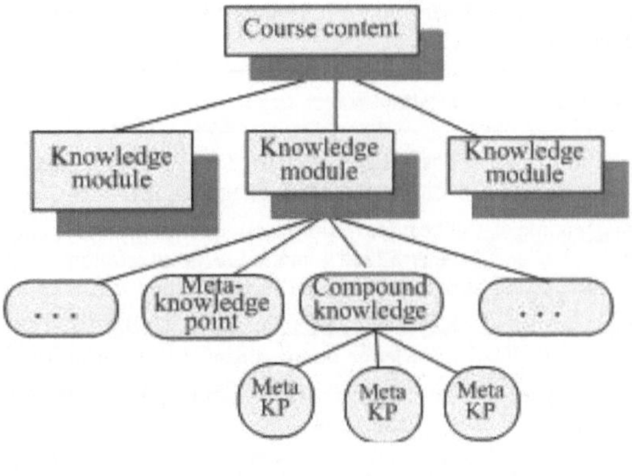


Figure 1. The hierarchical relationship of knowledge points.

3.3.1 *Analysing the type and relationship of knowledge points*

As mentioned earlier in this research, we mainly divided types of knowledge points based on the knowledge nature, namely divided them into factual, conceptual categories, programmes, and principles. Knowledge points-based course designing makes a course into a knowledge network. Therefore, the analysis for the relationship of the knowledge points is good for ensuring and analysing the correlation and its degree, and ensuring the logic and integrity of the course content in emphasizing the independence of a content. The relation of knowledge points includes hierarchy, precursor, correlation, and parallel.

(1) The hierarchical relation of knowledge points. The hierarchical relation of knowledge points is based on horizontal and vertical structure of the knowledge points. Horizontal structure refers to that the relationship between knowledge points is a peer hierarchy relationship and a brother structure. Vertical structure refers to that the relationship is a parent-child hierarchy relationship between knowledge points. Take micro-video courses as examples, the hierarchical relationship of the knowledge point as shown in Figure 1.
(2) The precursor relation of knowledge points. When learning a certain knowledge point, another knowledge point should be mastered first. The relationship between those knowledge points is precursor relations. Take the organic chemistry course at Khan Academy, and the micro-video course Nomenclature of Alkanes as examples, those knowledge points range from easy to difficult and have precursor relations.
(3) The correlation relation of knowledge points. The correlation relation reveals the network structure between knowledge points, and points out that knowledge is composed of a group of mutual connection and interaction nodes. The correlation relation of knowledge points can be divided into two categories: firstly, one to one correlation (1:1). This represents that one knowledge point only corresponds with another one. Secondly, one too many correlations (1:n). This represents that

Table 1. The description of ontology metadata for knowledge points.

| Element | Description |
|---|---|
| <KPid>ID | The unique identification of a knowledge point in the knowledge system |
| <KP>name | The name of knowledge ontology is a highly generalized knowledge point content, and represented by a shorter string |
| <kw>keyword | Keywords of knowledge point set, used for knowledge points retrieval and reusability |
| <type>type | Describing the type of knowledge points and different knowledge content corresponds to different cognitive structure. In the ontology database, the property of <type> can promote the establishing correlation between knowledge point ontology database and cognitive structure |
| <object> | Understanding, memory, and application. According to the nature of knowledge content, there are different learning objectives and requirements. Deciding the objectives and requirements can help learners use micro-video for autonomous learning. |
| <description> | Detailed description of knowledge points |
| <relation> | Relations and correlation between knowledge points, including hierarchical, precursor, correlation, and parallel relation. |

one knowledge point may have a relation with a number of knowledge points. Take the course sine function at Khan Academy as an example, the correlation relationship between knowledge points of definitions to sine function and sine.

(4) The parallel relation of knowledge points. Compared with the above relations, the parallel relation of knowledge points does not have a direct relation and has a weak link. Those knowledge points are parallel relations, which generally happens in the condition that the content is divergent and the logic of structure is not strong.

### 3.3.2 *Ontology metadata model of knowledge points*

In the computer field, ontology mainly refers to the explicit specification of concepts and relations between concepts in a specific domain, and indicates all concepts, relations between those concepts, and concept properties. Ontology of knowledge points is the expression of knowledge contents and the relations between knowledge points. Ontology metadata of knowledge points can be regarded as the description for attributes of knowledge points, which indicates nature, functions, and relations, etc. Therefore, the ontology metadata model of knowledge points for a micro-video mainly contains the following metadata elements: KPid, KP, kw, type, object, description, and relation. The specific descriptions are shown in Table 1.

The description of ontology metadata for knowledge points not only promotes the correlation of knowledge points which have been generated by decomposing a course's content, but also aggregate instructional micro-video and make these instructional micro-videos reuseable.

## 4 CONCLUSION

Decomposition and correlation design are important parts of designing a micro-video course. It can provide effective instructional micro-videos for learners' fragmental learning, and can also achieve the connection between knowledge and aggregation of learning resources by correlation design. This does not only ensure the integrity of the course structure, but also improves the application performance of online courses and ts sharing and construction.


## ACKNOWLEDGMENT

1. This paper is part of the achievements of Hainan production – study – research combination project (No: CXY20140003), National Twelve-Five Planning of Educational Information Technology Research Project (No: 146241926), and Research Fund project for the Doctoral Program of Hainan Normal University.
2. The corresponding author of this paper is Professor Huandong Chen (Email: chd@hainnu.edu.cn)


## REFERENCES


Annette Lamb. Theory to practice: Thematic Learning Environments [EB/OL]. http://eduscapes.com/ladders/themes/thematic.html, 2012-12-25.

About Khan Academy [EB/OL]. http://www.khanacademy.org/about, 2010-12-1.

Alexander McAuley, Bonnie Stewart, George Siemens, Dave Cormier. The mooc model for digital practice [EB/OL]. http://www.elearnspace.org/Articles/MOOC_Final.pdf, 2011-1-10.

Betty, Collis. Information technologies for education and training. In H. Adelsberger, B. Collis & J.M. Pawlowski (Eds). Handbook on information technologies for education and training [M]. New York: Springer, 2002: 1–22.

Jiang Yuefei. Intelligent tutorial planning based on knowledge structure graph [J]. Computer research & development, 1998, 35(9): 787–792.

Kekang He, Yongbai Zheng, Youru Xie. Instructional System Design [M]. Beijing: Beijing Normal University press, 2012.

Mi Wang. On Designing the Content of Micro-Video Courses in a Fragmented Learning Age [D]. Shanghai: East China Normal University, 2013.

Mi Wang, Bin He, Zhiting Zhu. Micro-Video Course: evolution, orientation and application domain [J]. China Educational Technology. 2013, (4): 88–98.

Rowland, G. Designing and Instructional Design, ETR&D, 1993, 41(1), 80.

Shengquan Yu, Xianmin Yang, Gan Cheng. The design and share of learning resource in Ubiquitous learning—the concept and structure of learning cell [J]. Open education research, 2009, 2(15): 47–53.
Wu Zhenghong. Research in Semantic Annotation on Video-Recordings of Classroom Teaching—Constructing a Semantic Model Based on Classroom Observation and Evaluation [D], East China Normal University, 2011.
Ying Peng. Research on Multimedia Web Courses and tools based on Knowledge System [D], Wuhan: Wuhan University, 2010.

*Frontiers in Computer Education – Wang (ed.)*
*© 2015 Taylor & Francis Group, London, ISBN 978-1-138-02797-8*

# First exploration of living space for old people with chronic disease, taking Diabetes Mellitus (DM) in the elderly as an example

Haiyang Gao & Jingyuan Li
*Faculty of Design, Kunming University of Science and Technology, Kunming, Yunnan*

ABSTRACT: With the growing of aging population in China, we pay more attention to the elderly chronic disease, and the care of elderly chronic diseases has gradually become the family and social problem which can't be ignored. This paper summarizes the design contents of living environment of chronic diseases in elder which in consideration of special psychological and physiological characteristics and the nursing care they needed from the perspective of human nature.

*Keywords*: Disabled elderly; chronic disease; diabetes mellitus; living environment

## 1 DIABETES MELLITUS

Diabetes mellitus (DM) is mostly diabetes mellitus type 2, referring to those people who are over sixty years old and in vivo due to insufficient insulin secretion or impaired insulin secretion, cause the human endocrine disorders, leading to human body material metabolism disorder. Thus, metabolic disorders of hyperglycaemia and high blood cholesterol as well as disorders of protein, water, and electrolytes occur. At the same time, diabetes mellitus type 2 patients have high incidence of hyperglycaemia, hypertension, high fat, hyperinsulinism, and also coronary disease, which is called X syndrome or metabolic syndrome.

Meanwhile, DM always coexists with many other geriatric conditions, and hypoglycaemia easily occurs along with accompanying infection, concurrent cardiovascular and cerebrovascular diseases, concurrent kidney diseases, concurrent eye diseases, and concurrent nervous system diseases.

This paper will take the psychological and physiological characteristics of DM patients as examples, and then analyse their environmental elements, in order to explore the environment design points for disabled elderly people.

## 2 DESIGN OF RESIDENTIAL SPACE OF DM

### 2.1 *Barrier to sensory system*

Among all the internal senses, vision is the most important feeling. Besides the apparent vision weakening, the concurrent pathology of the eyes of people

| Physiological function | Characteristics of function weakness |
|---|---|
| Sensory system | As age progresses, vision, hearing, tactile sensation, taste sense, and olfactory sensation all weaken; meanwhile, the ability to adapt to temperature change and a sensitivity to extreme climates weakens. |
| Motional system | Flexibility and the power of muscles gradually weakens. Activities are restricted and the risk of cataclasis increases. The ability to walk and understand weakens. |
| Central nervous system | The ability to react and the concept-forming ability weaken; flexibility and agility are reduced. Ability for keeping one's balance weakens. Energy and power are easily lost and depression easily occurs. |
| Urinary system | Nocturia occurs. Due to increased urination, sleep quality is affected and insomnia becomes more severe. Meanwhile, with the increase of urine in the bladder, uracratia is aggravated. |
| Respiratory system and vasculature | Activities are restricted; fatigue easily occurs; energy and power are easily lost; activities without confidence. |

Figure 1.1. Characteristics of physiological functions of DM patients.
(Source: drawn by the authors)

with DM includes the visual area, the dark adaptation ability the regulating function and the different degrees of color perception. For the mobility of the elderly, observation is the main way to feel the external environment.

The light environment in a residential space is very important. Indoor lighting forms include natural light and artificial light. DM patients will spend most of their time indoors. Therefore the design of the indoor lighting environment should not only satisfy the lighting requirements but also combined with the use of functional.

(1) *Natural lighting*
The most important part in the design of a lighting environment is natural light because the natural light is most consistent with the degree of adaptation to light eyes. Natural light can enable the elderly to feel the difference between day and night more clearly, which is beneficial to their physical and mental health. Meanwhile, ultraviolet rays in natural light do not only kill off harmful bacteria, but also promote the absorption of calcium

In order to provide a comfortable residential environment for the elderly, bedrooms should be arranged to face the best orientation in the whole design. A glass curtain wall can be used to guarantee sufficient light, but appropriate sunshade measures also should be adopted. At the same time, the indoor window glass should not be made of coloured glass that will hinder the vision of the elderly

The most common indoor lighting method in our country is unidirectional lighting. It should avoid dazzle and the depth of a room can not be too long. Meanwhile, increase the artificial lighting in the opposite window wall to balance the illumination or sunshade curtains can be used; when the natural lighting is too strong, some of the rays can be shaded so that the entire indoor lighting is comfortable. Sufficient indoor lighting can help prevent the elderly from falling down and also help them concentrate. But if the light is too strong, will cause confusion, excitement and even affect the walking balance.

The eyes of DM patients are slow in adapting to light. If the light between two spaces is different, a buffer space should be provided between them, in order to leave enough time for eyes to adapt to the change.

(2) *Artificial lighting*
Aiming at the vision barrier caused by DM and in order to satisfy the physiological and psychological needs of the elderly, indirect lighting systems should be used to provide sufficient lighting grades and avoid the strong light when the natural lighting is insufficient and at nights. Meanwhile, the design should try not to use dim lighting; using an all-optical spectrum and light source can reflect more accurate colours of people and objects, and can also reduce eye fatigue.

For the DM patients, because of its physiological function and flexibility significantly impaired eyesight low. In the lighting design, a warm light source should

Figure 2.1.1. Space with sufficient sunshine.
(Source: Authors' photographs)

Figure 2.1.2. Bright artificial lighting.
(Source: Own photographs)

be used in order to improve the illumination level and a bright residential environment helps open the mind of the elderly.

The lighting design of residential spaces should also avoid dazzle. The sleeping time of DM patients is longer therefore an indirect lighting method should be adopted. To avoid the elderly directly see the light source on the bed. Lamps and lanterns should be of diffused and reflective types, all the lamps should be installed with appropriate shades, and the height of lamps should be restricted. Indoor walls, suspended ceilings, and floors, all should all be decorated using non-gloss paints, in order to avoid creating dazzle.

Meanwhile, the use of auxiliary light sources can be a practical way to make adjustments for the entire indoor environment. Bedside lamps not only facilitate the individual activities of the elderly but also satisfy the lighting needs of nursing staff when caring for the elderly. Meanwhile, the bedside lights can strengthen the old man's personal space, enhance the confidentiality.

(3) *Colour of environment*
The living space of color suitable for harmonic color smart, low saturation, the interior space of the colors should be coordinated and unified. Among them, achieve the indicating effect local prominent in color, the common areas such as stairs, emergency exits, opening direction of fire door, floor signs, room signs, reminder signs, telephones, doorbells, lifts, and safety pilot lamps should be indicated with outstanding colours.

With weakening vision caused by DM, the feeling towards colour for the elderly will change accordingly.

Figure 2.1.3. Down outstanding colour contrast.
(Source: Authors' photographs)

Figure 2.2. Wooden floor.
(Source: Authors' photographs)

For example, it is difficult to identify the yellow. Therefore, a sign board with yellow characters on a white background should be avoided. Yellow on green should also be avoided, will the people of skin color reflection looks like a patient. Meanwhile, colours of a similar hue should be avoided because this combination can cause elderly confusion.

DM patients have difficulties in distinguishing space so when the space is changed, the height of the ground visually changes, and in some places which need a special reminder, an apparent border line should be provided.. If necessary, hues should be used for emphasis. The top tread of the stairs should be highlighted by colours; the front of the Stair tread should set a clear color antiskid warning.

(4) *Space temperature and humidity*
The ability of adapting to temperature change and sensation to extreme climates of DM patients weakens, therefore the change of indoor temperature and humidity will also influence the health of the elderly. Generally speaking, summer temperatures between 21–32°C and 24–26°C are the most ideal; winter temperatures between 16–20°C and 16–18°C are the most ideal.

## 2.2 *Barrier to motional system*

The flexibility and the power of the muscles of DM patients gradually weaken. Therefore, activities are restricted, the risk of fracture increases, and the ability to walk and understand weakens. Mild diabetes mellitus has symptoms of asymmetric progressive muscular weakness, along with muscle pain, in the muscle of cingulum extremitatis inferioris and the pectoral girdle, which is called diabetic amyotrophy.

Handrail materials should have a good durability, and be of high strength and fine tenacity. Handrails are usually made of strengthened plastics or plastic steel supported by lining metal materials, which has the advantages of being easy to clean, anti-corrosive, anti static and do not fade.

In an indoor environment, the ground is the space interface that people contact most; the tactile sensation that the ground reflects to people is the most direct. Within the residential space of the elderly, soft wooden floors and Elastic plastic materials are ideal materials, which have the ability for heat preservation and noise absorption, as well as elasticity and a soft tactile sensation. What is more, due to the walking difficulties and special physiological problems of the elderly, non-slip, dirt resistant, and easy to clean materials should also be considered.

Within the elderly's residential space, blankets are not suitable. Thick blankets always lead to tripping over; by contrast, light blankets do not have that defect. But no matter what kind of material is not easy to clean the carpet, at the same time as the bacteria reproduction of space.

Emergency alarm systems should be installed. Television sets, lamps, and curtains should be set up to be close at hand. In addition, when the elderly encounter a danger or the symptoms of illness occur, as long as they press the button of the emergency alarm system, doctors and nurses should arrive immediately.

## 2.3 *Barrier to central nervous system*

Ability to react and form concepts weakens; flexibility and agility reduces; ability for keeping one's balance weakens; energy and power are easily lost, and depression easily occurs.

Sound will influence one's mood and the attitude toward objective things. Too much silence will aggravate loneliness but too much noise will influence sleep and health. The biggest factor that influences the hearing of the elderly is noise.

Doors and windows are the main channels for spreading sound. Indoor doors and windows should do noise reduction and also be installed with double-glazed glass and hollow sounding insulating doors.

Within the residential space, the areas where furniture touches the ground should be equipped with soft mats. Interior can be used sound-absorbing performance good decoration materials, to weaken the effect of noise on indoor environment. In addition, should use the reduction of stone and suction noise effect of poor material.

A pleasant smelling environment is also very important. Odours and emotions have strong connections, which will directly affect people's mood and comfort. A pleasant smell will calm them down and release psychological pressure.

Figure 2.3. Soft and warm bedroom.
(Source: Own photographs)

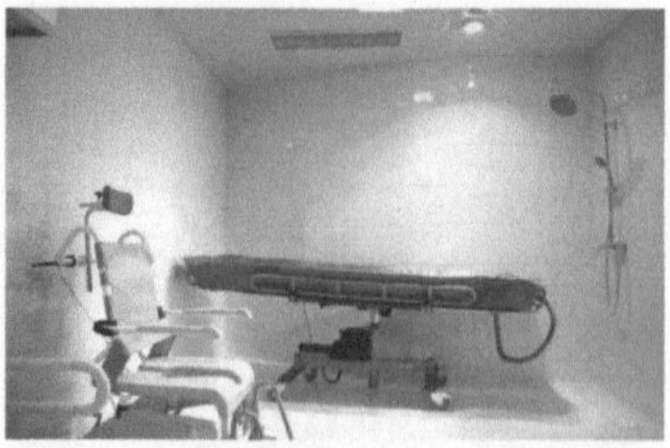

Figure 2.4. Down Modernized bathroom facilities.
(Source: Authors' photographs)

Green plants is quite important in improving air quality. Plants not only absorb carbon dioxide and release oxygen, but also get rid of harmful materials such as formaldehyde.

### 2.4 *Barrier to urinary system*

Nocturia occurs. Due to increased urination, sleep quality is affected and insomnia becomes more severe. Meanwhile, through the increase of urine in the bladder, uracratia is aggravated.

The time taken for DM patients to adapt to a dark environment is longer. When they go to the bathroom in the night, a prolonged light contrast will lead to visual barriers, therefore a night light or lamps of adjustable brightness can be used to solve the problem. In addition, if it is not convenient for the elderly to search for the switch at night, the switch should be set up in a convenient place, such as the front of the bed. A return circuit switch, an induction switch, or a night light of low brightness in the corridors can be adopted. In residential spaces, a perfect intelligent lighting system can help promote life quality.

In addition, a snow bright bathroom will perplex the aged, therefore dark walls and white sanitary ware will easily help the elderly find the toilet so as to reduce incontinence.

If urination at night increases, the consequent noise will disrupt other elderly people, so the drainage system should do denoising processing. Appropriate sanitation units should be chosen. If possible, use the quiet toilet

Unpleasant odours will cause breathing and heartbeat rates to become too fast, which can further influence health. Especially for those who spend most of their time in bed, a single room or one's own bathroom can reduce this influence. A ventilation system should be installed, which will help raise the air quality and create a pleasant smelling environment.

## 3 SUMMARY

Although nowadays the treatment for DM in our country is still limited, but we can investigate the key details of diabetes by elderly vision, hearing, touch and smell the living space environment, into the humane care, to provide a warm and comfortable environment for the elderly with chronic diseases, so as to improve their living feeling a positive.

## REFERENCES


An Qi, Zhou Bo, Zhang Linzi, Liu Hui. Environmental design for Disabled Elders [J] Settlement, 2013

Liu Xuyang, Liu Jia, Li Jing. Clinical Diagnosis and Treatment of Complications of DM [J] China Practical Medicine, 2007

Wang Jiangping. Planning and Design of External Residential Environment for the Aged [M] Beijing: China Electric Power Press, 2009

Wei Guiling. Medical Treatment of Senile Disease and Home Care [M] Beijing: China Medical Science and Technology Press. 2001

Xia Xiaoping. Geriatric Nursing [M] Beijing: People's Medical Press. 2004

Xu Leiqing, Yang Gongxia. Environmental Psychology [M] Shanghai: Tongji University Press, 2002

Zhang Kaiti, Sun Lujun, Mou Xinyu, Wang Haitao, Li Mingzhen. Researches on the Disabled Elders Conditions National Wide [J] Disabled Study, 2011

*Frontiers in Computer Education – Wang (ed.)*
*© 2015 Taylor & Francis Group, London, ISBN 978-1-138-02797-8*

# Teaching reform and practice of industry design specialty based on the Conceive, Design, Implement, and Operate (CDIO) engineering education pattern

Ge Lu
*Department of School of Art & Design, Zhengzhou Institute of Aeronautical Industry Management, Zhengzhou, China*

ABSTRACT: CDIO is an engineering educational concept. This paper analyses the feasibility that puts the CDIO engineering education pattern into the industrial design major. For industrial design major of the Zhengzhou Institute of Aeronautical Industry Management, the paper introduces the preliminary exploration and experience in guiding the mode into the teaching reform of industrial design specialty, in order to provide a teaching reference for the industrial design course at undergraduate colleges and universities.

## 1 MODE OF HIGHER ENGINEERING EDUCATION

CDIO stands for Conceive, Design, Implement, and Operate. It is currently a concept of talent cultivation of higher engineering education, its name inspired by the life cycle processes of product/system, which embodies the life circle from development stage to the modify and abandon stage of the modern industrial products. Based on the current situation, CDIO higher engineering education model takes the conception, design, implementation, and operation as the carrier to train students' ability for engineering, which includes not only academic knowledge, but also students' life-long learning skills, team communication ability, and system control ability.

CDIO emphasizes cultivating the comprehensive innovative capability of students through the project process (concept-design-implement-runs) to meet the employment needs of the industry, which coincides with the basic ideas of professional education. CDIO educational philosophy has a strong systematic rigor and operability, which is inspiring and instructive for the improvement of the professional training programme.

## 2 TEACHING REFORM AND PRACTICE OF INDUSTRIAL DESIGN

### *a) Curriculum integration*

According to engineering requirements for higher engineering education, engineering colleges and universities should offer four courses: personality education curriculum, engineering science and natural science courses, social sciences, a humanities science course, and an engineering course. Each course may include a core course (compulsory), a series of elective courses and the corresponding teaching practice (experimental courses, course-related training activities, internships, social activities, etc.). Industrial design courses present various professional courses and specialized courses for reintegration based on the specific needs of professional settings, forming a course which takes the product design as the core course and consolidates the basic knowledge. Practical teaching includes a corresponding curriculum (transport thematic design, furniture design), an experimental course, a graduation project (actual design or simulation design, engineering, and testing) and a proper engineering training. Graduation design should be prompted to strengthen students' integrated application of engineering knowledge and engineering capacity.

In the training programme of industrial design, in accordance with the intellectual content, an integrated curriculum can be divided into: modelling, engineering, engineering technology, design theory and expression, humanities and social sciences, and economics management. Project courses can be divided into: professional foundation, foundation design, improvement, and innovation. Practice teaching includes: internship, curriculum design, and graduation design.

A CDIO model places great emphasis on the cooperation of the project class, integration courses, and practice teaching. Figure I lists the correspondence relationship among the project class, the integration course, and practical teaching. CDIO-based industrial design is different from the traditional three-section structure (basic course- basic specialized courses-specialized courses). The former takes

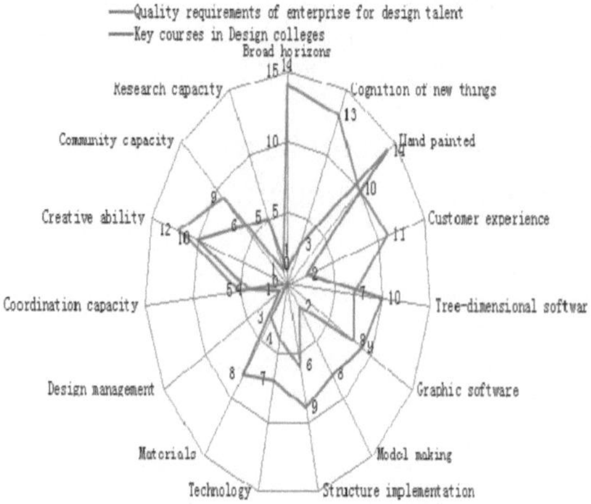


Figure 1. Differences between enterprise needs and college training.

capacity development as the core which develops students' comprehensive ability of problem finding, analysis and problem-solving, and communication and cooperation through the project course. In the meantime, it prompts students' self-study ability. This kind of structure may cultivate the students with a strong ability to learn, solve new problems, and adapt to new environments. Today, the social environment changes very quickly. New problems and new situations require practitioners to have strong abilities for problem-solving and innovation.

Through the CDIO approach in accordance with the project life cycle, related courses can be associated by teaching, learning and practicing, with the result of avoiding unnecessary duplication, and enabling students to master the course linkage between knowledge and the solved synthesis problems. A CDIO approach integrates teaching plans and curricula, breaks the barriers between teachers and courses, and better supports professional training goals.

*b) Project teaching practice*

Traditional teaching emphasizes knowledge specialization, and ignores knowledge of relevance and comprehensiveness. Therefore, industrial design education curricula at colleges and universities, and the enterprises' demand for industrial designers have certain deviations, as shown in Figure 1.

A CDIO mode of engineering education enables students to learn and practice according to the project cycle, which is fully in line with the teaching of industrial design education. Therefore, industrial design specialty arranges teaching for project teaching sequencing, taking projects as motivation to strengthen the skills, and expertise learning to implement project teaching.

Real designing is not to be found in the book. The product will encounter countless details likely to impact the final outcome from the bidding to the final sale. Throughout the production process, ranging from the early market demand and idea generation, to design, manufacture, sales, and feedback, a designer is involved in the entire process. In between, there are problems that can be solved using books and school experience, or details can be handled by accumulated experience. A CDIO project advocates "learning out of practicing", that is, the students' abilities are improved in the courses of practice. The philosophy of "learning out of practicing" lies in the practical programme, which helps students not only to gain theoretical knowledge but also to culticate team spirit though practice.

With school-enterprise cooperation teaching, students are required to discuss their idea with the enterprise while designing, to make proper sequence in accordance with work procedures while programme organizing, to implement in accordance with the conception, design, action, operation while carrying out programme, to know exactly about the production process. Students improved their capacity and accumulated experience through practice. For the implementation of a project-teaching concept, we joined the Industrial Design Association, the County of Henan province people's government, as well as companies such as Jiangsu and Zhejiang, to introduce the enterprise's production and management within schools, and supervise the teaching process and quality of the training base together. Colleges offer office and production sites, the training classes are on their sites, the equipment in the training class and the company's production equipment are fully compatible, and practice is in accordance with the company's production processes and the technical requirements. At present, the College of Art and Design Research Center covers an area of more than 3,000 square metres; instruments amount to 5.3 million dollars. The establishment of the Innovation Lab especially has provided a platform for exchanges and cooperation. During implementation, we introduce a teaching model of team cooperation. A team is usually made up of 6–8 students by free combination, captain election, and a clear division of responsibilities. After the completion of project ideas and project designs, a representative produces a design report, which will be evaluated and scored by all the members. The teams approved continue sequential projects while the teams unapproved have to redesign the current project. After the project, students are required to make a summary and share with each other, so as to develop the design further, and improve and consolidate the students' ability for team communication, coordination, and cooperation.

At the same time, students at different grades are trained to participate in industrial design competitions at home and abroad, in accordance with the differences of teaching plans and the mastery of professional skills. Exercising students' comprehensive practical abilities through competitions, effectively implements the concept of "learning by game" and builds their self-confidence for their future work. In recent years, substantially increasing are not only the number of students awarded but also the awarding levels in industrial design competitions at home and abroad, which witnesses the periodical success in the practice of teaching reform.

In the teaching courses, a teacher should establish small-scale practical projects according to the teaching needs. This should help students to understand more about the course content and application through curriculum development project training. The current curriculum projects mainly in the form of programme operations, such as a ceramic design in the course of Basis of Product Design. This project requires students to collect ceramics-related info, including brands, prices, categories, styles, functions, and even to bind the relevant paper info into booklets. Students will take the initiative to collect information during the implementation process, and get a more profound understanding of the properties and uses of the materials by comparison. Design is fundamental to the project. In terms of cultivating students' practical capability, colleges should take cooperation between schools and enterprises as the main line of the project, and make students learn by doing, which is supported by the competitions at home and abroad, and is based on the simulation course and the core curriculum.

*c) Deepening the undergraduate tutorial system*

Teachers are another important element in CDIO engineering education. Teachers are the keys in cultivating innovative talents and practicing the CDIO educational mode. The undergraduate tutorial system will improve college students' ability for innovation, and is a useful addition to high-quality talent training and teaching quality improvement. The CDIO model is implemented with extremely high requirements for teachers. To train students with the required capacity in CDIO, teachers should be well-prepared with a wealth of experience, a long-term cooperation with enterprises in all fields, and a deeper insight of the CDIO model. The long-term discipline-orientation in China resulted in the separation of engineering education and practice, and teachers lack comprehensive ability and practical ability training, which is a great challenge for us.

The CDIO engineering education reform, as far as teachers are concerned, provides certain opportunities and offers as a beneficial platform. With the "taking in, send out (inviting professionals out of college to give lectures and sending out teachers to be trained outside)" approach, teachers have been involved in the design of enterprise practice, improved their teaching ability, their engineering practice ability, and taken in the teaching contents and methods reform. In the practice of tutor systems for undergraduates, teachers can further contribute to effective implementation of the CDIO mode.

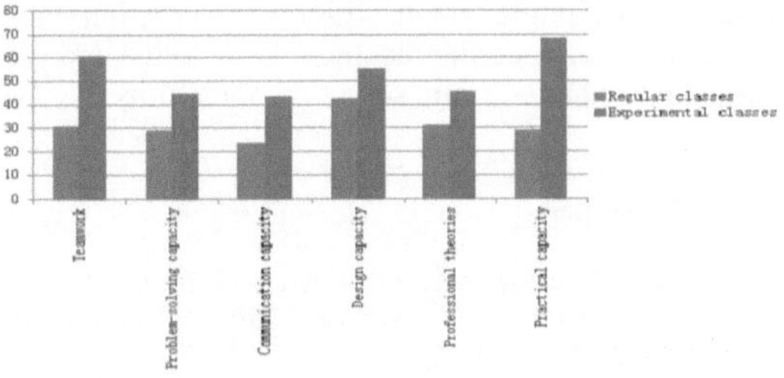


Figure 2. Learning Effect Analysis of CDIO Mode.

## 3 TEACHING EFFECT ANALYSIS OF CDIO MODE IN ENGINEERING EDUCATION

The CDIO mode derives from Europe and America. The reform involves various aspects of the reform of engineering education, and may have positive implications to colleges from different countries at different levels, and even with different circumstances. This study mainly selected the CDIO pilot courses as study objects, to clarify achievements and the existing problems through the investigation and analysis of students' learning outcomes and satisfaction levels.

The CDIO mode requires professional theoretical knowledge, practical abilities, design skills, teamwork, communication skills, finding and problem-solving abilities, and a sense of responsibility.

Thus, it is necessary to examine the learning effect, so as to examine the application effect of the CDIO mode. Through the survey found (as shown in Figure 2) that the improvement on students' teamwork skills, practical skills, and communication skills were quite obvious, while the improvement on their theoretical knowledge, designing capability, and problem solving capabilities was not so visible. Pilot classes were carried out in basic courses for freshers and sophomores, the CDIO project-based teaching is not yet fully expanded, and therefore, once the CDIO project-based teaching is applied over the next two years, the improvement of students' designing ability will be much more visible.

## 4 CONCLUSION

Through the CDIO mode, the reform and practice on the teaching of industrial designing has exerted positive influence on the traditional teaching mode. Through the integration of courses and practice of project teaching, the CDIO teaching training should fit the teaching contents, and improve teaching methods to enhance the ability to solve practical problems. Through constant searching and reforming in the courses of teaching practice, interdisciplinary industrial designers will be cultivated in accordance with the general background of "created in China" which comes out of the requirement on industrial transformation and upgrading development. That is, new industrial designing talents with design creativity and humanities, engineering technical knowledge and art, and ethical standards and social responsibility.

## REFERENCES


Chi Shiwei, Teaching Reform of Art and Design Majors in Higher Vocational Based on CDIO Mode [J]. Art Education, 2013, (6).

Gao Bingxue, Shi Xiaobin. Basic Structure of the Program of Industrial Design Specialty Based on CDIO Mode [J]. Art and Design, 2010, (3).

Han Yan, Feng Xingjie, Liang Zhixing, Zhang Jing. Teaching Effect Analysis of CDIO Mode [J]. Journal of Civil Aviation University of China, 2012, (8).

Tao Yongfang, Shang Cunhui. Inspiration of CDIO Framework to the Innovation of Higher Engineering Education [J]. Higher education Research in China, 2006, (11).

Wang Xuefeng, Cao Rong. Engineering and the Reform of Higher Engineering Education [J]. Higher Engineering Education Study, 2006, (4).

Wang Gang. Study and Thinking of CDIO Engineering Education Mode [J]. Higher Education Research in China, 2009, (5).

Wang Baozhong, Han Ju. Educational Reform and Practice of Packaging Engineering Major Based on CDIO Teaching Mode [J]. Journal of Packaging, 2013, (4).

Zuo Yuanzhi, Jiang Runhua, Yang Xiaoping. Innovative Design-oriented CDIO Engineering Education Model [J]. Journal of Dongguan University of Technology, 2010, (6).

*Frontiers in Computer Education – Wang (ed.)*
*© 2015 Taylor & Francis Group, London, ISBN 978-1-138-02797-8*

# Focusing on the mental health of rural left-behind children

Xiaoling Yang
*Baoding Vocational and Technical College*

ABSTRACT: The children left behind are a special group, the relevance ratio of their mental health problems is higher, and their mental health is worrying. Their psychological problems are the results of the interaction among many factors including school, community, parents, guardians and left-behind children, and forming a joint force from schools, families and communities is the main measure to solve the psychological problems of children left behind.

*Keywords*: left-behind children, mental health, psychological problems

"Left-behind children" refer to children who are left at home or staying with relatives due to their parents working outside hometown, and they lead lives without parents for a long time. Specifically, it refers to children left behind in rural areas and separated from one or both parents. According to the estimation of National Bureau of Statistics, the number of "left-behind children" in rural China is more than 58,000,000. With the transfer of surplus rural labor to urban area, the number of rural left-behind children is increasing. As these left-behind children are in the critical period of growing and developing, receiving compulsory education, and initially forming of enlightenment and view of life and world, the perennial lack of parental and family care and education would adversely affect their growth.

## 1 PSYCHOLOGICAL PROBLEMS OF LEFT-BEHIND CHILDREN

### 1.1 *Abnormal development of individual mental characteristics*

For personality traits, left-behind children often show two tendencies: introversion, social withdrawal, poor communication skills with people; irritability, impulsion or being too naughty, making some special action to attract the teacher to notice themselves.

### 1.2 *Deviant behavior*

Due to lack of timely and effective constraints and education, some children left behind are regardless of discipline and have deviant behavior. Many left-behind children do not follow guardians' instruction at home, and have poor moral character in schools, for example, non-compliance with rules and regulations, often having the behavior of being late, absenteeism, truancy, lying, fights, bullying and so on, some obsessing with billiard room, Internet bar and game room, and even hanging out with some bad adults in the society.

### 1.3 *Poor endurance for frustration*

In the face of their own discomfort or difficulties, left-behind children are often confused and overwhelmed, represented as slow action and weak willpower. Some children feel life is boring, and some even have the idea of suicide.

### 1.4 *Lack of leaning ability*

Most of "left-behind children" lack of initiative, self-awareness in studying, and are absent-minded and careless in coping with trouble. Part of "left-behind children" develop a new "uselessness of study" theory. Seeing parents who haven't been to school can still make money outside, the children in the rural areas begin to feel "there is no distinction between studying and not studying." For them, that "knowledge changes the fate" is just nonsense.

### 1.5 *Severe inferiority complex*

Most children have mind of rivalry and sense of pride, but as left-behind children have no reliance and protection because of the absence of parents, compared with the children with parents around, they prone to generate severe inferiority complex, and some even give up on themselves, lose confidence.

### 1.6 *Strong psychological dependence*

Children left behind are mostly raised by grandparents who often give their excessive care and protection, which cause the strong psychological dependence of children.

### 1.7 *Poor interpersonal skills*

When communicating with people, some left-behind children are timid, asocial and suspicious. When they are confronted with troubles, they would not tell classmates, not mention to communicate with the teacher.

## 2 REASONS BEHIND THE PSYCHOLOGICAL PROBLEMS OF LEFT-BEHIND CHILDREN

Modern psychology suggests that children's emotional capability is mainly learned from mother, and social capability of communicating with people is primarily learned from father. Therefore, in a family, the axis composed by both the father and mother is most conducive to the overall and healthy growth of children. Parents are children's "first teachers", and they play a vital role in their children's education, emotional training and growth.

The greatest need for love of human is unconditional love, that is mothers' love for their children. Humanity's greatest fear is to be abandoned and forgotten. The mother's love is unconditional inclusiveness, and this unconditional love will make people feel a deep "sense of security." Once people have a sense of security, feelings of confidence, stability, and free will then be produced. So, this need for love is the most basic need of human. In the growing process of children, it is necessary to listen to them and make them feel understood and accepted. And young people especially need to be listened to and be understood, because they have a lot of pressure in life and learning but have nowhere to express. Moreover, due to the hormonal influence, their emotions are rather unstable, so that if the parents can keep abreast of them and give care, support and encouragement, it will certainly be of great help to them. But parents of left-behind children work outside throughout the year, rarely can timely and effectively communicate with their children. When the children encounter confusion in the growing process, they have no right person to talk and ask, and various psychological problems arise because of this chronic lack of affection and care.

Piaget's "Cognitive Development Theory" puts forward the theory of moral cognitive development, the forming process of human morality is from heteronomy to autonomy. Heteronomy is not only a necessary stage, but also a necessary prerequisite to achieve moral autonomy of inner morality, it does not play its role naturally, and only through heteronomy training supported by external coercive power can people learn to consciously abide by the code of ethics. The moral behavior development of "left-behind children" is exactly at the stage of heteronomy, and the absence of parents and lack of timely and effective restraint and education from guardians lead to that they do not know the boundaries of words and deeds, have behavioral deviations, and some do whatever they want, if there is no one promptly shouting "stop", they will continue exploring, and they may test the boundaries of the law after entering into society.

## 3 MEASURES TO IMPROVE THE MENTAL HEALTH OF LEFT-BEHIND CHILDREN

### 3.1 *Parents of left-behind children should change their concepts, and attach importance to the overall and healthy development of children*

Many people believe that as long as enough money is left to the children, the children can accept higher education, and this is their responsibility for the child, and also the value of their hard work. This is a wrong one-sided concept which is often counterproductive. For each parent, the obligations and responsibilities taken for granted are to provide the necessary condition of material life, education and security for the children, and also to give their children family warmth, ensuring their healthy development in terms of emotion, morality, behavior and psychology. So parents should encourage children to communicate with their relatives, classmates and teachers, focus on the emotional exchange and psychological communication with children, and make children understand and respect themselves by sharing the experience of working, to help children form a correct view of life. In addition to telephone contact, parents can use the mode of letters, for experts advocates that writing letters between parents and children can better express feelings and opinions. Besides, parents should always contact with the guardian, urge the guardian to exercise supervision responsibility; contact regularly with the school, and truly pay attention to the child's growth.

### 3.2 *School should pay particular attention to the mental health of left-behind children*

For the left-behind children, the role of school education will be greater than that of the other children in general. A variety of methods should be used to standardize the daily behavior of children left behind, to achieve the combination of heteronomy and autonomy, which are the key of children education. Let the left-behind children "learn to be a good person , learn to deal with matters, learn to learn, learn to live and learn to care", which is the ultimate goal of left-behind children's education.

Mental health course should be offered by the school, mental health education seminars should be regularly organized, and "private message mailbox" should be set to help "left-behind children" solve problems that they have nobody to talk to, so as to unlock the "pimple" of mind.

Broadening the forms of education, and carrying out a variety of extracurricular activities, to create a good campus culture. Schools can offer activity classes once a week, focus on "self-study, self-discipline, self-reliance, self-care," teach from the contents like science and technology, reading, working, communication, art and so on, and correctly guide students to arrange their leisure time reasonably, to improve quality of life .

In order to foster a sense of cooperation of left-behind children and promote interpersonal relationships, they can be allowed to organize their own team activities. Children can organize into a team according to their own wishes, interests and hobbies instead of class, and the activities of this team can be presided by the one who devise the scheme that is acceptable to all team members, so that everyone has a chance to perform and exercise.

Carrying out "I'm perfect" activities. Students actively enroll in the activities, to demonstrate their learned skills in extracurricular activities, so as to develop their self-confidence, and make their depression and inferiority complex disappeared in the praise of others.

### 3.3 *Psychological support from volunteers*

Strengthening social propaganda and supervision, to create an atmosphere for a whole society's care to "left-behind" children. The media, units, organizations and individuals should try their best to make efforts for the "left-behind" children, establish a "committee of concerning about left-behind children", appropriately carry out activities for children to make up for their emptiness after leaving their parents. Cadres and workers of organs and institutions, village cadres, capable party members and people from all works of life in the community can help left-behind children as a pair according to the principle of "voluntariness and equality" and through the coordination of "committee of concerning about left-behind children" or voluntary claim, and then they become the "surrogate parent" of left-behind children and fulfill the responsibilities of "surrogate parent". "Surrogate parent" is required to, in accordance with the actual situation of left-behind children, take care of the child carefully, teach patiently in learning, guide patiently in mentality, and be the insiders of children's lives, the guider in learning, the protector of growth. Mobilizing veteran cadres and senior party members to be out of school voluntary counselor, to conduct ideological and moral education of left-behind children.

Rural left-behind children are a phenomenon that will exist for a fairly long period, and in order to solve the problem fundamentally, attention and cooperation from families, schools, communities and government are required.

## ABOUT THE AUTHOR

Yang Xiaoling (1972- ), Associate Professor, Department of Social Sciences, Baoding Vocational and Technical College, Engaged in Mental Health Education.

## REFERENCE

Zhang Yinhua, Opening Mind, Growing While Playing — Research on Leisure Counseling of "Left-behind Children" in Rural Primary Schools

*Frontiers in Computer Education – Wang (ed.)*
*© 2015 Taylor & Francis Group, London, ISBN 978-1-138-02797-8*

# A study on the implementation of the development policies of social work in Shenzhen

Chunyu Zhang
*North China Electric Power Univercity, Baoding, Hebei, China*

Yixiao Wang
*Xinlian College Of Henan Normal Univercity, Zhengzhou, Henan, China*

ABSTRACT: As an important means of social construction, social work has been attached great importance by the party and the government. The relevant policy documents become an important condition and guarantee to the development of social work. Based on the policy documents of social work, especially the development of social work of Shenzhen, a summary of the relevant experience of policies' implementation is presented in order to provide ideas for the development of social work in other provinces and cities.
*Keywords*: Social work, implementation experience, deepen reform

## 1 INTRODUCTION

China's economic development becomes rapid since the implementation of reform. But as an important task during the period of social transformation, the situation of social construction is increasingly serious. Under domestic experience, social work plays an important role in addressing social problems. In recent years, the social work began in the country especially in the Pearl River Delta region . As the forefront of China's economic and social development, Shenzhen is one of the cities in China's earlier exploration of social work. Based on the development of social work in Shenzhen, we study lessons of policy implementation, so as to develop other provinces and cities nationwide and to provide ideas in social work.

## 2 RELEVANT POLICY DOCUMENTS

### 2.1 *Relevant policies of the central government*

In October 2006, the Sixth Plenary Session of its Sixteenth Central Committee of Chinese Communist Party clearly stated, "building great social work personnel." It is focused on the construction of social work personnel becomes the basis of social work development policy in China. During 2007–2009, policies for the development of social work in China is to start. In terms of experimental work, job evaluation, social collaboration, academic education, and training of cadres were actively explored and achieved some success. During 2010 to 2012, China began to promote the development of social work policy, and showed it a high degree of attention, multi-sectoral joint collaboration, and focus on the characteristics of long-term planning.

Under the guidance of the central policy document, different countries began to implement the theory and practice of social work. The representative is Shenzhen mode.

### 2.2 *The interpretation and implementation of the central policy in Shenzhen*

In the central policy guidance, Shenzhen began to develop social work as a breakthrough in the public administration reform in 2007, it promulgated and implemented a municipal relevant policy documents, establishing the social work system with Shenzhen characteristics. The policy response to the situation of Shenzhen can be summarized as follows:

First, the common understanding. A delegation led by the Shenzhen Municipal learned theoretical knowledge, learned from experiences around, the Shenzhen municipal government agreed that as a kind of institutional arrangement, social work will improve the social welfare system, and it has an important significance in the innovation of the social management system.

Second, an innovation system. In order to ensure the authority of decision-making, and promote the effectiveness of the work, the establishment of the Shenzhen municipal government implemented a procedure of coordinating bodies and strengthening the system of combining department functions. In 2007, the Shenzhen Municipal established a new work system, which is "catch total sector-led organization, specifically responsible for civil affairs departments, actively cooperating with relevant departments."

Lastly, establishing a system. In October 2007, on the basis of learning the experiences of Hong Kong, Taiwan, Shanghai, and other places, the Shenzhen municipal government issued policies such as "Opinions on Strengthening Social Work qualified personnel". As the programmatic document to promote the development of social work, the policy made institutional arrangements to the social work management system and the specific operation.

Through the interpretation and implementation of central policies, Shenzhen formed the social work development pattern of "the unified leadership of Party committees, government guidance, the folk organization operation, and broad public participation", which has been hailed as the "Shenzhen model".

## 3 SHENZHEN POLICY IMPLEMENTATION EXPERIENCE

Shenzhen social work related to the implementation of the central development policy provides a rich and profound inspiration for other provinces, which is a milestone in the professional development of social work in mainland China.

### 3.1 *The party committee and government attention is the prerequisite*

In order to co-ordinate economic and social development, the Shenzhen municipal government takes "people-orientation" as the starting point and ruling goal and attaches great importance to social work. To develop social work in China one must get party committees' and governments' attention and support. This is decided by the characteristics of China's social and political system, but also by the phasic requirements of China's social work development.

### 3.2 *Fiscal input is the guarantee*

The success of Shenzhen's promotion of the development of social work is the due to the huge financial investment in government purchase of social work services. The government must seize the opportunity to vigorously promote the public finance, establish and improve the "government purchasing social work service" mechanism.

Through the implementation of a fair bidding system,we can attract more social organizations to undertake public service projects commissioned by the government.

### 3.3 *The innovation under the central document of policy is the fundamental*

Shenzhen took advantage of the region being adjacent to Hong Kong and Macao, following the courage to reform the traditional, and explored the development model which is "under the leadership of the government and social organizations development". "Shenzhen model" is required from the regional development reality to realize innovation, which is the efficient execution of policy.

## 4 SHENZHEN'S INADEQUATE IMPLEMENTATION OF POLICIES

Specific policy implementation exist certain irrationality, hindered the further development of the localization of social work, mainly reflected in the following aspects:

### 4.1 *The intensity of policy implementation still needs to be improved*

In policy implementation, the various stakeholders' roles, responsibilities, and locations are not clear. Meanwhile, due to the people's understanding of social work exist a big gap, resulting in some departments not respecting the basic concepts and working methods of social work as well as social work intervention in some departments.

### 4.2 *Absence of legislation limiting the development of social work*

At present, China has not yet established a dedicated social work law, especially in the tax incentives and exemption regime, which have a significant impact on the resources absorbing of the social work field. This will result in an effective donation bill not being issued, thus affecting businesses, foundations, and other donors' enthusiasm. In addition, the absence of law leads to a lack of legal status of social workers,low salaries ,a lack of social recognition, non-standardized rewards and punishments assessment, which will hinder the process of the construction of social work personnel.

### 4.3 *Assessment and regulation has not been effectively incorporated into the social work assessment*

First, the current evaluation is focused on the evaluation results, rather than the process supervision , leading it difficult to make timely appropriate treatment-related violations. Second, social service agencies is short of rewards and penalties, not only lead to the institutions and social workers' lack of initiative,but it is difficult to form the deterrent force because the penalty is not timely. Furthermore, the government has yet to establish a supervision system of social organization, causes more irregularities.

### 4.4 *Mechanism of government purchase of social services is to be further optimized*

In the process of the implementation of government purchase, there is no detailed purchase criteria, resulting in the purchase behaviour itself having a certain degree of randomness, which is detrimental for the orderly competition and healthy development of social service agencies. In addition, the ability to raise their own resources and professional services capabilities is relatively weak, social service agencies are dependent on the government as the most important source of funds, so that the formation of social service agencies rely on government funding inertia, weakening the ability to integrate its own social resources.

## 5 FURTHER REFORM OF THE IMPLEMENTATION OF POLICY IN SHENZHEN

Social work is the inevitable outcome of China's economic and social development. In the modern society it is there to meet the needs of the people, basic services, and the means of solving social problems. Thus, the field of social work should be focused on promoting the social construction. Combined with difficulties and problems occurring in the development of the Shenzhen Social Work experience and implementation of the development of social work in the implementation of central policies, we recommend trying the following dimensions from the direction and content of the reform policy implementation in order to achieve the sound development of social work development.

### 5.1 *Continue to improve the social work organization system*

We should continue to improve the social work organization system, and further improve the social work organization system policy by an executive body of competent authorities, service usage units, service providers, industry associations, research institutes and other components of the assessment, giving full play to their capabilities, strengthening coordination, and the formation of a work force. Whilst strengthening co-ordination between the party and government departments, a clear division of responsibilities between departments is needed, in order to build a complete system of an effective social work service.

### 5.2 *A sound legal system of social work*

We should speed up the legislative process of social work, standardized processes determine the social worker job duties, professional title, evaluation, salary, reward and punishment evaluation, selection, and recruitment. At the same time, we should formulate comprehensive government purchase service regulations and mechanisms to further regulate the government purchase service of social work and public service. We should improve donation duty-free concessions law, implement the preferential development of tax-free donations, to maximize the mobilization of social forces to participate in social work services.

### 5.3 *Strengthen the guiding role of regulatory assessment of social work development*

First, we should strengthen the specification, guidance and supervision, according to the government to purchase a certain proportion of service funds to set up special evaluation regulation funds. We would bring the evaluation into social work system, optimize the social service agencies, the board of supervisors and other internal governance structures, establish a public complaints system, and increase the transparency and credibility of strong social work service mechanism. Second, improve the structure of third-party professional evaluation mechanisms, the implementation of a comprehensive assessment of social work services and social work service performance will be evaluated based primarily on the results of the audit of government procurement of service. Finally, we should formulate the behaviour system of social institutions and social worker.

### 5.4 *Optimization mechanism for government purchase of social work services*

First, the government should set a reasonable policy purchase objectives of the social work service, reasonably determining the government's financial payment mechanisms to effectively guide the social service agencies, in order to ensure their welfare and achieve proper services. Second, the government need to scientifically establish the criteria of the government purchasing funds. On the basis of criteria, the government should establish a dynamic adjustment mechanism. Besides, according to the quality of service of social work agencies, classifing the government purchasing social work service funds, in order to achieve healthy and orderly development of competition.

## REFERENCES

Liu Shujuan, Shenzhen Development of Social Work and Implications for the province [J]. Northeast Agricultural University (Social Sciences). 2010, 6 (8)

Liu Zheng. Situation, Problems and Solutions of Social Work Development [J]. Changsha Social Work College. 2009 (1)

Mu Yang, Social work development mode of Shenzhen [J]. Chinese Social Work. 2013,(4)

Wang Sibin, New opportunities and new orientation of China social work development [J]. Chinese Characteristics of Social Work Forum, 2008 (9)

Yuan Guangliang, China's social work legislation thinking [J]. Theory Journal, 2011 (7)

*Frontiers in Computer Education – Wang (ed.)*
*© 2015 Taylor & Francis Group, London, ISBN 978-1-138-02797-8*

# Policy on education equity in different places at college entrance examination

Chunyu Zhang
*North China Electric Power University, Baoding, Hebei, China*

Yixiao Wang
*Xinlian College of Henan Normal University, Zhengzhou, Henan, China*

ABSTRACT: On the orders of the central government, the policies being successively issued by local governments about the children of migrant workers could take the examination locally. Meanwhile, there have been heated debates in society. This paper, from the perspective of education justice, analyses the problem of migrant students taking college entrance exams in cities, evaluates the existing the questions, highlights the relevant policies promulgated by the government over the years, and compares them with each other to present some problems that need to be addressed. In so doing it puts forward the countermeasures of perfecting the system.

*Keywords*: Migrant students sitting college entrance exams in cities, Education justice, Reform of college entrance examinations

## 1 INTRODUCTION

Since the restoration of college entrance examinations in 1978, the college entrance examination system has been accompanied by an equal and fair value concept, and it is under continuous innovation and development. With the development of society's economy, the rural surplus labour force has been flowing into the cities in recent years. The number of the children of migrants is increasing, the problem of their education has become a social problem and an educational conundrum. The reform of the national college entrance examination system is becoming a hot topic in the field of education. After years of research and exploration, the migrant students taking the college entrance exam in cities policy of China has finally been put into action, many provinces have already launched the policies, such as Henan, Hebei, Liaoning, Jilin, Beijing, Shanghai, Shandong, Xinjiang, etc. These policies appear to have solved the immediate problems, but behind the policies, the true justice problem have not been addressed. This paper, from the perspective of education justice, evaluates the existing migrant students taking college entrance exams in cities system, in order to promote further reform to achieve relative fairness.

## 2 THE FAIR PRESENT SITUATION

Following the principles of equality and fairness, our migrant students taking the college entrance exams in cities system has passed through a succession of developments and has realized the transformation from theories to practice. Some eliminate the barrier about the household registration restrictions on education fairness, and ensure the education rights of the children of migrant workers, show the equity and justice of college entrance examination. But we have to admit that in our current system there still exist some unfair and unjust problems, which would have a bad influence on society.

### 2.1 *The migrant students sitting the college entrance exams in cities policies are unfair in different regions*

The ministry of education cannot solve the problem on their own anyway. This issue needs collaborative efforts of all social parties. In the face of practical problems, the ministry of education will have to promote the migrant students taking the college entrance exams in cities system. With years of reform, the government followed the principles of fairness and efficiency. Along with the ministry of education progressively liberalizing rights, it not only has brought vitality into the state and society, but it has also reduced a lot of responsibility of the central government. The ministry of education asks the local government to issue new college entrance examination policies which adapt to circumstances. But without a clear guiding reform plan, it just places the awful burden on local governments, which is unfair on the local governments.

### 2.2 *The starting point of the migrant students sitting the college entrance examination system is not fair*

The migrant students taking college entrance exams in cities policies are mainly targeted at the migrant children which don't have a local household registration. The policies shown that children of migrant workers can enter senior high schools and take college entrance exams locally if certain conditions are met: their parents must have a steady job, a stable household, a steady revenue, should pay all sorts of insurance, and should be permanent residents. Students must have gone to school locally, and have a school roll at least one year, etc. Under this assumption of differences between cities and the countryside and between different regions, the local governments set access conditions for children wishing to sit college entrance exams. This has a certain objective rationality. But the migrant students taking college entrance exams in cities policies tried to promote more educational equality, the conditions cannot be too hard. In fact, the migrant students taking college entrance exams in cities policies by the local governments are clearly unfair.

### 2.3 *The unfair distribution of the migrant students sitting college entrance exams in cities policies*

After the local government implemented policies, admissions and resources will be reallocated. And as things stand for now, some provinces have been clearly defined that admissions criteria about migrant children are the same as local candidates, but other provinces haven't defined this. If the migrant students recruited by the local university, it will have impact on the existing local recruiting resources, and this is unfair for the local candidates. If the migrant students recruited by the household school,it will cannot get effective ranking due to the examination papers are different, thus influence the examinee admission, and this is not fair to the candidates from other places.

No matter how different college entrance examination policy design is, it will trigger a new unfair phenomenon . Children of migrant workers to sit college entrance exams locally policy must achieve, these new unfair phenomenon both to the university entrance examination system poses a challenge.

## 3 THE PATH TO REFORMING AND PERFECTING THE SYSTEM

Rawls in his theory of justice said, "justice is the primary value of society... Some laws and systems, regardless of how they are efficient and orderly, as long as they are not just, must be reformed and abolished." [1]

[1] John Rawls. A theory of justice China [M]. Beijing: Social Science Press, 1988.1

The essence of the different college entrance examinations should embody justice, the core value should be fair. Based on the idea of fairness and justice, we put forward three aspects of thinking on China's different college entrance examinations reform.

### 3.1 *To ensure a fair starting point for off-site college entrance examinations*

Education equality is a citizen's right regardless of nationality, race, sex, occupation, property status or religious belief. Once there is a guarantee of a fair starting point, then children of migrant workers to sit college entrance exams locally policy may be effectively achieve a settlement in migrant workers in the education of their children.

First, establish the values of fairness and justice reform. Education equity is the foundation of social justice, college entrance examinations reform must establish a fair and just value first. Only by following the fair rules, all sectors of society will be realize the positive interaction. Fairness and justice is the that social interest relationship are properly coordinated and internal conflicts and other contradictions are dealt with properly. Only fairness and justice of the children of migrant workers to sit college entrance exams locally policy, it will be coordinate the relationship between outflow and inflow of interest to the population, and will be promote reform and development.

Second, establish a long-term enrolment mechanism. China's open remote college entrance examinations, are an essential requirement of achieving educational equity, but are made by the different provinces, these reform ideas within the existing framework of the college entrance examination system is flawed. This idea in the influx of a small, highly competitive college entrance examination populous is feasible, but for Beijing, Shanghai and other places, is not effective. The "national college entrance exam migrant", is leaving the low enrolment rate of domicile to the admission of non-domicile to take the exam, the only driving force is the difference in the provinces enrolment targets. Root cause of these things is that our college enrolment management at the provincial implementation of the quota system. The implementation of off-site college entrance examinations needs to break the original provincial enrolment system and establish long-term enrolment mechanisms. To break the provincial planned centralized admission system, once the national key university enrolment targets are no longer allocated to the provinces, then the conditions for the entrance to bind household college entrance examination system will lose the meaning of existence, this may be a driving force to promote college entrance examination reform.

Lastly, reform the household registration system. On the one hand, the household registration system to confirm citizenship had been unable to adapt to current growth trends of population flows. The realization of different college entrance examinations, must be under the environment of the joint college entrance examination system reform and the household registration system reform, thus realize the equality of education. On the other hand, remote college entrance examination is not only solve the education issues, but also social issues. As the implementation of policy reforms, related policies need to be synchronized in order to realize the value of the policy.

### 3.2 *Remote college entrance examination to ensure a fair process*

Off-site college entrance examination is the starting point for the fair value reflected in policy making, but the fair process is the core implementation. Therefore, to ensure the process fairness to different college entrance examination policy, in order to make the children of migrant workers benefit from the policy, in order to effectively solve different problems of the floating population in the entrance examination.

First, the establishment of scientific "access conditions". The offsite college entrance examination system is to address children or to take the exam in the non-household system, thus setting the conditions should be set around students.. Children of migrant workers to sit college entrance exams locally policy for parents conditions need to be improved in two aspects. On the one hand, parents should weaken the social background conditions affecting their children to take the exam. On the other hand, we should set a "controlled" standard, the "stable house", "stable job" are no uniform standards to measure, it can be changed to require parents with "stable income." Because a steady source of income means a candidate's parents have a stable job and a roof over their heads.

Second, strengthen the control of central government. The Chinese Communist Party's 18th Congress makes this clear: fairness and justice are inherent requests in building socialism with Chinese characteristics. It is the responsibility of the central government to maintain social equity and justice, and to provide equal education opportunities. For "Migrant Students sitting college entrance exams in cities" reform at any one time, the protagonist is central government reform. To promote the "off-site entrance" reform, to achieve true educational equity, the central government should be led by local governments to strengthen monitoring and evaluation of off-site entrance policy implementation, and continuous reform and improve relevant policies and systems.

### 3.3 *To ensure the equality of results*

As a solution to the problem, the migrant students sitting college entrance exams in cities policies must be feasible, the policies could be the key to the problem.

First, ensure fair admission. After publication of the college entrance examination policy, we could see that less than half of those policies have set admissions criteria. To solve this problem, the policies must set admissions criteria that the migrant children are the same as local candidates, so that policies can be implemented effectively.

Second, the implementation of compensation for the weak. After the local government implemented policies, the import places' interest become the focus of general attention. And ignore the outflow places of examinee's interests. Whether the success or failure of the policies will hardly affected they participate in college entrance examination, but there may violate their rights and interestsl. So act on fair principle, we should make up for the lack outflow places.

The migrant Students sitting college entrance exams in cities system is improving, the scientific policies and the government's push are necessary to realize the migrant students sitting college entrance exams in cities. Multiple stakeholders are involved, society's public resources will be reallocated; this is an issue that needs collaborative efforts of all social parties. We should therefore follow the principles of equality and fair, to set policies and allocate resources, solve the education of a transient population, supply and perfect the college entrance examination system, to further guarantee the right of citizens to receive education.

## REFERENCES


Chen Bin. Different college entrance examination the interests of the game, the plight of analysis and Countermeasures [J]. Education and Test, 2012(3)

Guo Caiqin. Theory of educational equality [M]. China University of Mining and Technology press, 2004:55

John Rawls. A theory of justice China [M]. Beijing: Social Science Press, 1988.1

Liu Yao. Remote predicament and path of high test [J]. Henan Education, 2012 (9)

Zhong Ling. Different college entrance test the ice when the three big problems [J]. Life Education, 2012 (19)

*Frontiers in Computer Education – Wang (ed.)*
*© 2015 Taylor & Francis Group, London, ISBN 978-1-138-02797-8*

# A study of the personalized physical education teaching of sports courses

Lan Yu
*Jiangxi University of Finance and Economics, Nanchang, China*

ABSTRACT: A yearly Physical Education (PE) sport teaching arrangement, which includes various teaching units, is designed towards the objectives of skill learning and physical quality promotion. Considering the difference among individual students, a personalized PE sport teaching plan is formulated for each student. At the same time, correspondent teaching modes, methods, and organizing forms are discussed.

In a physical education (PE) sport teaching class, the interests, physical qualities and skill levels of different students are distinct. The uniform teaching plan, teaching mode, teaching method, and organizing form are not suitable [1,2]. Therefore, a personalized teaching plan based on the characteristics of each student should be formulated to maximize the teaching effectiveness, and proper teaching mode, method, and organizing form should be adopted [3–7].

## 1 PERSONALIZED PE SPORT TEACHING PLAN

Figure 1 presents the unit component of the personalized teaching plan, and Table 1 gives a thirty-six hours example of yearly teaching plan.

Next contents of all units are introduced.

### 1.1 *Theory unit*

There are two sub-units in the theory unit.

(1) Sub-unit 1: including exercise and health, sports history, theory of skill and tactics, exercise safety, exercise prescription theory, and the relation between physical quality and sports.
(2) Sub-unit 2: competition rules, design and evaluation of exercise plans, formulation and evaluation of exercise prescription, relationship between exercise and profession, nutrition and health care.

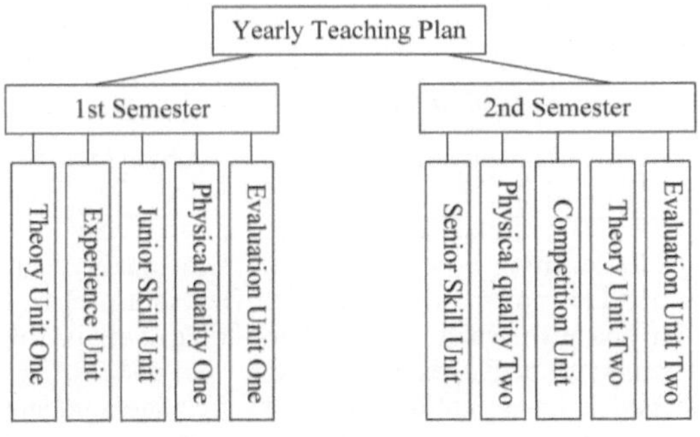


Figure 1. Constitution of teaching plan.

### 1.2 *Experience unit*

Experience unit includes three modules: (1) Games related to a specific PE optional course; (2) Skill show; (3) Exercise experience of different physical qualities.

### 1.3 *Skill unit*

There are two sub-units, named junior and senior sub-units. A personalized skill exercise plan, which is determined by the current skill and physical quality levels of a certain student, is carried out in this unit.

As we all know, most skills of a specific sport often have several learning tips. For example, shooting is one of the basic basketball skills. There are many patterns of shooting: free throw, long shot, fadeaway, and pull shot etc. The difficulties and the effectiveness for physical quality promotion of these shooting patterns are different. Therefore, shooting skills could be divided in details. Students with poor foundation are required to

Table 1. Common teaching methods

| Teaching Unit | Hours | Teaching Unit | Hours |
|---|---|---|---|
| Theory Unit One | 4 | Senior Skill Unit | 18 |
| Experience Unit | 4 | Physical Quality Strengthen Unit | 4 |
| Junior Skill Unit | 20 | Competition Unit | 6 |
| Physical Quality Promotion Unit | 4 | Theory Unit Two | 4 |
| Evaluation Unit One 1st Semester | 4 | Evaluation Unit Two 2nd Semester | 4 |

learn junior shooting patterns and students with certain level foundation can learn senior patterns.

In addition, different exercise forms could be used to meet distinct physical quality requirements of different students. Take shooting as the example again, we may arrange standing shoot, multiple spots shooting, and half-court shooting and full-court shooting to meet different endurance and strength promotion requirements.

However, since a specific sport has its preferred physical quality, the effectiveness of skill unit for physical quality promotion is limited. The physical quality unit is designed to further promote the physical quality.

### 1.4 *Physical quality unit*

Similar to the skill unit, the physical quality unit is divided into two sub-units: one is called physical quality promotion sub-unit and the other is called physical quality strengthening sub-unit. The exercise plan in the physical quality unit is also designed for an individual. Objectives of physical quality unit include: (1) Learn the theory of exercise prescription. (2) Make up the difference between the skill teaching plan and physical quality promotion requirements.

### 1.5 *Competition unit*

Competition unit teaches students how to apply their skills and tactics through teaching matches. It has two phases: one is performed before the physical quality strengthening sub-unit and the other is performed after the physical quality strengthening sub-unit. The other objective of the competition unit is to let students experience the pleasures of matches.

### 1.6 *Evaluation unit*

Evaluation contents include physical quality, skill, and theory. In the skill module, basic skills are more important. Physical qualities include endurance, strength, and flexibility. Theory evaluation includes basic theory, theory of PE sport, and theory of exercise prescription.

## 2 PERSONALIZED PE SPORT TEACHING MODE

### 2.1 *Experience unit*

There are three teaching modules in the experience unit: sport games, skill show, and exercise experience. The relationship among the three modules is described in Figure 2. That is to say, the skill show is implicit and arranged in the sport games module so that students will not have burdens. The exercise experience module is arranged after the sport games module.

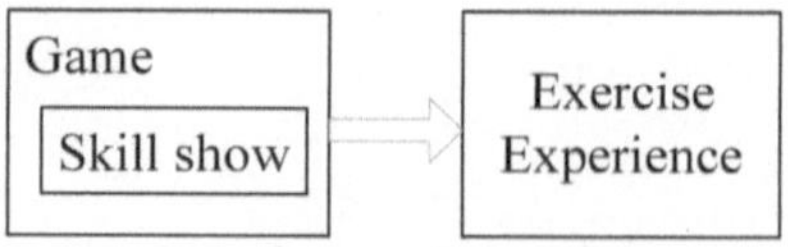


Figure 2. Relationship among three modules.

The sport games module includes three phases: preparation, execution, and conclusion. In the preparation phase, the teacher is responsible for explaining the rules of games and arranging group division. In the execution phase, two or three activities which can show students' skill level are arranged and executed during the game. The teacher should be responsible for arrangement and paying attention to the enthusiasm of students. In the conclusion module, the teacher must illustrate the relationship between games and the specific sport to improve the learning interest of students. The relationship between sport skill and physical quality is also required to illustrate and impel students to take part in exercise.

The exercise experience module aims at the promotion of endurance, strength, and flexibility. At least one to two exercise methods for each physical quality should be introduced. The objectives of this module include: (1) Make students realize the importance of the three physical qualities. (2) Learn the difference among the three physical qualities. (3) Learn the distinct effectiveness of different exercise methods. (4) Understand the theory of exercise prescription further, which includes exercise item, strength, duration, and frequency. In this module, the teacher should emphasize exercise safety.

### 2.2 *Skill unit*

The basis of the skill unit teaching is skill learning law, exercise prescription theory, and the promotion of skill and tactic exercises for physical quality.

The teacher should divide the students into several groups. Students in each group often have similar skill exercise plans.

The unit is divided into two phases: one is a skill learning phase and the other is an integrated exercise phase. The former phase has several lessons for different skills. Each lesson has two links: one is uniform learning and the other is personalized exercise. In the personalized exercise link, each student has their own exercise form to meet their physical quality promotion requirements.

### 2.3 *Physical quality unit*

The basis of the physical quality unit teaching includes the physical quality level of a student, teaching plan of the skill unit, and exercise prescription theory.

The teaching process of the physical quality unit is divided into three phases. The first is a preparation phase in which the teacher should explain the teaching objective, execution mode, and notices. The second is an execution phase in which the students exercise personally and the teacher must pay attention to the status of each student so that he can help the student solve emerging problem. The third is a conclusion phase in which the students discuss their feelings first and the teacher then draws conclusions. During the whole process of the unit, the teacher must guide the students' action according to the physical quality promotion plan.

### 2.4 *Competition unit*

Different match forms should be designed in this unit.

The task of the first sub-unit is combat exercises. The procedures include illustration, competition, conclusion, competition again, and conclusion again etc. After multiple matches and conclusions, the skill and tactic levels of students could be improved, and the students could have an objective knowledge of their skill level and physical quality level.

The teaching process of the second sub-unit is divided into several phases which include preparation, competition, and conclusion links. A main objective of this sub-unit is to let students feel the pleasures of matches. In a conclusion link, the teacher must encourage the students to share competition experiences and arouse their enthusiasm for skill and physical quality exercises.

Table 2. Common teaching methods.

| Type | Teaching method |
|---|---|
| Explanation | explanation; questions and answers; discussion |
| Experience | demonstration; protection |
| Exercise | decomposition exercise; whole exercise; loop exercise |
| Competition | game; competition; scene |
| Exploration | discovery; question exploration; small group learning |

## 3 PERSONALIZED PE SPORT TEACHING METHOD

Table 2 presents the teaching methods that are often adopted in physical education. Each unit of personalized PE sport teaching consists of several modules or phases. Table 3 shows the adopted teaching method and organization form of each module or phase.

## 4 ORGANIZING FORMS OF PERSONALIZED PE SPORT TEACHING

Generally, common organizing forms of a PE classroom are class teaching and group teaching. Group teaching is divided into four forms: (1) Big group

Table 3. Teaching method and organizing form adopted in different phases of various units.

| Unit name | Module or phase | Teaching method | Organizing form |
|---|---|---|---|
| Theory unit | Sport exercise plan and exercise prescription arrangement | Discussion, Explanation, Exploration | Class |
| | The other modules | Explanation, Questions and answers, Discussion | |
| | Competition rules<br>Match appreciation | Demonstration | |
| Experience unit | Sport game<br>Skill show | Game, Scene | Class |
| | Exercise Experience | Scene | Big group |
| Skill unit | Warm up and Review | Whole exercise | Class |
| | Explanation and Demonstration | Demonstration, Explanation | Class |
| | Uniform learning | Decomposition exercise, Whole exercise, Exploration | Big group/ small group/ double |
| | Teacher Evaluation | Explanation, Questions and answers | Class |
| | Personalized exercise<br>Integrated exercise | Decomposition exercise, Whole exercise, Loop exercise | Small group/ double/ individual |
| Physical quality unit | Preparation | Exploration | Small group/ double |
| | Execution | Decomposition exercise, Whole exercise, Loop exercise | Small group/ double/ individual |
| | Conclusion | Explanation, Questions and answers | Class |
| Competition unit | Preparation | Explanation | Class |
| | Competition | Match | Big group |
| | Conclusion | Explanation, Questions and answers | Class |

(eight and more than eight persons); (2) Small group (three to seven persons); (3) Double (two persons); (4) Individual. Each organizing form has its suitable teaching phase, which is described in Table 3.

## 5 CONCLUSIONS

The design and execution of a personalized PE sport teaching example are presented in this paper. First, a yearly PE sport teaching scheme is designed, which consists of a theory unit, experience unit, skill unit, competition unit, and evaluation unit. Second, the corresponding teaching mode is designed for each teaching unit. Third, the teaching mode applied in different phases of different units is presented. Finally, organizing forms in PE classroom are described.

## ACKNOWLEDGMENT

This work was supported by the 11th five-year plan of Jiangxi Education Research (No. 09YB282), the 11th five-year plan of Jiangxi Social Science Research (No. 10JY64), and Science and Technology Project of Jiangxi Education Department (No. GJJ11091).

## REFERENCES

Fangyuan Wang. College physical education sport course theory tutorial. Shanghai: Fudan University Press, 2009.

Nuyao Wang, Zhaobin Du, Yuqin Sun. College's basketball align specific sports and fitness research. Shandong Sports Science & Technology, 2007, 29(1): 59–61.

Rongfang Cao, Li Guo, Lei Li. Experimental research on personalized physical education teaching method. Education and Profession, 2007(5): 146–148.

Rongwei Wang. Personalized teaching: melody of college physical education teaching reform. Heilongjiang Researches on Higher Education, 2004(11): 117–118.

Wenwen Qi. Influence Factor analysis on physical quality personalized teaching. Teaching and Management, 2011(3): 110–111.

Zhenming Mao. Theory of physical education teaching. Beijing: Higher Education Press, 2011: 118–119.

Zhiwei Deng. Theory of personalized teaching. Shanghai: Shanghai Education Press, 2002: 52–55.

*Frontiers in Computer Education – Wang (ed.)*
*© 2015 Taylor & Francis Group, London, ISBN 978-1-138-02797-8*

# The teaching reform and practice of hydraulics based on the Problem-Based Learning teaching mode

J.C. Liu, Y.W. Shi & J. Zhao
*School of Water Resources and Environmental Engineering, Changchun Institute of Technology, Changchun, China*

ABSTRACT: Problem-based learning is the learning and teaching mode. The paper introduces the PBL teaching mode and it's application in hydraulics. PBL can reform the classroom teaching mode, encourage teachers' classroom changes, and is helpful for students to construct new knowledge. It resolves the lack of traditional teaching mode. The paper takes the cohesion and energy dissipation of downstream drainage buildings as an example, this fully demonstrates the application of PBL teaching mode in hydraulics teaching, it also provides food for thought for other courses on the reform of water resources and hydropower engineering.

*Keywords*: Hydraulics; teaching mode reform; PBL teaching mode

## 1 INTRODUCTION

Hydraulics is an important professional foundation course in the water conservancy specialties of water resources and hydropower engineering, hydrology and water resources engineering, and agricultural water conservancy engineering. The hydraulics content is complex and huge which consists of water statics, fluid movement of the beam theory, energy dissipation of downstream drainage buildings, seepage, and high speed water flow. On the one hand, it involves many calculations of building of water conservancy projects, the application of many theories and formula, and a lot of half experience and half theory formula. On the other hand, PBL haves clear engineering point, the energy dissipation of downstream drainage buildings, seepage, high speed water flow, etc all can be used it to solve the practical problems.

At present, the hydraulics courses mainly adopt the traditional teaching method, such as classroom teaching, laboratory experiments, and the combination of multimedia and blackboard. The traditional teaching method is based on the textbook content, teachers are always positioned at the centre of organizations, and they organize and monitor the whole process of teaching activities, while students can only passively accept, digest and understand the content. For the traditional teaching method, students don't search the course content which is related to the literature and engineering case beforehand, when the actual problems appear, they have no idea what to do, so it is difficult for them to adapt to the needs of the development of the university education system, and their innovative spirit and practice ability is difficult to put to use.

What kind of teaching method can be used to improve the applied scientific research innovation ability of students in institutions of higher education? The diversified teaching method reformation is reported in literature from domestic and overseas institutions, however these methods cannot change the dominant role of the teacher and the passive role of students in teaching. As for the learning teaching mode, its kernel is to place the learners at the centre, the learners are the main body of information gathering, processing, and knowledge of autonomous refactoring instead of passively accepting state of external stimulus.

The role of the teachers is transformed from that of a person imparting knowledge to that of the organizers, helpers, and promoters of a student's learning process, maximizing a student's enthusiasm, initiative, and ultimately inspiring learners' creativity. This method which integrate the PBL teaching idea has been applied to the hydraulics teaching reform attempts in the class of 2010 students in our school, the education reformation achieved a very good result.

## 2 CONNOTATIONS OF THE PBL TEACHING MODE

In 1969 the United States professor of neurology HowardBarrows at McMaster University in Canada founded the PBL teaching mode, as a kind of brand-new teaching mode, which was based on the constructivism learning theory. Dewey's pragmatism education theory and Bruner's discovery learning theory. The kernel of PBL is case teaching, it takes the student as the main body, and teacher as the leader. We can appreciate from B. Hs and Y.B. Du's literatures.

### 2.1 *The definition of PBL*

PBL takes a set of unique and interrelated tasks as the prerequisite, by efficiently using the existing resources, the goals of particular efforts are achieved; it is also a method which uses a variety of resources to solve the multiple tasks associated with each other problems. PBL gains a more complete and specific knowledge by the practice experience, internalizes absorption, explores innovation, forms the special skills, and makes the learning process more fully. We can appreciate these information from J.F. Liu's literature.

### 2.2 *The components of PBL*

PBL mainly consists of four elements, that is the content, the activity, the situation, and the result. J.F. Liu's and Q. Feng's literatures are described. The content is a series of complex, non-predictive and multidisciplinary problems.

The activity refers to an action of inquiry where students make use of certain technical tools and research methods to solve the problem. In fact, it actually gives students a close connection with the process of the real work of a learning experience, allowing students to construct and generate their own knowledge, at the same time, forming the corresponding work ability or skill.

The situation refers to an environment where students carry out an exploration; it can be a physical entity learning environment, a virtual environment.

The result refers to a work which is formed by the students through the inquiry activity, it can be software, a plan, or a conclusion, etc.

### 2.3 *The characteristics of PBL*

Characteristics of PBL mainly have five aspects. We can appreciate these information from Z.W. Ai's and L.X. Wen's literatures.

(1) The real and specific learning situation.
(2) Comprehensive and open learning content.
(3) Diverse and collaborative learning way.
(4) Digital and networked learning methods.
(5) Diversity and effective of learning mode to the evaluation of students.

## 3 THE APPLICATION OF PBL IN HYDRAULICS TEACHING—TAKING "THE COHESION AND ENERGY DISSIPATION OF DOWNSTREAM DRAINAGE BUILDINGS" AS THE TEACHING EXAMPLE

According to the situation of the water conservancy engineering laboratory, PBL learning mode in hydraulics is as follows: the study should be closed to the students' learning content, the practical problems in engineering design and innovation of science and technology to establish the learning program. Under the guidance of teachers, the students should be divided into study groups based on the characteristics of the students. At the same time, teachers should guide students towards completing the project and exploring the method to solve the problem. On the basis of the connotation of the PBL teaching mode, combined with the actual situation of the curriculum of hydraulics, the PBL teaching is as follows: design project → the division of groups → make a plan → collaborative inquiry → report the demo → conclusion and evaluation.

### 3.1 *Designing projects and establishing the situation*

The design project is a very important link, the rationality of the project directly determines the success or failure of the follow-up teaching activities. According to the nature, training objective, the properties of hydraulic factors such as social needs and the characteristics of students of the hydraulics course, the teachers can let the students do what they really need by taking the design of a hydraulic engineering problem as a starting point. For example, teachers design the projects and create the problem situation: "the cohesion and energy dissipation of downstream drainage buildings". The questions can be what is the water? How much is the unit weight of water energy under the case of the building dam? Why is the width of the drainage buildings smaller than the original river, what results will it produce? How much energy should be eliminated in unit width bed every second? How to connect to the downstream river channel which is down from upstream of the water? etc? These problems are the real situations which are always encountered in the design of water conservancy projects. The students' attention and enthusiasm can thus be attracted and maintained.

### 3.2 *Dividing the groups and determining the tasks*

In order to fully mobilize students' learning motivation, teachers should divide the whole class students into groups and make sure each group has 5–7 people based on each student's interests and hobbies, business ability, professional competence, etc. Finally, we can assign respective tasks.

There ware 42 people in the excellence of grade 2010 of water resources and hydropower engineering class, in the project of "the cohesion and energy dissipation of downstream drainage buildings", we divided the class into 6 groups according to the actual situation, and each group had 7 people. A captain for every group who is responsible for coordinating the team's entire tasks, and a recorder for every group who is responsible for all the problems and solution of the project. One people responsible for computing, three people responsible for modeling, testing, and the production of the relevant model, one person for reporting production, demonstration, interaction, and presentation.

### 3.3 *Planning and analysing the problem*

In order to better explore the activity, teachers should fully respect the group opinion in planning. So it is important to note your role transformation and playing a good supporting role and avoiding too much intervention. By team group discussion, the framework is built up to solve the problem, listing the detailed work plan and forecasting possible difficulties, solving the problem of the implementation of the plan, implementing the personnel and the completion time in the scheme, and ensuring the project is completed on schedule.

In the project of "the cohesion and energy dissipation of downstream drainage buildings", students can put forward a series of problems, such as how to calculate the discharge energy size, how do you determine the contraction section depth, and how to use the computer to visually show the output. At the same time, Excel calculation and Matlab programming method is considered to solve the problems.

### 3.4 *Collaborative inquiry and problem solving*

In the inquiry activity, team members should always aim at project goals, collaboration, taking advantage of access to information resources and technical tools, collecting relevant data and selecting, processing the useful information, and sharing resources between team members. They also should construct new knowledge on the basis of discussion, such as: the cause of the upstream channel siltation, the causes and the prophylactic measures of downstream river channel scour, the optimization of engineering problems. Team members use their spare time and holidays to carry out the project, including: collecting information, processing information, field visits, numerical calculation and model simulation and model testing, etc. In project inquiry, students may experience many technical and life problems, once they encounter problems, they may be in an irritable mood, and even have the idea of giving up the project. At this stage, teachers should pay attention to guide students, on the one hand, to prevent the project deviating from the topic, on the other hand, to encourage students and help students build up self-confidence.

In the discussion of the "the cohesion and energy dissipation of downstream drainage buildings" project, we mainly divide the contents into three parts that is the basic theory, numerical calculation and physical model. Every part is equipped with an instructor. In the weekend and the students' free time, teachers should instruct the students by turns. We can also put a project weekly meeting system in place according to the needs of the project, because that can facilitate the timely communication between teachers and students.

### 3.5 *Demonstration report and results*

After finishing the project, the whole class should get together and engage in a group discussion.

In the inquiry of "the cohesion and energy dissipation of downstream drainage buildings" project, students' achievement reports mainly include three parts: 1) the form of the report, each team presents the whole research process of the project, as well as the problems encountered in the study of and the measures to solve the problems; 2) based on the laboratory platform, each group summarizes phenomena which appeared in the process of testing, and the reasons for their occurrence; 3) based on the computer workstation, teams show their numerical calculations and visualization of the entire project implementation process.

### 3.6 *Summary evaluation*

Through the self-assessment and mutual reporting between each group, teachers and students should summarize their topic.

In the inquiry of "the cohesion and energy dissipation of downstream drainage buildings" project, through summarizing and exchanging between the students, the design content of this part can be well understood. Method such as the high order implicit algorithm, iterative method, numerical method, such as hydrolysis can also be well mastered. At the same time, combining Excel, Visual Basic and Matlab with professional design can also inspire the students' enthusiasm and extend the students' knowledge.

## 4 CONCLUSION

It is necessary for students to gasp the corresponding knowledge of the hydraulics. In this paper, the PBL teaching mode is introduced into the teaching of hydraulics to make up for the disadvantages of the traditional teaching mode. Because the PBL teaching mode is different from the traditional mode, it pays attention to the performance of the students learning process and embodies the guiding role of the teacher and the students as leading actors. There are practical projects throughout the course of the learning process which can effectively stimulate the students' intrinsic motivation. This teaching model can also provide students with decision-making opportunities according to their own interest, so that the students can carry on the study and explore independently, thereby, students' autonomous learning ability and innovation ability can be effectively promoted. By this reformation, students' knowledge analysis ability and the ability to solve practical engineering problems have been significantly improved, that is in accordance with the high quality water engineering applied technological talents 'cultivation according to the outstanding engineers training plan for the needs of social and economic development.

## ACKNOWLEDGMENTS

This work was financially supported by Jilin Province education scientific planning.

## REFERENCES


B. Hs, T.R.M. The portable patient pack: a problem-based learning unit [J]. *Journal of Med Edu,* 1977, (12): 1002–1004.

Y.B. Du. The Reform Practice of "Precision Machinery Manufacturing Technique" Course Based on PBL Teaching Mode [J]. *Value Engineering*, 2010, (13): 214–216. (in Chinese)

J.F. Liu, Z.X. Zhong. A Study on The Model of Project-based Learning [J]. *Studies in Foreign Education*, 2002, 29(11): 18–22. (in Chinese)

Q. Feng, C. Guo, C.Y. Wu, et al. Application of problem based learning method combined with virtual realistic technology in wastewater treatment teaching process [J]. *China Academic Journal Electronic Publishing House,* 2011, (4): 88–92. (in Chinese)

Z.W. Ai, H. Wu. Application of problem-based learning teaching model in the experiment of liquid surface tension coefficient measurement [J]. *Physics Experimentation,* 2013, 33(7): 20–22. (in Chinese)

L.X. Wen, F.Y. Wei, C.Y. Luo. Discussion on teaching reform of case-based PBL teaching method in pharmacology of traditional Chinese medical formulae [J]. *Journal of Hunam Univ. of CM,* 2013, 33(8): 98–100. (in Chinese)

# *Communication and intelligent control*

*Frontiers in Computer Education – Wang (ed.)*
*© 2015 Taylor & Francis Group, London, ISBN 978-1-138-02797-8*

# Design of a Variable Frequency Drive (VFD) controller based on the Memobus Remote Terminal Unit (RTU) protocol

BenZhen Guo, ShangFu Hao & Xiao Zhang
*School of Information Science and Engineering, Hebei North University, Zhangjiakou, China*

XiaoJing Wang
*School of Humanities and Social Sciences, Zhangjiakou University, Zhangjiakou, China*

ABSTRACT: The traditional Programmable Logic Controller–Variable-frequency Drive (PLC-VFD) control method is inflexible and expensive. This paper develops a new type of VFD controller based on the Memobus Remote Terminal Unit (RTU) protocol. The hardware circuit and Memobus RTU protocol software were designed and a prototype was produced. The experiment results shows that this controller transports data reliably and runs stably. The control method and technology are worthy of using abroad.

*Keywords*: dsPIC30F6010A; VFD; Memobus RTU

## 1 INTRODUCTION

With the industrial automation speedily developed, the Variable-frequency Drive (VFD) is widely used in all types of motor-driven making control more and more flexible and energy efficient [1].

There are two ways to drive a motor by the VFD: external terminal control or communicational control. The first way uses an external relay sequence circuit or PLC output terminals to achieve the VFD stop-forward-reverse and other operating modes [2]. The second uses communication to achieve VFD control. An external sequencer circuit control has poor reliability and a low intelligence level. Using this method, it is difficult to achieve human-computer interaction and it has other shortcomings. A VFD that is controlled by the PLC is very common, but the cost can seem pretty high. In addition, a human-computer interface and an optional encoder are standard modules, which further increase the cost. Designing the VFD controller with a microcontroller is cheaper and the controller can easily be expanded to include modules such as human-computer interaction and other intelligent features.

This paper develops a new type of VFD controller based on a Memobus Remote Terminal Unit (RTU) and a ds-PIC30F6010A. A Memobus RTU communication protocol is achieved between a dsPIC30F6010A and an A1000 VFD which are produced by Yaskawa. The experiment results shows that this controller transports data reliably and runs stably. The control method and technology are worthy of using abroad.

## 2 MEMOBUS RTU COMMUNICATION PROTOCOL

The Memobus RTU communication protocol uses a master-slave response communication. In this communication protocol, the initiating communication device is called the communication master (main station) The device which receiving Communication requests and make the appropriate response called the communication slave (slave station). The entire communication network allows only one master station. The master station communicates with only one slave station at the same time. Master communication data packets contain the slave address, the slave command to query their own information and respond by identifying address. Memobus RTU communicate data in RTU model. Compared with the ASCII mode, the RTU mode has continuous data transmission, high byte transmission efficiency, and is more suitable for real-time transmission of data, which is necessary in automation and control systems.

## 3 THE HARDWARE CIRCUIT DESIGN

### 3.1 *Overall design diagram*

Figure 1 shows the overall block diagram of the VFD controller which is developed with dsPIC3-0F6010A. The controller includes an RS232 bus interface, RS422/485 bus interface, SPI bus interface, quadrature encoder interface, 8 digital inputs and 8 digital

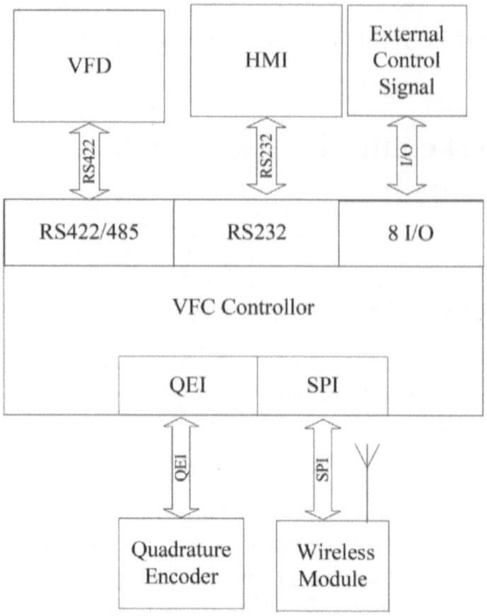


Figure 1. Block diagram of the overall design.

outputs. The powerful controller can easily communicate with the common human-computer interface via RS232 and it supports Memobus RTU communication via an RS422/485 bus interface. The quadrature encoder feedback motor speed to the controller. In addition, eight digital inputs can achieve external control buttons or trip switch control, an eight-way switch output can directly drive contactors action.

## 3.2 *Host controller*

As can be seen from the overall design diagram, dsPIC30F6010A is the core of the whole system. dsPIC30F6010A is a 16 fixed-point digital signal controller chip (DSC) which was launched by Microchip.

The high speed operational ability of the DSP and MCU control capabilities is combined in the dsPIC30F6010A with a high anti-jamming capability. The maximum operating speed is up to 30 MIPS. A high-performance Reduced Instruction Set Computing (RISC) Central Processing Unit (CPU) core and a modified Harvard architecture reduce the difficulty of development.

An enhanced instruction set of a special C compiler optimization strongly supports DSP. Large-capacity memory is integrated in this DSC, including 144kByte ROM/FLASH, 8kByte SRAM and 4kByte data EEPROM [3]. An integrated motor control Pulse-width Modulation (PWM)module has four pairs of PWM output pins which can be set to a variety of operating modes. An integrated optical encoder interface makes motor control applications easy.

## 3.3 *RS422 circuit*

The VFD A1000's Memobus RTU electrical standards meet the standard of the RS422 bus. Compared with RS232, the RS422 drive is stronger due to its high input impedance of the receiver and its transmitter driver; RS422's differential form of transmission signal [4] makes its anti-interference good. Figure 2 shows the RS422 driver circuit, SP490E is a low-power differential transceiver which meets the RS-485 and RS-422 electrical standards, data transfer rates is up to 10Mbps. dsPIC30F6010A UART1 transmit data signals U1TX_P is converted to RS422 differential signal RXD- and RXD + via SP490E, RXD- and RXD + connect to the A1000 receiving terminal; A1000's differential data transmission terminal TXD + and TXD- is converted to single-ended data signal U1RX_P via SP490E. Communication between the controller and the VFD uses DB9 connectors.

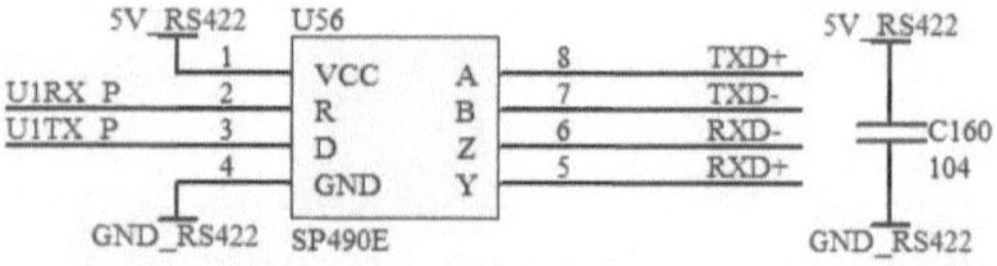


Figure 2. RS422 circuit schematics.

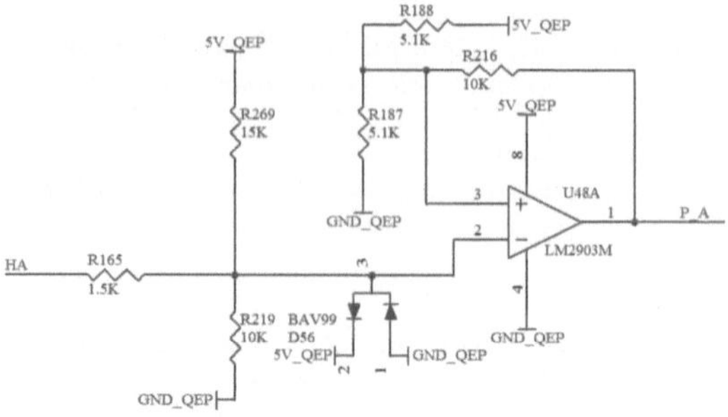


Figure 3. Hysteretic comparator circuit.

## 3.4 *Quadrature encoder circuit*

The Quadrature Encoder Interface (QEI) is internally integrated in the dsPIC30F6010A. To improve the electromagnetic compatibility and encoder pulse signal transmission reliability, a hysteretic comparator circuit and optical isolation circuit are designed. A dual voltage comparator LM2903 with peripheral circuit structures achieves the hysteretic comparator. An optical coupler 6N137 makes the signal isolated electrical. The signal is treated by the hysteretic comparator and optical isolation connected to the quadrature encoder interface (QEI) of the dsPIC30F6010A. Figure 3 shows the encoder output signal which is processed by the hysteretic comparator circuit.

## 3.5 *Electric reliability design*

The electromagnetic work environment of the VFD is complex. In order to improve overall system reliability and anti-jamming performance, the data transmission path between the microcontroller and the VFD such as the RS232, RS422/485 interface and the quadrature encoder interface is electrically isolated by an optical coupler. The I/O output circuit uses relays to achieve

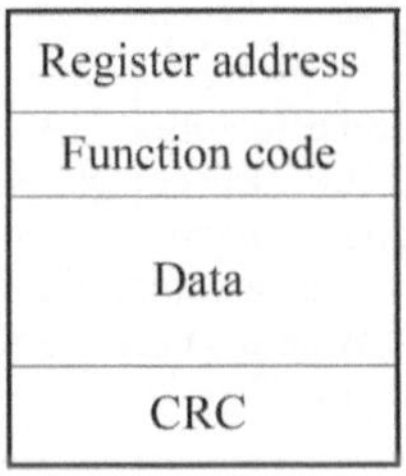


Figure 4. Memobus RTU data formats.

electrical isolation. The AS is shown in Figure 4, the optical coupler chip 6N137 is a high-speed optical coupler, its conversion rate is up to 10 MBit/s, with a temperature, current and voltage compensation function, isolation voltage is up to 2.5 kV. One end of the optical coupler is powered by an inboard power supply VDD_5V, the other end is supplied by the DC-DC power modules [5].

## 4 IMPLEMENTATION OF THE MEMOBUS RTU PROTOCOL SOFTWARE

### 4.1 *A1000 VFD master-slave communication requirements*

The Memobus RTU's communication between the controller and the VFD A1000 is done by way of a master and slave protocol. The VFD as a slave station launches communication with the master station is time-sensitive, after receiving the information from the master station, it is not allowed to send the same kind of instruction information to the same slave station for some time, otherwise the drive will generate a communication error and an overshoot alarm will be raised. The specific waiting time varies depending on the different commands, the minimum waiting time of the read operation instruction and monitoring setting is 5 ms, written arguments need 50 ms, written by determining the change of command parameters need 3–5s [6].

### 4.2 *A1000 Memobus RTU protocol data format*

Figure 4 gives the Yaskawa A1000's Memobus RTU data communication formats. A data frame is composed of four data segments: slave station address, function code, data, and error detection (CRC). The VFD slave address is from 0 to 31, by setting the station address register between 0 to FF, when setting the slave address to 0, the master transmits data to all slave stations in a broadcast mode.

A function code is used to distinguish the master station command type. The Yaskawa A1000 VFD has three types of function codes: 03H indicates that the master reads the contents from the station registers; 08H, indicates that the loop test Memobus RTU communications is being performed; 10H, indicates that the main station would like to write more data to the register of the slave station.

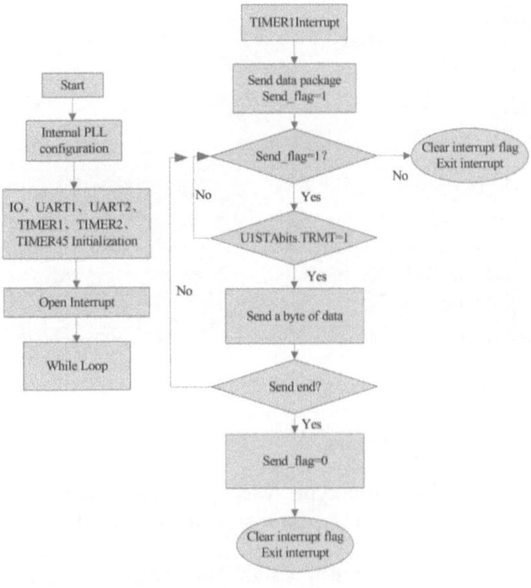


Figure 5. Software flow diagram.

The data frame includes a data segment, the slave register address, and the number of data during the master reading operation. During the writing register operation, the data frame data include the register address, the number of data, and total amount of bytes and data segments. A1000 Memobus register is two bytes long, the high byte of data is sent first and then the low byte of the data in data transmission.

### 4.3 *Software design*

The control system software consists of the main program and some ISR. The main program completes the master controller core clock configuration, IO pin, timer and UART modules and other peripherals. Initialization, then open interrupt switch and enters the main loop in the end. The system uses three timer interrupts to complete the data transmission and unpacking work. Timer 1 interrupt period is 200 ms, the VFD sends a reading register request when a Timer 1 interrupt occurs. Timer 2 interrupt period is 150 ms, mainly to complete unpacking received data. Two buffers VFD_REV_BUF and LCM_REV_BUF are defined. The values of the register returned from the VFD are stored in VFD_REV_BUF, variable values and various control commands sent by industrial serial screen are stored in LCM_REV_BUF. Timer 45 interrupt period is defined as 200 ms, register data of the VFD that is stored in VFD_REV_BUF is sent to the industrial serial screen to display every 200 ms. Data communication between the controller and the VFD uses a dsPIC30F6010A's UART1 interface, and industrial serial screen communication using UART2 interface. Data is stored in VFD_REV_BUF and LCM_REV_BUF by the UART receive interrupt service program. Figure 5 shows the main program, TIMER1 interrupt subroutine flow diagram.

Figure 6. Prototype physical photo.

## 5 EXPERIMENTAL VERIFICATION

A controller prototype was produced based on the Memobus RTU protocol and was verified by experiment.

The experiment results show that data communication between the controller and the VFD is reliable and stable. The operating power, current, voltage, and other parameters can be displayed in an industrial serial port panel screen. It can also control the VFD such as start-stop and run reverse by the panel screen. Figure 6 is a photograph of the controller prototype.

## 6 CONCLUSION

In this paper, using dsPIC30F6010A as the core design of the Yaskawa A1000 VFD controller. In addition, the controller is equipped with an industrial serial screen to achieve a friendly man-machine interface and convenient control. This system has very good practicability, dynamic, and extendibility.

## REFERENCES


B&B Electronics. *RS-422 and RS-485 Application Notes.* http://www.bb-elec.com.

Microchip Technology Inc. 2005. *ds-PIC30F6010A DataSheet.*

Si-Xue Cheng. 2007. *Study on integrated control method and its application based on PLC and frequency converter.* Shenyang: Shenyang University of Technology.

Wen-Hong Lv. 2008. *The principle of electromagnetic compatibility and Application.* Beijing: Tsinghua University Press.

Yaskawa. 2008. *YASKAWA VFD A1000 Datasheet.*

Zhong Hua Gong. 2011. *AC servo motor with frequency conversion technology and Application*, Beijing: People's Posts and Telecommunications Press.

*Frontiers in Computer Education – Wang (ed.)*
 *ISBN 978-1-138-02797-8*

# A condition assessment model of power transformers based on fuzzy information fusion

Y.Y. Zhang & R.M. Bi
*Guangxi Key Lab of Power System Optimization and Energy Technology, Guangxi University, China*

Z.H. Zhong
*Archives of Guangxi University, Guangxi University, China*

Y. Sun
*Guangxi Power Grid Power Dispatching & Control Center, Guangxi, China*

ABSTRACT: A power transformer is one of the most expensive pieces of equipment in power delivery; it is therefore important to assess the condition of a transformer. In this paper, a transformer condition assessment model is proposed based on fuzzy information fusion. First, an index system is established based on part division of a transformer. A Cauchy membership function is then employed to simulate uncertainty of the transformer condition. A fuzzy synthetic evidential reasoning method is then proposed to combine the fuzzy information of the transformer. Finally, case studies are conducted to demonstrate the effectiveness and feasibility of the model. Results show that the model can provide useful suggestions relating to transformer maintenance for electric power companies.

*Keywords*: Power Transformers; Fuzzy; Information Fusion; Evidential Reasoning; Condition Assessment

## 1 INTRODUCTION

Recently, with power systems' future orientation towards ultra-high voltage, large capacity and smart grids, power equipment is required that is of a higher level of safety and economic operation. Since a transformer is the key piece of equipment in power delivery, the high costs and the critical nature of transformers make it extremely important to prevent failures as early as possible [1]. Therefore, obtaining the accurate condition of a transformer is essential for performing necessary and effective maintenance on the transformers [2].

Since Dissolve Gas Analyses (DGA) is an effective tool for fault diagnosis of transformers, genetic programming [3] and a support vector machine [4] are applied to obtain DGA information in order to evaluate the power transformer's condition. Furthermore, a particle swarm optimization-least squares support vector [5], a weakening buffer operator and a least square support vector machine [6] are applied to forecast the power transformer's condition based on the DGA. However, these achievements are often limited in their representation of the fault condition. In fact, the condition of a transformer is often somewhere between normal state and faulty state [7]. Therefore, the DGA method alone is not enough for precisely assessing the condition of a transformer and it is hard to get accurate results when assessing the transformer condition [8], [9]. Professor Liao [10] presented a method of condition assessment for transformers based on fuzzy evidential theory, offering a novel way of thinking about transformer condition assessment. However, it does not include bushing and any accessory of the transformer in its index system, and it just focuses on the weight relationship between indexes in the process of evidence fusion but does not take evidence conflicts into account.

In order to solve these problems, in this paper, a new assessment model, based on fuzzy information fusion, is developed to assess the condition of a transformer. Some advantages are obtained such as the fuzzy synthetic properties and the capabilities of evidence combination. An assessment index system is first established. Both the information of index weights and the conflict of evidences are taken into consideration in this study, which can improve the accuracy rate of the evaluation. The results show that the proposed approach is practical and effective.

## 2 FRAMEWORK FOR TRANSFORMER CONDITION ASSESSMENT

### 2.1 *Assessment system of transformer condition*

According to relevant standards [11], items of a transformer's main body, bushing and accessory are

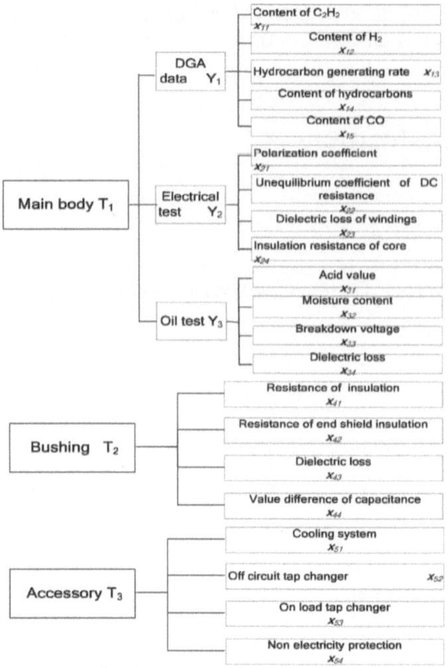


Figure 1. Assessment index system.

selected as the index system. The parameters of $T_1$ and $T_2$ are defined as quantitative indices while $T_3$ consists of qualitative indices. The index system is illustrated in Figure 1.

As shown in Figure 1, supposing that there are a set of indices $X, X = \{x_1, \ldots, x_m, \ldots, x_M\}$, and that there are three parts defined as $T = \{T_1, T_2, T_3\}$ in the index system. Unlike $T_3$ and$T_2$, $T_1$ has its own factor level $Y = \{Y_1, Y_2, Y_3\}$, which consists of DGA indices, electrical test indices, and oil test indices. Furthermore, T, which is called the target layer, is the evaluation outcome of the transformer's condition [12].

## 2.2 *Determination of the assessing grade*

Generally, it is difficult to compare inferiority degrees among indices because the index data are in different orders of magnitude. In order to resolve this problem, this paper employs Equations (1) and (2) to unify the dimension of the index values. The unified index values range within [0, 1], with 0 and 1 representing the initial state and the final state of insulation deterioration of a transformer respectively. For indices (e.g. insulation resistance of core) of which big values are desired, their values of the relative inferiority degree can be calculated by (1); for others like $H_2$, whose content are small values, are defined by (2):

$$x_m = (e_m - e_0)/(e_1 - e_0) \quad e_0 \le e_m \le e_1 \tag{1}$$

$$x_m = (e_0 - e_m)/(e_0 - e_1) \quad e_1 \le e_m \le e_0 \tag{2}$$

where: $e_0$ represents the index's value given by the manufacturer of the equipment, namely the normal value; $e_1$ represents the index's value when the insulation has seriously deteriorated, namely the cautionary value; and $e_m$ is the index's current value. Following that, the state of a transformer can be classified into five levels: good, normal, attentive, poor, and serious, which can be expressed by a matrix $g = \{g_1, g_2, g_3, g_4, g_5\}$.

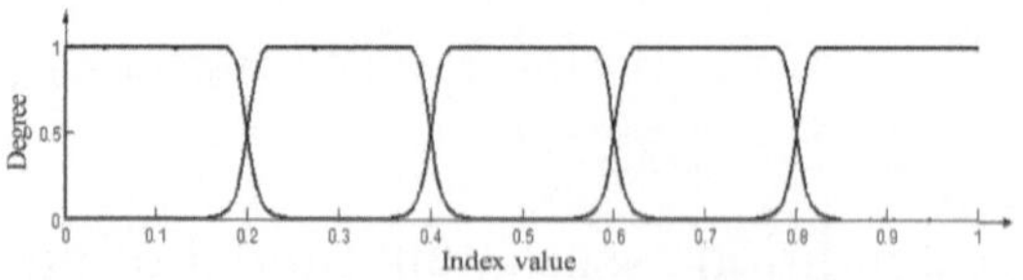


Figure 2. Curves of Cauchy membership function.

## 3 AN INTEGRATED METHOD BASED ON FUZZY INFORMATION FUSION

### 3.1 *Fuzzy method*

According to fuzzy theory, the border of neighbouring assessing grades should be an interval. In order to obtain the membership degree, a Cauchy membership function is established on the basis of the relationship between state and membership degree. The function and its curve are illustrated in Equation (3) and Figure 2 respectively.

Parameters of the Cauchy membership function are obtained by experts' experience: $c_1 = 0.15$, $c_2 = 0.25$, $c_3 = 0.35$, $c_4 = 0.45$, $c_5 = 0.55$, $c_6 = 0.65$, $c_7 = 0.75$, $c_8 = 0.85$, $b = 8$, $a = 1/0.05^8$.

$$f(x) = 1/[1 + a(x - c_k)^b] \quad k = 1, 2, 3, 4, 5, 6. \tag{3}$$

### 3.2 *Index level judgment by fuzzy evidential reasoning*

Evidential Reasoning theory has the advantage of handling problems with fuzziness and uncertainty.

Supposing that all independent possible solutions or hypotheses of a proposition constitute a finite set $\Theta$, $\Theta$ is defined as the frame of discernment of the proposition, and there exists a basic hypothesis $G_n$, which fits $G_n \subseteq \Theta$. All possible hypotheses of $\Theta$ constitute a set $G = \{G_1, G_2, \ldots, G_N\}$, which is presented by the power set $2^\Theta$. For example, in this paper, the state of a transformer is classified into 5 grades. Furthermore, an extra grade $G_6 = G(\Theta)$ is introduced to reflect the uncertainty of each grade, and these six grades can be defined as a frame of discernment.

The effects of various indices are different in the assessment system: some are notable while others are weak, therefore different weights may be assigned to different indices. Indices and factor weights calculated by the Analytic Hierarchy Process (AHP) method are shown in Table 1 [13]. The weights of Main body $B_1$, Bushing $B_2$ and Accessories $B_3$ are 0.5297, 0.3378, and 0.1325, respectively.

Table 1. Constant weights of assessing indices.

| Item | weight | weights of the item |
|---|---|---|
| $B_1$ | 0.3974, | {0.1150,0.2375,0.3566,0.1750,0.1150} |
| | 0.3665, | {0.2381,0.2057,0.2287,0.3275} |
| | 0.2361 | {0.1718, 0.1974,0.4295,0.2014} |
| $B_2$ | 1 | {0.2381, 0.2287, 0.4264, 0.1722} |
| $B_3$ | 1 | {0.2163, 0.1793, 0.3520, 0.2524} |

Considering the weight differences, a revision of the original assignment value must been done before combining evidence. The revision equations are given by (4)–(13):

$$\bar{w}_m = w_m / w_K \tag{4}$$

$$\alpha_m = \alpha_K \bar{w}_m \tag{5}$$

$$x_m(H) = \alpha_m X_m(H) \tag{6}$$

$$x_m(\Theta) = 1 - \alpha_m \tag{7}$$

where $w_m = \{w_1, \cdots, w_M\}$ is a uniform weight vector, $w_r$ is the relative importance of $x_m$ and the sum of $w_m$ equals to 1, $w_K = \max\{w_1, \cdots, w_M\}$ denotes the most important index in the index level $\alpha_m (m = 1, \ldots, M)$ denotes a belief degree coefficient, and $\alpha_K$ is referred to as a prior coefficient representing the importance of the role of the most important factor in the assessment. According to repeated studies, in the index level, $\alpha_K$ can be defined as 0.95. $X_m(H)$ is the original basic assignment probability of index $x_m$, and $x_m(H)$ is the revised basic assignment probability, naturally, $x_m(\Theta)$ is the uncertainty of belief assignment [10].

Suppose that con is the measure of partial conflicts between the two evidences ($con$=0.5 in this paper), if $0 < con < 0.5$, the classical D-S evidence approach is available. On the other hand, if $0.5 \leq con$, it denotes that the conflict of the two evidences is too large, the D-S approach fails. Rectifications of evidences must be tackled with the following process:

$$con_h = \max_{\mu \neq f, v=(1,2,\ldots,n), v=(1,2,\ldots,n)} m_h(T_\mu)\ m_{h+1}(T_v) \tag{8}$$

$$\beta_h = \sqrt{\frac{\gamma}{con_h}} \tag{9}$$

where $con_h$ is the partial conflict between the $h$th index evidence and the $(h+1)$th index evidence. The value of $\beta_h$ can be calculated by (9). Suppose that the $h$th index evidence $m_{h,n}(G_n)$ is combined with the $(h+1)$th index evidence $m_{h+1,n}(G_n)$, the rectified evidences of the $h$th and the $(h+1)$th grades are shown as Equations (10) to (13); the result of the combination is shown in (14):

$$m^1_{h,n}(G_n) = \beta_h \times m_{h,n}(G_n)\ (n = 1,\ 2,\ \ldots,\ 5) \tag{10}$$

$$m^1_{h,n+1}(G_\Theta) = m^1_{h,n+1}(G_{n+1}) = 1 - \beta_h + \beta_h \times m_{h,n+1}(G_{n+1}) \tag{11}$$

$$m^1_{h+1,n}(G_n) = \beta_h \times m_{h+1,n}(G_n) \tag{12}$$

$$m^1_{h+1,n+1}(G_\Theta) = m^1_{h+1,n+1}(G_{n+1}) = 1 - \beta_h + \beta_h \times m_{h+1,n+1}(G_{n+1}) \tag{13}$$

$$m^1_{(h)} = m^1_h(G_n) \oplus m^1_{h+1}(G_n) \tag{14}$$

After combining the index evidence one by one, the original basic probability assignment of each factor's evidence can finally be calculated. After combining all the factors' evidential information, the total assessment of one part of the transformer is obtained by Equation (15):

$$F^1_{(s)} = f^1_s(G_n) \oplus f^1_{s+1}(G_n) \tag{15}$$

Finally, the D-S evidence decision-making criterion for evaluating condition grades is defined as Equation (16):

$$\begin{cases} m_s(G_{N0}) > m_s(G_\Theta) \\ \quad m_s(\Theta) < \varepsilon \end{cases} \tag{16}$$

where $m_s(G_{N0}) = \max\{m_s(G_n)\}$, $G_n \subseteq G$, n= 1,2,..., N}; $\varepsilon$ is set to be the threshold value based upon experts' field experience and standard techniques ($\varepsilon$=0.04 in this paper). The less the value of $\varepsilon$ is, the more precise and reliable the assessment result becomes. The assessing grade of the given example can be judged according to $G_{n0}$

## 4 CASE STUDY

A 220 kV transformer in an electric power company in China was used for evaluating the presented approach. On the basis of the test data, an assessment of the transformer can be performed. Specifically speaking, for factor layer $Y_1$, the relative inferiority degree can be represented as follows: $\{x_1 = 0,\ x_2 = 0.0647,\ x_3 = 0.4586,\ x_4 = 0.0584,\ x_5 = 0.1600\}$. The judgment matrix of the relative inferiority degree, obtained by (3), of each part is given in Equation (17):

$$X_1 = \begin{bmatrix} Y_1 & 0.0000 & 0.0647 & 0.4586 & 0.0584 & 0.1600 \\ Y_2 & 0.1150 & 0.1800 & 0.4375 & 0.0000 & \\ Y_3 & 0.5056 & 0.5714 & 0.0320 & 0.0900 & \\ T_2 & 0.0000 & 0.0000 & 0.0200 & 0.4004 & \\ T_3 & 0.0600 & 0.0000 & 0.0600 & 0.0400 & \end{bmatrix} \tag{17}$$

Then, taking Main body $T_1$ as an example, using $X_1$, the membership degree between the indices and

Table 2. Calculation of membership degree.

| Index | Assessing grade $g_1$ | $g_2$ | $g_3$ | $g_4$ | $g_5$ |
|---|---|---|---|---|---|
| $x_{11}$ | 1.0000 | 0 | 0 | 0 | 0 |
| $x_{12}$ | 1.0000 | 0 | 0 | 0 | 0 |
| $x_{13}$ | 0 | 0 | 1.0000 | 0 | 0 |
| $x_{14}$ | 1.0000 | 0 | 0 | 0 | 0 |
| $x_{15}$ | 0.9911 | 0.0089 | 0 | 0 | 0 |
| $x_{21}$ | 1.0000 | 0 | 0 | 0 | 0 |
| $x_{22}$ | 0.9394 | 0.0606 | 0 | 0 | 0 |
| $x_{23}$ | 0 | 0.0111 | 0.9889 | 0 | 0 |
| $x_{24}$ | 1.0000 | 0 | 0 | 0 | 0 |
| $x_{31}$ | 0 | 0 | 1.0000 | 0 | 0 |
| $x_{32}$ | 0 | 0 | 0.9745 | 0.0255 | 0 |
| $x_{33}$ | 1.0000 | 0 | 0 | 0 | 0 |
| $x_{34}$ | 1.0000 | 0 | 0 | 0 | 0 |

the grades can be obtained by Equation (3) and the results are presented in Table 2.

By using the weight information of factors relating to Main body, an assignment matrix is acquired, as shown in Equation (18):

$$G(T_1)=\begin{bmatrix} g_1 & g_2 & g_3 & g_4 & g_5 & g_\Theta \\ 0.5557 & 0.0004 & 0.3544 & 0 & 0 & 0.0895 \\ 0.9757 & 0.0007 & 0.0147 & 0 & 0 & 0.0089 \\ 0.8304 & 0 & 0.0986 & 0.0011 & 0 & 0.0699 \end{bmatrix} \quad (18)$$

The output of the transformer's condition assessment is shown in the matrix in Equation (19). After combining evidences of all three parts, the total condition of the transformer can be obtained and it is displayed by the matrix in Equation (20):

$$G_{body} = [0.9265\ 0.0001\ 0.0520\ 0.0000\ 0\ 0.0214] \quad (19)$$

$$G_{total} = [0.8632\ 0.0004\ 0.1082\ 0.0003\ 0\ 0.0280] \quad (20)$$

Finally, the assessment can be conducted by applying Equations (15) to (16), obtaining the conclusion that the transformer is on grade $g_1$, which indicates that there were no faults in the transformer. Based on the conclusion, the period of the preliminary test could be prolonged. Further analyses suggest that the condition of the Main body and the Accessories are very good, while the Bushing is on the attentive level and its maintenance and test should therefore be performed ahead of schedule. Compared with the maintenance log, this result is reasonable. In fact, as maintenance of the transformer, only one operation, cleaning of insulation waste on the bottom of oil tank, was carried out. It can be concluded that the integrated model is capable of delivering reliable results of a transformer's condition assessment.

## 5 CONCLUSIONS

This study presents an integrated approach of transformer condition assessment (220kV). The results of the reported work and analyses can be summarized as follows:

(1) This paper established an index system (including the main body, the bushing, and the accessories) of transformers' condition based on a transformer related procedure and experts' experience.
(2) A comprehensive evaluation method based on a fuzzy theory and evidential reasoning integrated model is proposed.
(3) The proposed method has been verified through case studies. It proved to be an efficient tool to assess the overall state of a transformer and capable of giving guidance for condition-based maintenance of a transformer.

## ACKNOWLEDGMENT

The project is funded by "Power Transformer Reliability and Economical Life Assessment and its Maintenance and Decommissioning Strategy Research", "National Natural Science Foundation of China (Grant No. 51277034)", "Study of Power Flow Control Form of Electromagnetic Ring Circuit in Guangxi Grid" project (K-GX2013-104), and "The National Natural Science Foundation of China" (Project for Young Scientists Fund, Grant No. 51107103).

## REFERENCES

Bacha, K. Souahlia, S. Gossa, M. 2012. Power transformer fault diagnosis based on dissolved gas analysis by support vector machine. *Electric Power Systems Research* 83: 73–79.

IEEE Guide for the Evaluation and Reconditioning of Liquid Immersed Power Transformers. 2007. IEEE Std. C57.140–2006 *IEEE New York Monitor*, USA.

Liao, R. Zheng, H. Grzybowski, S. Yang, L. Particle swarm optimization-least squares support vector regression based forecasting model on dissolved gases in oil-filled power transformers. *Electric Power Systems Research* 2011, 81, 2074–2080.

Liao, R.J. Bian, J.P. Yang, L.J. Grzybowski, S. 2012. Forecasting dissolved gases content in power transformer oil based on weakening buffer operator and least square support vector machine-Markov. *IET Generation Transmission & Distribution* 6: 142–151.

Liao R. Zheng, H. Grzybowski, S. et al. 2011. An Integrated Decision-Making Model for Condition Assessment of Power Transformers Using Fuzzy Approach and Evidential Reasoning. *IEEE Transactions on Power Delivery* 26: 1111–1118.

Liao R. Zhang, Y.Y. Yang, L.J. et al. 2014. A cloud and evidential reasoning integrated model for insulation condition assessment of high voltage transformers. *International Transactions on Electrical Energy Systems* 24: 913–926.

Li, J. He, Z. Wang, Y. Lv, J. 2012. A Two-Dimensional Cloud Model for Condition Assessment of HVDC Converter Transformers. *ENERGIES* 5: 157–167.

Muthanna, K.T. Sarkar, A. Das, K. 2006. Waldner, K. Transformer insulation life assessment. *IEEE Transactions on Power Delivery* 21: 150–156.
Saaty, T.L. 1980. *The Analytic Hierarchy Process*. New York: McGraw-Hill.
Shintemirov, A. Tang, W. Wu, Q. H. 2009. Power Transformer Fault Classification Based on Dissolved Gas Analysis by Implementing Bootstrap and Genetic Programming. *IEEE Transactions on Systems Man and Cybernetics Part C [Applications and Reviews]* 39: 69–79.
Singh, J. Sood, Y.R. Jarial, R.K. Verma, P. 2008. Condition monitoring of power transformers—Bibliography survey. *IEEE Electrical Insulation Magazine* 24: 11–25.
Wang, M. Vandermaar, A.J. Srivastava, K.D. 2002. Review of condition assessment of power transformers in service. *IEEE Electrical Insulation Magazine* 18: 12–25.
Tang, W.H. Spurgeon, K. Wu, Q.H. Richardson, Z.J. 2004. An evidential reasoning approach to transformer condition assessments. *IEEE Transactions on Power Delivery*. 19: 1696–1703.

*Frontiers in Computer Education – Wang (ed.)*
*© 2015 Taylor & Francis Group, London, ISBN 978-1-138-02797-8*

# Spatio-temporal TEN model and the dynamic expression of moving object based on Clifford algebra

Ke Xu
*School of Information Management, Dezhou University*

Dongyan Cao
*College of Physics and Electronic Engineering, Dezhou University*

Xuedou Yu
*Science and Technology Department, Dezhou University*

ABSTRACT: The tetrahedral grid TEN model is an important data model in 3D Geographical Information System (GIS). It has the limitation that it describes the dynamic characteristics of geographical things, such as pollution cloud, ocean and so on. This paper proposes a kind of Spatio-temporal TEN model, which defines corresponding data organization structure, and storage structure, and algorithm used to generate it. Finally, it puts forward the railway patrol system, makes a geometrical measurement analysis and dynamic simulation of a dynamic object by using Clifford algebra.

*Keywords*: Spatio-temporal TEN model, Clifford, railway GIS

## 1 INTRODUCTION

Geographic information system (GIS) has been widely applied in many fields, such as geology, oceanography, city planning, and environmental monitoring. The relevant scholars have put forward various 3D GIS data models such as 3D FDS model, SMM model, and OO3D model. However, Pilouk proposed irregular tetrahedron grid (TEN model) about the irregular boundary objects like the pollution cloud, ocean, geological structure, metal mine etc. 3D data model based on TEN has many advantages such as simple topological relations which are convenient for three dimensional analysis and display.

At the present, the development of GIS demands that GIS can effectively support spatial and temporal integration of expression, modeling and simulation of the continuous geographical phenomenon[1]. The original TEN model doesn't express dynamic characteristics of geographic objects, so the construction of spatio-temporal TEN model can describe the dynamic characteristics of geographic objects such as a cloud of pollution, ocean etc. This paper presents a spatio-temporal TEN model which defines the corresponding data organization structure and storage structure, and gives the algorithm for its generation. Finally, it puts forward the railway patrol system, makes a geometrical measurement analysis and dynamic simulation of a dynamic object by using Clifford algebra.

## 2 TEN MODEL

Spatial point-set TEN[2] is a set of tetrahedra which are mutually adjacent but non overlapping and satisfy the Delaunay criterion. Each tetrahedron's vertices is concentrated nearby four points, the Circumscribed sphere of the tetrahedron does not contain any point in point set. TEN which conforms to Delaunay standards and 3D Voronoi graph is a dual, 3D Voronoi diagram is space expansion of the 2D Voronoi diagram.If the space point set is composed of n points, the spatial point is Corresponding to the n 3D Vorinoi map and each point is the center of one of 3D Voronoi map[2].

In the 3D GIS, a tetrahedron is the basic element which describes 3D space objects. Through the subdivision of tetrahedrons, an arbitrary 3D space object is divided into a series of adjacent but non overlapping sets of tetrahedrons. The constitution principle of the TEN model is that node is a boundary of the arc, arc is a boundary of the triangular and the triangle is a boundary of the tetrahedral[2].

The development project of science and technology of DeZhou No. 2011B12

## 3 CLIFFORD ALGEBRA

Clifford algebra is also known as geometric algebra, is an algebra based on the dimension calculation, the operation rules realized the uniform of scalar calculation,vector calculation, dimension calculation and geometric calculation. Clifford algebra consists mainly of two kinds of operation of an inner product and an outer product. The outer product mainly reflects the mutual construction of the relationship between different levels of geometric forms. And the inner product can express the measurement features of distance, angle etc. For the two vectors A and B, Clifford product can be expressed as $ab = a \bullet b + a \wedge b$. $a \bullet b$ as the inner product a and b, the results of the operations is a scalar quantity, $a \wedge b$ as the outer product, the result is a vector. When $a \bullet b = 0$, non zero vector a, b orthogonal; outer product is raising dimension operation, and when $a \wedge b = 0$, a, b in parallel. The outer product operation and inner product operation is also provided based on two kinds of expressions of the measurement relationship and the dimension relationship, which makes the expression of the geometric object contains the dimension relationship, measure relation attributes[3].

## 4 SPATIO-TEMPORAL TEN MODEL

### 4.1 *Data model based on tetrahedron*

The spatio-temporal tetrahedral data model is based on the space-time tetrahedron as the basic element, a spatial-temporal tetrahedron consists of four basic elements:the node of spatio-temporal (Tnode), the edge of spatio-temporal (Tedge), the face of spatio-temporal triangle (Ttriangle) and the tetrahedion of spatio-temporal (Ttetrahedion). You can use a relational database table to store the relationship among four basic elememts. The relationship between Tedge and Tnode is stored by edno table;the relationship among Ttetrahedron, Ttriangle and Tedge is stored by tte table.

### 4.2 *Data structure*

To make the topological relationship among the geometry objects, in order to make the model better establish the space body model with a spatial location, topology relation and time information, more detailed data structures must be designed. the tetrahedral data structure is as the follows:

(1) The data structure of the node.

```
Struct Tnode{
IntVertexID / / point identification
IntVertype / / type
Float x, y, z / / point coordinate
Float t// time t
TetreVer [n] / / point belongs to the body
TriVer [m] / / point belongs to a triangle
Pointer to a StructVertex * next// point
Edgever [p] / / edge of point
}
```

(2) The data structure the of the triangle edge.

```
Struct TEdge{
IntEdgeID / / boundary identification
IntEdgetype / / edge type
Float t// time t
EdgeVer Va, Vb / / edge at both ends
IntTriNo [2] / / edge adjacency triangle
IntTetraNo [2] / / edge adjacent tetrahedra
}
```

(3) The data structure of the triangular face.

```
Struct Ttriangle{
IntTriID / / triangle number
Float t// time t
IntTriTetra [2] / / triangle where the tetrahedron
TriVer [3] / / three vertices of triangle
TriEdge [3] / / three edges of a triangle
}
```

(4) The data structure of the tetrahedral.

```
Struct Ttetrahedion{
IntTetraID / / tetrahedral number
Float t// time t
TetraVer[4] / / four tetrahedral vertex
TriTetra[4] / / tetrahedral four triangles
TetraTet [4] / / tetrahedral adjacent tetrahedra
```

The simple data structure reduces the data redundancy and the occupied space, it is also very good at maintaining the topological relationship of entities, which are reflected in the data structure.

### 4.3 *The generation algorithm of the spatio-temporal TEN model*

(1) A set of spatio-temporal data, in addition to the time dimension, inevitably corresponds to convex boundary constituted by a triangle[4]. We first find the point on the convex boundary, then examine all triangles using Delaunay rule to get all triangle satisfying Delaunay condition.

(2) For a selected arbitrary triangle in a convex boundary triangulation network, let vertices A, B, C be the base triangle, rearrange the order of vertices, in order to normal vector pointing to the inner of convex boundaries. The triangles is written into the data structure.

(3) Along the normal vector direction, we find a point D satisfying Delaunay standards, make the circumscribed sphere of the ABCD four points exclude other points of the convex boundary.

(4) Write the tetrahedral ABCD, and the time attribute together, write into the data structure, establish topological relations, external and makes normal vector of each triangle pointing to the outer of the tetrahedral, and the three new triangles is the new base triangles.

(5) For each base triangle repeat step (2), and (3), form a new tetrahedron and a new triangle, and write into the data structure and update topological relationships. Finally finish the spatio-temporal TEN model.

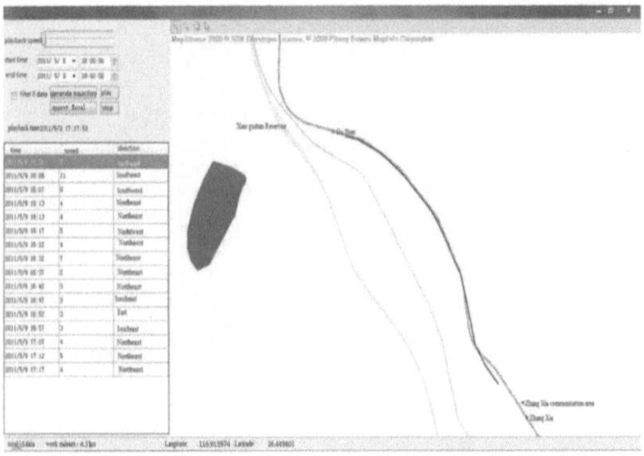

Figure 1. Track playback graph.

## 5 DYNAMIC SIMULATION

Clifford algebra has advantages in dynamic geometric measurement analysis of a moving object[5]. For example, in a railway patrol system, patrol personnel hold a positioning system, we can select the patrol personnel in four buildings under surveillance, we can analyse the synchronous geometric measure relation of the the four buildings and patrol personnel.

(1) A sampling frequency of satellite monitoring vehicle as the interval, transmitsts the inspector's current coordinate information immediately.
(2) We construct the synchronous calculation model of the geometric measure relationship by using Clifford algebra with a centre point of the monitored person and four buildings that are controlled as the basic geometric object.
(3) Synchronous computational model based on the step (2), constructs a synchronous calculation matrix of the patrol and distance.realize the synchronous calculation and distance between buildings and patrol.
(4) We reconstruct the motion process of personnel monitored by using the relevant data and versor interpolation operators in Clifford algebra, then output the related data for visualization.

The simulation analysis of the movement scene between patrol personnel and four buildings as follows in Figure 1.

The simulated results reproduce the characteristic parameters of motion of the inspection personnel, giving the dynamic change between personnel and building controlled.

## 6 CONCLUSION

Clifford geometric algebra has a wide application in calculus, statistics theory, signal processing, and other fields. Clifford algebra is an algebra of the dimension calculation based, the operation rules realized the uniform of scalar operations, vector calculus, dimension calculation and geometric operation. Clifford algebra consists mainly two kinds of operation of an inner product and an outer product. The outer product mainly reflects the mutual construction of the relationship between different levels of geometric forms.The inner product can express the measurement features of distance, angle etc. This paper presents a spatio-temporal TEN model which defines the corresponding data organization structure and storage structure, and gives the generating algorithm for the model. Finally, it puts forward the railway patrol system,as an example, making a geometry measurement analysis of a dynamic object by using Clifford algebra and dynamic simulation.

## REFERENCES


[1] Zhaoyuan Yu. Multi-dimensional Unified GIS Data Model based on Geometric Algebra [D] May, 2011
[2] Han Wenzhong, Wu lixin, Liqingquan, Wang Yanbing. Models and Algorithms for Three Dimensional Spatial Information System [M]
[3] Linwang Yuan, Guonian Lu, Zhaoyuan Yu. Multi-dimensional Unified GIS Data Model based on Geometric Algebra [M]
[4] D.Z. Sui, R.C. Maggio. Integrating GIS with hydrological modeling: practices, problems, and prospects. Computers, Environment and Urban Systems. 23 (1999) 33–51
[5] P.A. Burrough. Dynamic modeling and geocomputation. Geocomputation: Aprimer, John Wiley and Sons, Chichester (1998), pp. 165–191

# Author index